"Michael Berdine has performed a great service in giving us a full biography based on the Sykes Papers and an array of other archival sources. It is comprehensive, exact and reliable. On the famous Sykes–Picot agreement of 1916 and a range of other major issues, this is an indispensable work."
     – Wm. Roger Louis, Kerr Chair in English History and Culture and
          Distinguished Teaching Professor, University of Texas at Austin

"Mark Sykes is one of those figures in the annals of British imperial history that many a budding student will come across, but seldom will they receive a thoughtful and engaged consideration as to who he was. Dr Michael Berdine now makes it easier for both student and scholar, analyst and observer, to interrogate and understand the context of the man and the role he played in this rather problematic and regrettable period of Britain's forays beyond its borders. Dr Berdine's efforts are all the more unique as compared to other historians' work in shedding light on this era because Sykes's own exploits were mostly classified, and access to records and sources that would give us more insight was cut off. Dr Berdine does the reader a great service by going into those sources and delivering to us, with a true historian's penchant for detail and interest, a narrative that is not merely conjecture, but based on genuine discovery. As the Arab world continues to feel the ramifications of those early years of the twentieth century, more than a hundred years later, it behoves us to understand more, not less, about this poignant past – and we have Dr Berdine to thank for furthering the ability for us all to do so."
     – Dr H.A. Hellyer, senior non-resident fellow at the RH Centre for the
          Middle East, Atlantic Council, Washington DC, and professor at the
               Centre for Advanced Study of Islam, Science and Civilisation at
                    the University of Technology Malaysia, Kuala Lumpur

"Michael Berdine has written the professional biography of a crucial figure in the Middle East's emergence from World War I and the creation of the region's modern nation states. Sykes's name is well known, but his participation in Britain's Middle East policy making was long classified as secret. Now Berdine has brought his deeds to light in a detailed analysis covering the years 1911–1919, using an impressive array of government papers, memoirs and private papers from the leading political actors of the period, plus Sykes's own unpublished and published works. Anyone interested in the creation of the modern Middle East, World War I in the region, or Britain's role in it must read this book."
     – Linda Darling, Professor of History, University of Arizona

"This is the definitive book which everyone interested in the history and formation of the Middle East should read. I recommend it wholeheartedly."
     – Christopher Catherwood, author of *Winston's Folly*

REDRAWING THE MIDDLE EAST

Sir Mark Sykes, Imperialism and the Sykes–Picot Agreement

MICHAEL D. BERDINE

I.B. TAURIS

LONDON • NEW YORK • OXFORD • NEW DELHI • SYDNEY

I.B. TAURIS
Bloomsbury Publishing Plc
50 Bedford Square, London, WC1B 3DP, UK
1385 Broadway, New York, NY 10018, USA

BLOOMSBURY, I.B. TAURIS and the I.B. Tauris logo
are trademarks of Bloomsbury Publishing Plc

First published 2018
Paperback edition first published 2020

A catalogue record for this book is available from the British Library.

A catalog record for this book is available from the Library of Congress.

ISBN: HB: 978-1-7883-1194-6
PB: 978-1-8386-0467-7
ePDF: 978-1-7867-3406-8
eBook: 978-1-7867-2406-9

Series: Library of Middle East History, 78

Typeset in Garamond Three by OKS Prepress Services, Chennai, India

To find out more about our authors and books visit
www.bloomsbury.com and sign up for our newsletters.

To Nolly

CONTENTS

LIST OF ILLUSTRATIONS

Map

Plates

Plate 10 Sir Mark Sykes, Unknown and François Georges-Picot, Unknown

Plate 11 Sharif/King Husayn – Lowell Thomas Papers, Arabia, Graphic Materials, James A. Cannavino Library, Archives & Special Collections, Marist College, USA

Plate 12 Lord Curzon – George Grantham Bain Collection, Prints & Photographs Division, United States Library of Congress

Plate 13 Chaim Weizmann – National Portrait Gallery

Plate 14 Nahum Sokolow – George Grantham Bain Collection, Prints & Photographs Division, United States Library of Congress

Plate 15 Arthur J. Balfour – Imperial War Museum Photograph Archives

Plate 16 Sir Mark Sykes and family. From left: Everlida, Angela, Richard, Lady Sykes, Daniel, Christopher and Freya (c. 1917) – Historic English Archive

Plate 17 General Sir Edmund Allenby – National Portrait Gallery

Plate 18 Sir Mark Sykes brass image on Eleanor War Memorial at Sledmere – The Revd Gordon Plumb

ACKNOWLEDGEMENTS

This book could not have been written without the help and support of others. First of all, I would like to thank Cambridge Muslim College for giving me the research fellowship that enabled me to begin my research. Thanks also to the staff of the Hull History Centre, who were helpful in assisting me in the research of the Papers of the Sykes Family of Sledmere and The Papers of Sir Mark Sykes, 1879–1919, as well as the British Online Archives for making the papers available online.

Many others were also of great assistance in making this book possible. These include the staff at the Churchill Archives Centre in researching the Churchill Papers, Hankey Papers, Amery Papers and George Lloyd Papers. So, too, were the staffs at The National Archives at Kew, the Asian and African Studies Reading Room at the British Library, Cambridge University Library and the Sudan Archive at Durham University. Also, thanks to Andy Aldridge at The National Archives, Emily Dean at the Imperial War Museum, Kate Woolhouse at the Hull Daily Mail, Alexis Valentine at the Library of Congress in Washington, DC and the Revd Gordon Plumb for providing the map and photos for the book. Special mention also goes to Dr John Ansley, Director, Archives and Special Collections at Marist College, Poughkeepsie, New York, for assistance with photos from the amazing Lowell Thomas Collection at the Marist College Library.

Special thanks are due to Emeritus Professor Richard Cosgrove of the University of Arizona, my supervisor, mentor and friend, who read and edited the original manuscript through a long and often challenging creative process. Also, very special thanks to Carole Pearce, editor extraordinaire, for the extra time and effort she took to improve every facet of the manuscript and bring it to its final form.

To my publisher, Iradj Bagherzade, founder, chairman and senior editor of Middle East Studies at I.B.Tauris, who believed in the project and supported it from the beginning, go special thanks. Also, thanks to Lizzy Collier, Jo Godfrey and Angelique Neumann at I.B.Tauris for their patience with my many questions and to commissioning editor Dr Lester Crook for his guidance and support.

Last of all but not least, I could never have written this book without the help and support of my wife Nolly, to whom this book is dedicated. From the beginning she assisted me in so many ways that made it possible to do the research and writing necessary to complete the book. She was always available to encourage, spur me on, read and re-read the manuscript through its many incarnations over the four years it took to research and write, and was always there to offer sound advice. For all this and so much more, I cannot thank her enough.

Despite all the assistance I received from others that made this book possible, any and all mistakes in the telling of this story are mine and mine alone.

NAMING CONVENTIONS

Arabic names and titles

The spelling of Arabic names and titles in English has varied over the centuries, but has recently been standardised by the *International Journal of Middle East Studies* (*IJMES*) Transliteration System,[1] which I have followed. In researching this book I came across a variety of English spellings of Arabic names and titles and have retained the original spellings in quotations, but otherwise used their modern equivalents. I have retained the name of Constantinople in quotes, but otherwise used the city's modern name of Istanbul in the text, which is closer to the Turkish "Stambul" (meaning "in the city"). However, with Mecca, Jeddah and Yemen have I continued to use these archaic spellings in the text in order to save confusion, because their usage generally persists even in the Middle East. I have included some titles and their meanings as an aid to understanding the importance of some individuals mentioned in the narrative.

Archaic	Refers to	Modern (*IJMES*)
Abu Bekr	Islam's first caliph	Abu Bakr
Beyrout	Capital of Lebanon	Beirut
Caliph	Sunni leader	Khalif
Caliphate	Government of the caliph, leader of Sunni Muslims	Khalifat
Constantinople/ Stamboul	Ottoman capital	Istanbul
Emir	Prince, or ruler; or male member of ruling house	Amir
Feisal/Faisal	Man's name	Faysal
Hedjaz/Hejaz	Red Sea coastal region on the Arabian peninsula	Hijaz

Hussein/Husain	Man's name	Husayn
imam (lower case)	Sunni prayer leader, mosque religious leader	imam
Imam (capitalised)	Shiʿa religious leader, descendent of Ali, the 4th caliph and son-in-law of Prophet Muhammad	Imam
Jaffa	Palestinian port city	Haifa
Jeddah	Main port of Hijaz	Jiddah
Kerbala	City in Iraq	Karbala
Kerkuk	City in Iraq	Kirkuk
Koraysh/Koraish	Tribe of the Prophet Muhammad	Quraysh
Mecca	Holiest city of Islam	Makkah
Medina	Second-holiest city of Islam	Madinah
Mesopotamia/Irak		Iraq
Mohammed/ Mohamed	The Prophet of Islam; a common Muslim male name	Muhammad
Mohammedan/ Moslem	Follower of Islam	Muslim
Saudi	Kingdom of Ibn Saʿud, Saʿudi Arabia	Saʿudi
Sayid	Descendant of the Prophet Muhammad through his grandson, Hassan, son of Ali	Sayyid
Shaikh/Shaik/Sheikh	Sunni religious leader, or scholar; older man	Shaykh
Sherif	Descendant of the Prophet Muhammad through his grandson, Husayn, son of Ali	Sharif
Sherifate	Meccan government under Sharif Husayn	Sharifate
Shiʿa/Shiʿism	Minority sect of Islam, followers of Ali, the 4th caliph and son-in-law of Prophet Muhammad	Shiʿa/Shiʿism
Ulema	Body of Sunni scholars	Ulama
Wejh	Coastal town in Hijaz	Wajh
Yanbo	Coastal town in Hijaz	Yanbu
Yemen	Country at southern tip of the Arabian peninsula	Yaman

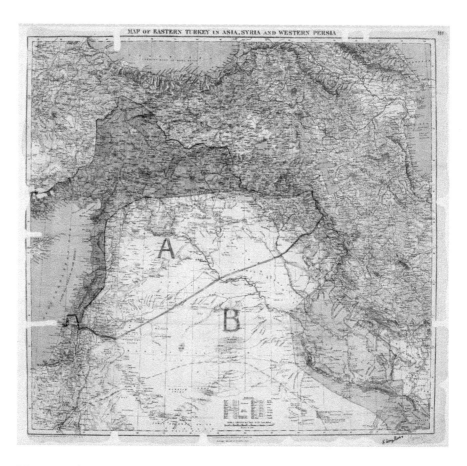

Map 1 Sykes–Picot Map (c. 1916) – The National Archives

INTRODUCTION

What a relatively unknown British aristocrat did during World War I in the Middle East greatly affected its outcome and, as a result, contributed much to the troubles in the region today. While it is only part of a much larger and complicated story, for a time Lt. Col. Sir Mark Sykes, Bart, MP, played a pivotal role in the war in the desert and made a major difference at critical junctures during the war and in its aftermath. It is time for the full story to be told.

Between 1915 and 1916, Sykes was Lord Kitchener's agent at home and abroad, operating out of the War Office until the war secretary's untimely death at sea in 1916. Following that, from 1916 to 1919 he worked at the Imperial War Cabinet, the War Cabinet Secretariat and, finally, as an advisor to the Foreign Office. Tall, charismatic and energetic, Sykes was convinced of his importance to the Middle East war effort because of his frequent travels in the region since boyhood and the books, articles and reports he had written, and the maps he had submitted of the area to the Foreign Office and War Office over the years. He was hard to ignore or put off when Middle East matters were discussed. For a time, he was the government's Middle East expert, involving himself in virtually every facet of policy and strategy concerning the region. Included among these were the Sykes–Picot agreement, the Arab Bureau, the Arab revolt and the Balfour Declaration.

However, it was the name of T.E. Lawrence and his stirring exploits in the Middle East during World War I that came to symbolise the Middle East war. Today this name instantly brings to mind actor Peter O'Toole in the 1962 movie *Lawrence of Arabia* as the eccentric Lawrence racing across the Arabian desert on a camel with his native robes flying behind him, leading bands of Arab Bedouin against the Turks and blowing up trains. Over the past century Lawrence's exploits in the Middle East war have been lauded in numerous books, articles and movies, while Sir Mark Sykes's involvement in the desert war is virtually unknown.

Unlike Lawrence, Sykes's activities took place mostly in government offices in London and were long classified as secret. He also did not survive the peace, dying of the Spanish flu at the Paris Peace Conference in February 1919. As a result, until recently there have been only two biographies about him: Shane Leslie's *Mark Sykes: His Life and Letters* published in 1923[1] and Roger Adelson's *Mark Sykes: Portrait of an Amateur* published in 1975.[2] Other than these, Sykes and his Middle East activities during World War I are mentioned only occasionally, primarily in academic books and articles. However, both men played major roles in Britain's war against the Turks. While Lawrence became Britain's hero and poster boy in promoting the war in the desert, Sykes was its *éminence grise*, working behind the scenes and often at cross-purposes with Lawrence and without his knowledge.

Since the end of the war, despite Lawrence's much-publicised wartime exploits in the Middle East, the sands of time have long since erased virtually any evidence he was even there. Instead, what remains is a legacy found mostly in books and articles, countless photographs, newsreel footage, archival records and legend. The same cannot be said of Sykes. His wartime activities had major, long-lasting consequences in the Middle East and today we are living with the results. It is Sykes's legacy, not Lawrence's, that lives on today in the region. The purpose of this book is to acquaint readers with the generally unknown activities of Sir Mark Sykes in the halls of power during World War I and offer some reasons why the Middle East is the way it is today.

Other than these two biographies of Sykes, both of which were limited by their lack of access to secret government documents and private archival material unavailable at the time, to date there has been no single source that reveals the extent of his wartime activities. With these resources available, and seeing the consequences of many of Sykes's actions over a century ago in the turmoil of the Middle East today, I decided to write this book.

In my research I found Leslie's and Adelson's biographies helpful in providing otherwise unavailable personal information and background material. Both authors had access to the Sykes family and their recollections, as well as to papers at the family estate at Sledmere in East Riding, Yorkshire. Leslie had the support of Lady Edith Sykes, who asked him to write a book about her husband after his untimely death. Over fifty years later, Adelson had the support of Sykes's eldest son and heir Sir Richard Sykes, the 7th Baronet, to aid in his research. I was not so fortunate, and had to rely primarily on these two books and Sykes's own works as sources for his private life and letters. However, this did not prove much of a problem. Most of Sir Mark Sykes's papers relating to his World War I activities, which are the subject of this book, were donated by the family to the University of Hull and are available to the public at Hull's History Centre and the British Online Archives.

Research of the Sykes Papers, and the official documents, papers, memoirs and books written by those who knew and worked with him reveal, the surprising extent of Sykes's involvement and influence in British Middle East wartime policy and strategy and the key role he often played between 1915 and 1919. Among these, in addition to negotiating the Sykes–Picot agreement, he proposed the creation of what became the Arab Bureau that was to make Lawrence famous. He played a key role at a critical time early in the Arab revolt. He was also involved in writing the Zionist declaration and ensured its subsequent passage as the Balfour Declaration by the War Cabinet, which was to lead to the creation of the state of Israel and so much more.

CHAPTER 1

THE "MIDDLE EAST EXPERT"

He was clearly marked out for service in the East. He became an invaluable factor in all that intricate and remarkable policy which split the Arab from the Turk, divided the Moslem world at a most critical juncture, and eventually furnished important forces on the desert flank of Allenby's armies.

Winston Churchill[1]

On 27 November 1911, the new MP for Central Hull, Mark Sykes, made his maiden speech in the House. It was later observed that he touched "in a brilliant manner ... on every national danger from Tunis to Travancore. As for Turkey, whatever was the British policy, it had alienated the Young Turks. Should the importance of this be lost on his listeners, he warned that the recent war in Tripoli between the Turks and Italians for control of that Ottoman province could prompt "a spark" among Indian and Arab Muslims elsewhere.[2] Deploring the government's lack of support for the Turks, which had led to the loss of influence in Istanbul, Sykes "gave historical examples of the British being vulnerable in the East when war came elsewhere, citing the way the Indian Mutiny and the Crimean War were linked." He "concluded by repeating his general position: if war came in the West, the British must send troops to the East."[3] With that, the heir to the baronetcy of Sledmere in East Riding, Yorkshire took his seat.

His words were timely and well received, following as they did on Foreign Secretary Lord Grey's report to the House "on Germany, the Moroccan crisis, and the balance of power in Europe." This was after a lengthy debate "in which Liberal and Labour members reiterated their distrust of the Foreign Office." As was customary, the speaker, following a member who had given a maiden speech, would congratulate him. In Sykes's case, no less a member than Prime Minister Asquith followed him and praised Sykes, "for 'as promising and successful a maiden speech as almost any I have listened to in my long experience.'"[4]

Throwing himself energetically into his parliamentary duties and Tory-related activities, Sykes joined the important Conservative clubs, including the Carlton Club, the 1900 Club, the Beefsteak and Prince's. He also spent much of the rest of 1911 addressing meetings around the country and involving himself in party fundraising and promotion. By the end of the year, Sykes had given speeches at thirty-one meetings. "He had also attended 27 social events, spent 50 days in Westminster, taken part in 88 divisions, asked 16 questions and made one speech."[5]

In 1912, his mother died and the following year his father died. Now the Sixth Baronet of Sledmere, Sir Mark Sykes was fully involved in a busy political career as well as managing the 30,000-acre estate, one of the largest in the north of England, which he had inherited from his father. In late September 1913, six months after his father's death, he took time from his busy schedule to travel to the Ottoman Empire. It was a place of fond memories from the many times he had travelled there both with his parents and alone with his father.[6]

This time Sykes sought to mix business with pleasure, as this trip was to assess the effects of the recent Balkan Wars on Turkey. These two successive military conflicts in 1912 and 1913, the second of which ended in August 1913 with the signing of a peace treaty between the Ottomans and Bulgaria, had deprived the Ottoman Empire of almost all of its remaining European territory. He had referred to the seriousness of this in a speech on 12 August, as a matter that "in time would come to trouble the War Office":

> The break-up of the Ottoman Empire in Asia must bring the powers of Europe directly confronting one another in a country where there are no frontiers because the mountains are parallel to the littoral, and because there being only three rivers, one moving in a circle and the others running side by side over a level plain, it is very difficult for any power to find a frontier.[7]

Very soon Sir Mark Sykes would find himself having to resolve this very matter.

He wrote of his trip to Constantinople and through Anatolia, Kurdistan[8] and the Arab Ottoman lands in detail in *The Caliph's Last Heritage*, his third and last book on the Ottoman Empire, which was to be published in 1915. This was an area he knew well from his travels over the years and one he had written about previously. Now Sykes dramatically described how it had changed since the deposition of Sultan Abdul Hamid II in 1909:[9] "What a strange mood [Constantinople] is in to-day — after five years of progress, of folly, of squalid intrigue, of violent negotiations, of senseless destruction, of ignominy, of instability, of wars and devastating fires."

He had been back to the city four times since 1908, when the Young Turks established themselves as the new rulers of the Ottoman Empire:

> Outwardly, the change is trivial. The streets are cleaner, the roads smoother, the dogs have gone ... There are fewer turbaned heads, fewer horses, fewer soldiers; more officers, more newspapers, more ruins. Gone is the palace, the retainers, the swarms of eunuchs, the gay equipages of the ex-Sultan's favourites.

Moreover:

> [T]here is at the root of things a deep change ... The fall of Abdul Hamid has been the fall, not of a despot or tyrant, but of a people and an idea. The Sultan meant something to his subjects, his people something to him. Good or ill, he represented not only a system but life, a scheme of things, an idea, a tradition, a faith, a species of continuity.

A wistful nostalgia permeates the entire chapter. His depiction of Abdul Hamid's reign as rule by "ill — in blood, confusion and terror," one in which the sultan "stood for the old order, and fought for it craftily but childishly, bravely but narrowly, pertinaciously but despairingly" showed this. "One able, frail, sickly, uneducated old man could not bear the weight. He willingly surrendered to the spirit of the age. With him fell things good and evil, as they must on a Day of Judgment."[10]

Sykes's resentment at the loss of the old and familiar was almost palpable. "In the place of theocracy," he wrote:

> [i]mperial prestige and tradition, came atheism, Jacobinism, materialism and licence. With the old order went the palace, its spies and intrigues, its terrorism and secrecy; with the new order came secret societies, lodges, oaths of brotherhood, assassinations, courts-martial, and strange, obscure policies.

For Sykes:

> In an hour, Constantinople changed; Islam, as understood by the theologians, as preached in the mosques, as the moral support of

the people, as the inspiration of the army, died in a moment; the Caliphate, the clergy, the Koran, ceased to hold or inspire."[11]

Contrasting the old with the new, Sykes noted that, in appearances, "the new Turkey was progressive," but in fact, "a mock Parliament made mock laws; mock ministries and mock ministers rose and fell; a mock counter-revolution served as a pretext for shattering even a semblance of authority, and set up a mock Sultan." As a result of all these and other changes, its territories "of Bosnia, Herzegovina and Eastern Roumania were lost, then Tripoli was snatched away, then the unnatural Balkan Confederation was made possible, and Turkey-in-Europe was lost."[12]

He saved his strongest criticism for the new Turkish soldier. Far from the once fearsome "Terrible Turk" of the past, his German-trained modern counterpart was not "the bold, sullen type of Abdul Hamid's day." That was a time when "the men were fanatical, truculent and rude; they terrorized the people, insulted European women and made life unpleasant; there was something of the pampered Pretorian guard about them, of the overbearing janissary, and yet there was something more." He believed the missing element in the modern Turkish soldier was

the innate idea that military service was the holy duty of a Moslem, that the years with the colours were sanctified in the sight of God, that each man was to die for the faith and the Caliph. Now, the khaki-clad, bewildered levies who slouch about with puzzled pathetic faces have no such idea.[13]

If this were not bad enough, the soldiers were ignorant peasants whose officers themselves "disregard all discipline, intrigue and quarrel among themselves, control their generals and assassinate their commanders-in-chief." Speaking as a military officer himself, a disgusted Sykes further described them as "[h]ysterical, pedantic, idle and vicious," asking rhetorically, "what could be expected from the leadership of such creatures, what influence could they expect to have on the men they are supposed to prepare for battle?" His harsh assessment of the modern Turkish military was that they were "a mere horde of helpless, leaderless villagers, misunderstood and misunderstanding, with no more enthusiasm or hope than a chain-gang."[14]

In the years to come during World War I, Sykes and Whitehall would learn just how wrong this biased, prejudiced and shallow assessment of Turkey's new German-trained military was. Sykes's vivid imagination was in harmony with British assumptions. The romantic in him made him see things through the lens of his youthful memories, which contrasted sharply with his resentment of and disillusionment in the modern Turkey. He was not alone in this; a fact that would prove costly in the early years of the war – particularly in the Dardanelles and at Kut.

In spring 1914, Sykes gave two parliamentary speeches on Eastern affairs in which he dwelt on his hopes and fears for the Ottoman Empire. He told the members of the House that "he worried about Germany gaining control of the Turks. Great Britain must not allow this to happen, despite the present Government's unwillingness to initiate any measures to avoid it." He had much to say about the German presence in Turkey, particularly its military influence.[15]

The 28 June 1914 assassination of Austria's Crown Prince Franz Ferdinand and his wife in Sarajevo caught most of the country and Europe by surprise, including Sykes. This was followed by the German invasion of Belgium six weeks later and Britain's declaration of war on 4 August. What was described at the time as the "war to end all wars," a variation on the title of a booklet written and published in October 1914 by H.G. Wells, entitled *The War that will End War*,[16] had begun. However, it did not end war in the twentieth century and what became World War I proved to be a major catastrophe. It was a "conflict that ... mobilized 65 million troops, claimed 3 empires, 20 million military and civilian deaths, and 21 million wounded."[17] Although Sykes had preached for years about the need for military readiness of Britain's armed forces in the event of a coming war, when it did happen he was caught off-guard like everyone else.[18]

As soon as war was declared Sykes rushed to Yorkshire from Wales, where he had been training with other Territorial officers. A long-time serving officer in the Territorials ("Terriers"), he had been commissioned a lieutenant in 1897 in the Princess of Wales's Yorkshire Regiment, a voluntary militia Territorial battalion known as the Green Howards. In November 1899, he was called up with his regiment to go to South Africa to fight in the Boer War and after two years Captain Mark Sykes returned home to Sledmere on 16 May 1902.[19] In 1911 he was promoted Lieutenant Colonel and appointed Colonel-Commanding of the 5th Yorkshire "Green Howards" Regiment.[20]

Before he became engaged in readying his Terriers for war, Sykes went to France with his private secretary, now Sergeant Wilson, for a quick tour. Over a period of several days, the two men visited several French villages, as well as battlefields and eight supply centres. Sykes wanted to learn as much as he could from what he was able to see of the war. Everywhere the "smell of war was omnipresent: 'the stink of dead things, [and] dead horses.'" Moreover, he "saw a good many wounded ... and poor refugees trudging back in the wake of our advance ... [women], children and weary old men.'"[21]

It was the condition of the "seriously wounded" that shocked him the most. Sykes wrote home of seeing "one officer 'muttering at the point of death'; a soldier 'without a face − just a red pad of lint' being led by a couple of other wounded men." He estimated there was only one doctor for about 500 wounded people in a "station which was filthy with dust and oil. There were no hospital trains or beds − just cattle cars filled with straw." Once he was back home, Sykes and his wife sought to do something about the lack of medical facilities and care on the front. They paid for the Metropole Hotel in Hull to be converted into a military hospital, which they also used to store supplies. Responding to "an appeal from the French government and the Union of French Women to support hospitals in France," Lady Sykes left for France in early October with several nuns to assist at a 150-bed hospital at a château twenty-five miles from the Front. By November

> she set up a 35-bed hospital at a villa in Dunkirk to serve as an emergency stopping-place for seriously wounded soldiers. In co-operation with an effort begun by the Hospital Auxiliary of the French Red Cross, [Lady Sykes] brought 5 doctors and 25 nurses, as well as orderlies and drivers from the East Riding to her hospital at Villa Belle Plage.

In her appeals for support at home, Lady Sykes "explained that she and her husband were equipping the hospital with a full X-ray department and two motor ambulances as well as paying half of the running expenses estimated at £900 per month [equivalent to over £94,500 in 2018]." Meanwhile, she ran the hospital with her sister Eva Gorst until the summer of 1915, when regular army medical personnel were able to take over the emergency treatment of the wounded.[22]

With his wife busy in France taking care of the sick and wounded, on his return from touring the front Sykes resumed training his Terriers in Newcastle upon Tyne with a renewed intensity and newfound sense of purpose.[23] "We'll be home before Christmas" was the optimistic prediction made by soldiers as they headed off to war in August 1914. As "Churchill's daughter Mary Soames wrote later: 'The general feeling throughout the country was one of elation and excitement, and a confident certainty (shared even in some well-informed quarters) that it would 'all be over by Christmas.'"[24] However, by the end of the year the high rate of casualties and shocking acts of violence against civilians made it clear that this was not to be the case.

The war proved not to be short and more of a stalemate, where success was measured not in miles, but in inches, often given up the following day. Cannon and machine gun fire quickly made cannon fodder of charging men with bayonets and officers on horseback waving their swords. The lesson of Sebastopol about modern warfare had not been learned sixty years later. In the French countryside it had become a war of attrition and battles of inches, as each side crawled towards the other through smoke, muck and mire under unceasing volleys of bullets through barbed wire, from their own trench to the enemy's and back again, as land gained one day was given back the next. New strategies had to be invented.

On 24 August less than three weeks into the war, the First Lord of the Admiralty Winston Churchill received an astonishing handwritten note from Sir Mark Sykes, MP. In it Sykes offered his services and those of his battalion to Churchill for service in the Middle East against the Turks.[25] Churchill's father Lord Randolph was a close family friend and frequent household guest of Sykes's mother at Mayfair while Mark was growing up.[26]

Although World War I had not yet broken out in the Middle East, Sykes was certain it would and was anxious to play a part when it did so. He told Churchill the Germans were doing all they could to involve Turkey in the war on the side of the Central Powers, which would have serious ramifications for the Allies – in particular, Britain in India and Russia in the Caucasus. He reminded Churchill, "I know you won't think me self-serving if I say all the knowledge I have of local tendencies and possibilities, are at your disposal."[27] Sykes suggested:

> [I]if operations are to take place in those parts I might be of some use on the spot, than anywhere else. My Batt^n is practically willing for foreign service, i.e., 85 per cent of its possible antagonists in the regions I mention, I could make it serve a turn, raise native scallywag corps, win over notables or any other oddment."[28]

Acknowledging Sykes's letter two days later, Churchill made no mention of Sykes's offer of himself and his battalion, but wrote instead, "My many thanks: but I hope we shall avoid a rupture with Turkey; tho! The situation is not good with those people. What a conflict!"[29] For a time there was no further correspondence on the matter between the two men. Although they knew each other well, belonged to some of the same clubs and would talk after sessions in the House, no doubt Churchill ignored Sykes's offer because he was probably already overwhelmed with offers and advice from his admirals, the War Office and its generals and Cabinet ministers. By comparison, Sykes was merely an amateur without general command experience. It would, however, not be their last communication on war matters.

When on 5 November news reached Sykes in Newcastle of the declaration of war on the Ottoman Empire, the confident, self-assured

"Middle East expert" was more anxious than ever to be where he believed he should be. Instead, he found himself stuck in Newcastle, doing a job he felt any other competent officer could be doing and not using his Eastern expertise. "It maddens me," later he wrote his wife, "not to be where I could be most useful, i.e., in the Med[iterranea]n .... Is it not ridiculous the haphazard way we do things!"[30]

As one who had devised war games for his family and guests to play on the grounds of the family estate at Sledmere, and who had served on the board planning the Royal Naval and Military Tournament of 1913,[31] Sykes enthusiastically rose to the challenge. His keen and inventive mind came up with "an elaborate scheme for improving the defence of trenches. He made drawings and elevations to illustrate his scheme, had them photographed and dispatched to the War Office." However, there is no record of how his ideas were received at Whitehall. Meanwhile, Sykes was sent home with a serious throat infection to recuperate at the beginning of 1915, while his battalion was readied to leave for the front. He was still unable to join them when they left in April for Ypres, and so his second-in-command, Major Mortimer, led them.[32]

While recovering from his throat infection at Sledmere, Sykes was champing at the bit to get involved in the war in the East. His notes to Churchill had received polite responses but no offers of service and his inquiries at the War Office met a similar lack of response. But, not being one to be put off, whenever he was in London on parliamentary business, he took every opportunity to meet Churchill and others who might help him with his "Eastern" aspirations.

On 27 January 1915, the day before the War Council was meeting, Sykes again wrote to Churchill. This time, in a six-page typed letter he outlined a detailed and elaborate plan, which he characterised as "a point of view a little away from the ordinary," with extensive advice and numerous suggestions on how to conduct the war in the East. Noting the hopeless "deadlock" in which the Allies currently found themselves on all fronts in Europe, Sykes observed that if the situation were not to change, "the war may continue indefinitely and as attrition and exhaustion are mutual the chances of forcing a decisive peace grows less certain."[33]

To end this stalemate, he proposed a different tack:

As it stands now we are planning a safe certain game, taking no risks . . . and in so doing are following a line of action which certainly would not have commended itself to either Saxe,[34] Frederick,[35] or Napoleon, since we are fighting the enemy on his chosen ground, or to put it another way we are fighting the enemies [sic] best troops at their best.

Coming to the point of his letter, Sykes told Churchill, "I am sure that to end this war we must sooner or later take risks (I would not say this to any one but you, because you are the only man I know who would take risks.)"[36]

The risk to which Sykes alluded was making a concerted effort to take Constantinople,[37] where the Germans have "an immense influence both East and West, [and, as such,] it is a diplomatic stronghold for the Balkans, a place whence you may make offers both financial and territorial." Furthermore, as "a military stronghold Eastward, there is hardly an officer or Pasha in Asia who has not got some hoard or interest in Constantinople." By taking the Ottoman capital, he told Churchill, "the whole fabric of Turkey would come rumbling to the ground,- the terrorist committee, the army headquarters, the gold, the ideal all would be gone, and the thorn in Russia's side would be extracted, the menace to Egypt dispelled, the Panislamic danger which is ever present, no matter what we may pretend, would be gone for ever."[38]

Sykes believed that Germany's Achilles heel was the German south. In support of this, he expressed the rather peculiar opinion that cultured and Catholic Austria was "soft, easy going, peaceful, and by religion and tradition and antagonistic to the Prussian" with its centre "in Vienna, where everyone is refined, courteous, inefficient, and soft, [it is] the very antithesis of the North German." So, if "by June [you could] be fighting towards Vienna, you would have got your knife somewhere near the monster's vitals, and perhaps might achieve the line Mulhausen [Alsace], Munich [Bavaria], Vienna [Austria], Cracow [Poland], before next winter." This made the southern approach the best way to relieve the Allies on the Western and Eastern fronts and thereby end the current impasse. To do this, he maintained, "the base for this campaign is Constantinople, [and] the road lies through Tiflis, Sofia, Belgrade, and Budapest – your second line of supplies is from the Adriatic."[39]

Taking Constantinople, Sykes believed,

> would produce an entirely new situation, and ... cannot be an impossible task. The Gallipoli peninsula is open to attack from the Western side and the Eastern batteries are also within range of the heights of the Island of Mudros, which is Greek.

Here he detailed how this could be done, again reemphasising the strategic importance of taking the Ottoman capital. "Once you get Constantinople you are in a position to bargain and diplomatise with the Balkan states." As he saw it, this was crucial, because "in return for these considerations you will purchase (1) a right of passage through Bulgaria (2) Passive co'operation of Roumania (3) The goodwill of Greece and Italy."[40]

If none of this would induce Churchill to consider his scheme, Sykes no doubt thought the next remark would. He was well aware that in executing such an operation Churchill already faced with "troop shortages and competing schemes for offensive operations" on the Western Front and in the Baltic. Therefore, he "was reluctant to support military operations in a southern theatre because they would have spread the available troops too thinly."[41] Addressing these issues in turn, Sykes assured Churchill that "in Constantinople you will find immense stores of war material, guns cartridges, rifles, and a well equipped arsenal, besides large quantities of gold in private hands." Not only that, but once in the city the Allies could use it as a staging area to prepare its army for a northward advance on Vienna, speculating that a combined military force could be assembled composed "of British, Russian and Indian elements ... of say 100,000 Indians, 50,000 British, and 600,000 Russians, bringing with the [Serbians] the force on the South Eastern front to a million men." Thus, "with this force on the [Serbian] Danube the Austrian situation becomes worse – the Russian pressure on the Carpathians more irresistible as the danger of being taken in reverse increases." For the Central Powers, an Allied

crossing of either the Danube or the Carpathians would be tantamount to the loss of Hungary and opening an immediate prospect of separate negotiations with the Hungarian government – the basis of which would be of course the right of passage.[42]

This would open a third front on Austria's southern flank.

"Every step you advanced towards Vienna would make whole masses of the German confederation more eager for peace, and if the terms of peace were more generous to the Southerners then to the Prussians, the more likelihood of splitting the ranks of the enemy," Sykes assured Churchill. "[T]he Austrians and the South Germans have no interest in sea command – care nothing for Alsace Lorraine, or Belgium, and must be tired of being [pressured] by the Prussians." Therefore, "South Germany is not vitally antagonistic to our desires per se but per accidens. So you would have the twitch on the people most likely to give way, and once they gave way the rest of the fabric must be weakened."[43]

Sykes ended his letter advising Churchill

[w]ith Russia on our side we have numbers but in modern war numbers only count if you can deploy, if you cannot deploy mobility is gone and you sink into trenches and the fatal scheme of attrition so deadly to the interests of civilisations and the future vitality of this country.

No doubt Sykes believed he had presented a strong case. However, while the First Lord of the Admiralty must have received this letter, unlike his previous correspondence with Churchill, there is no evidence in either the Sykes's archives in Hull or in the Churchill Archives in Cambridge that Churchill ever replied to it.[44]

In early February, Sykes went to London where he met Churchill and also made visits to the War Office, "where all the talk was about planning the Dardanelles expedition. [He] immediately wrote to Churchill in support of the venture." Apparently, this letter "has not survived."[45] While Churchill may have sought his advice on an attack on the Ottomans, which prompted Sykes's support of a combined naval and military attack, he apparently heeded the counsel of others. This Churchill described in detail in *The World Crisis*, in a chapter entitled "The Genesis of the Military Attack."[46] He was in general agreement with Sykes, however, as were others at the time, that the Ottoman Empire was the weakest link in the coalition of the Central Powers. Thus, an attack would take pressure off the Allies on both the Western and Eastern fronts, where their advances had either ground to a halt in a deadlock or were in danger of collapse. In retrospect, Sykes's advice and the counsel Churchill received elsewhere would prove to be based on flawed intelligence about the Turkish military and Turkish defenses. As a result, the Gallipoli campaign was one of the worst military disasters of the war and almost ended Churchill's political career.

In another letter to Churchill on the Dardanelles campaign, dated 26 February 1915, Sykes got straight to the point. Apparently encouraged by recent events, in light of his earlier letters to Churchill, he began on a positive note:

> I see by the papers that there has been liveliness in the vicinity of the Dardanelles, though what it portends I know not, but as you bore with me the last time, I venture again to write of certain things passing through my mind.[47]

Sykes asked if the British forces in the Mediterranean had "sound political advisors, such as accompany our troops in extra-Indian expeditions." If not, he named two men who were then in London and were available "whose knowledge would be invaluable to your people in the Agean [*sic*]." Returning to the theme of his previous letters about the importance of taking Constantinople, Sykes noted that, given the course the war was taking in Europe, Constantinople was becoming "more and more important ... [and] that the blow there should be hard, decisive, and without preamble." Writing as one who believed he knew the Turks from all the time he had spent there,

Sykes warned Churchill: "Morally speaking, every bombardment, which is not followed by a passage of the Dardanelles is a victory in the eyes of the mass of Turkish troops around the Marmora." Moreover, "the Turks are accustomed to thinking in terms of passive defence ... [which] make Turks think a long resistance or repulsed attack all that can be wished for." With this in mind, he further warned Churchill that "Turks always grow formidable if given time to think, they may be lulled into passivity, and rushed, owing to their natural idleness and proneness to panic, but they are dangerous if gradually put on their guard."[48]

By this time Churchill was getting advice from all sides about the Gallipoli campaign, and in the final stages of planning prior to its being launched on 15 April. Aware of this and still hoping to make an input, Sykes pressed home his points:

> I suggest there are two ways [to succeed in the Eastern campaign] (1) To take the Gallipoli peninsular [sic] and begin negotiating with Bulgaria, or (2) to play the great stroke and take Constantinople by a combined attack by sea from the North and South.

He emphasised the importance of this approach, which "should be done as near as possible in one bound." This was to secure the backing of Bulgaria, whose support he deemed essential in attacking Germany from the south. "Bulgaria will not move until she sees some reason for doing so – at present, she knows of no real allied success on land – why should she even negotiate with people who cannot touch their enemies [sic] territories."[49]

In summarising, Sykes once again emphasised his main point:

> [T]he whole panorama becomes quite different in the face of an occupied Gallipoli Peninsular [sic] and a Turkish capital at the mercy of an invader – Then there will be some inducement to move and daily I become more and more sure that the war will not end until the Balkan States are mobilized against Austria.

Finally, appealing to the politician in Churchill, Sykes wrote, "as time goes by without any results, the people of this country will grow sick and irritable, and there will be sore danger of a feeling of suspicion and friction growing up between the British and the French." If this was not persuasive enough, he pointed to Wellington's successful campaign in the Iberian Peninsula during the Napoleonic wars as a prime example of one that opened up a southern front against France and effectively drained its resources to fight elsewhere as being "no mistake." In Sykes's view, "[H]ere is another Peninsular [sic] far easier of access with far greater prizes in it." He ended his three-page typed letter with a handwritten note: "Perhaps my letter is ridiculous in view of what tomorrow's [papers] may contain, I write as one in the dark to one who is in the light, still you will forgive me."[50]

On receiving his note, Churchill responded by telegraphing Sykes "his thanks for it,"[51] and followed up the next day with a letter in which he wrote, "Many thanks for your most interesting letter. I expect we shall succeed in forcing the Dardanelles though of course we have not yet come to the crux of the enterprise." To this he added, apparently in reference to Sykes's earlier letter to which he had not replied, "I was more in sympathy with y[ou]r letter of a fortnight ago than I dared to tell you when we met!"[52]

It will probably never be known whether Sykes's advice in these letters had any influence on Churchill. However, it is possible they may have had some effect, given his apparently high opinion of Sykes's knowledge of the area, and their longstanding family connections and interactions in the Commons, as indicated by his prompt personal responses to Sykes's earlier letters and this last one. It also seems likely the last two letters may have been prompted by private conversations between the two men, and a request by Churchill to Sykes to put his thoughts in writing.

In his two-volume history of World War I, *The World Crisis 1911–1918*, written during his so-called wilderness years when he was out of public office in the early 1920s, Churchill makes no mention of Sykes, or of any discussions or correspondence between them on this subject or any other at this time.[53] However, given their relationship and in light of Sykes's subsequent influence on British strategy and policy in the Middle East, which began shortly afterwards, the letters may be the first evidence of Sykes's influence in wartime affairs, even though it was private, personal and behind the scenes.

Both letters reveal a self-confident and, at times, arrogant Sykes, typical of the British upper class of the time. So sure of himself and what he thought he knew, and so obviously caught up in the excitement of the war and his desire to play a part in it, he exhibited a biased and prejudicial view of the Ottomans. The letters also show a tendency on Sykes's part to reach fanciful and far-fetched conclusions based more on personal opinion than fact. Fortunately for him, because his correspondence was private, Sykes would not be painted with the same brush as Churchill after the tragic events of Gallipoli unfolded. While Churchill was ostracised from power and positions of influence, Sykes's encouragement of the Gallipoli campaign remained unknown and his reputation intact. Shortly after the last of these letters was written, Sykes, and his "knowledge" of the East, became highly sought-after and he would play a critical role in several key events affecting the Middle East and the aftermath of the war.

An interesting postscript to these exchanges and the events that followed, in particular Churchill's subsequent ouster from the Cabinet and his post at the Admiralty over the Gallipoli disaster, was recounted by Churchill's biographer Martin in *In Search of Churchill: A Historian's Journey*. The book details Martin's 1969 visit to the Gallipoli Peninsula and his subsequent meeting with Ismet Inönü.[54] At the time of the Dardanelles Campaign, Inönü had been Mustafa Kemal's (Ataturk's) lieutenant, and was later his successor as president (Kemal was the Turkish frontline commander at Gallipoli). As Gilbert recalled, Inönü

had two points he wanted to make about Churchill: that the naval attack [of 18 March 1915] at the Dardanelles had come to within an ace of success, and that he, Inönü had thought at the time that the Turkish naval forces would have been decimated in the Sea of Marmara, exposing Constantinople itself to the vastly superior naval forces of the Entente.

He made a further point, one

Churchill had made at the time, in vain, that in the immediate aftermath of the setback of 18 March 1915, had the Admiral tried again [instead of deciding not to do so for fear of the mines believed to be in the Straits], he would have had a good chance of success. The Admiral was John de Robeck, whom Churchill quickly nicknamed "Admiral 'de Row-back.'"[55]

CHAPTER 2

KITCHENER'S MAN AND AGENT-AT-LARGE

Sir Henry {McMahon} was England's right-hand man in the Middle East till the Arab Revolt was an established event. Sir Mark Sykes was the left hand: and if the Foreign Office had kept itself and its hands mutually informed our reputation for honesty would not have suffered as it did.

T.E. Lawrence[1]

Eventually, Sykes's comments, ideas and knowledge of Turkey and the Middle East came to the attention of Lord Kitchener, the new secretary of state for war, who had recently returned to London from Cairo, where he was high commissioner of Egypt and the Sudan. Under intense pressure from the press, politicians and public alike, Prime Minister Asquith, who had temporarily assumed the post of secretary of war in March 1914, resigned from the War Office on 5 August and offered the post to Kitchener.[2]

Field Marshal Horatio Herbert Kitchener, 1st Earl Kitchener, KG, KP, GCB, OM, GCSI, GCMG, GCIE, ADC, PC, was a much decorated senior military officer and popular hero. Other than his work as chief of staff in South Africa in the Boer War (1900–1902), he had spent almost his entire career in the east and spoke fluent Arabic.[3] His postings included Palestine (1874–1878), Cyprus (1878–1879), Anatolia (1879–1883), Egypt and Sudan (1883–1899), and India (1902–1909), before the "Hero of Omdurman" returned to Cairo as British agent and counsel general (1911–1914).[4] As a result, he was "Eastern" in outlook, and liked what he heard from Sykes.

Kitchener tentatively supported Churchill's plan in the Dardanelles, but for reasons other than those of Sykes. Constantly being pressured by demands for troops on every front, Kitchener was relieved that in the East "Britain's commitment to Russia would be shouldered by the Admiralty."[5] So, while

Churchill focused on a naval invasion of the Gallipoli peninsula to take Constantinople, Kitchener wanted to land forces on the shores of the eastern Mediterranean and take Alexandretta in northern Lebanon, so he could use it as a base. Once he was in Alexandretta, he could carry through with his own plans, as he had unfinished business in the east. In early 1914, just before Kitchener returned to England from Cairo on leave, he had been approached by Amir Abdullah, the son of Sharif Husayn of the Hijaz, for assistance in fighting against the Turks. Because Britain was not at war with the Ottoman Empire at the time, Kitchener was unable to do anything and had turned him down. However, anticipating the imminent outbreak of war with the Turks, he was reluctant to let such an opportunity pass and directed his subordinates in Cairo to continue communications with Mecca.

Meanwhile, anticipating the break-up of the Ottoman Empire after the war and with Sharif Husayn in mind, Kitchener asked Oriental Secretary Ronald Storrs and Brig. Gen. Gilbert Clayton, director of military intelligence in Cairo, to come up with "details of a plan for a separate Arab kingdom in the postwar Middle East." They suggested "a North African or Near Eastern viceroyalty that would include Egypt and the Sudan and stretch all the way from Aden to Alexandretta" to be ruled by Kitchener from Cairo. It also included a plan "to detach Syria, or part of it, and form a separate entity controlled directly or indirectly by Britain."[6]

When Kitchener put forward this idea in November 1914 it was opposed by the Foreign Office, which held that France had a special interest in Syria that included Alexandretta on the Mediterranean coast. Foreign Secretary Sir Edward Grey bluntly told Kitchener, "We cannot act as regards Syria."[7] Nevertheless, Kitchener held on to his idea of an Arab state and Arab caliphate under British auspices. Three days before the 17 March 1915 War Council meeting he repeated his position in a memorandum:

> Should the partition of Turkey take place . . . it is to our interests to see an Arab Kingdom established in Arabia under the auspices of England, bounded in the north by the valley of the Tigris and the Euphrates and containing within it the chief Mahommedan Holy Places, Mecca, Medina and Kerbala.[8]

With this, things reached an impasse and Middle East policy remained undecided.

In early February 1915, Lancelot Oliphant, a friend of Sykes's from their time together at the British Embassy in Constantinople and now in the Political Affairs department of the Foreign Office, introduced Sykes to Lt. Col. Oswald Fitzgerald, Kitchener's personal military secretary. Sykes quickly took the opportunity to share his views on Constantinople with Fitzgerald, who asked for a memorandum on the subject, which he promised to pass on to Kitchener.[9] On 2 March Sykes sent his memorandum to Fitzgerald, entitled "Considerations on the fall of Constantinople." He followed this up two days later with a letter to Fitzgerald in which he raised the possibility of his working with the War Office:

> As regards myself I want to make it quite clear that I could only leave my battalion on a direct order & not as a volunteer. My personal duty is to my regiment, but an order makes it different. Personally I think if you are sending out any troops for work in those parts that as a battalion I could be of good value and I could give it special training.[10]

In early April, recuperating from the throat infection that made him unable to join his battalion, which was being sent to France, Sykes went to London from his battalion's training headquarters in Newcastle and met Kitchener, Churchill and Oliphant to discuss the war effort. On his return to Newcastle a few days later he wrote to his wife Edith about the meeting and described the war secretary as being "very genial, & seemed in great spirits." He added, without further elaboration, "there was again some talk of sending me, but I said [it] must be a command no more no less – and there it lies."[11] Shortly afterwards, he received orders to report to the General Staff at the War Office in London. This change in Sykes's situation was to have immediate and long-lasting repercussions in the Middle East.

In early 1915, the Asquith government had agreed in principle to partition the Ottoman Empire. However, there was no consensus on what its interests were in the region. After a general discussion in a meeting on 10 March the War Council agreed that "the nation's territorial claims would be tailored to safeguard the Empire after the war against possible rivals." However, there was no agreement on what this meant or how it was to be done. Kitchener was very clear in his opinion, stating, "The security of Egypt and the strategic Suez Canal route to India, East Asia, and the newly acquired oil supplies of southern Persia were essential to Britain." He further argued that "in a postwar world, Russia, and to a lesser extent France, posed a greater threat to Britain than Germany." It was his belief that the Entente powers would not long survive Germany's defeat and, thus

> with Russia in Constantinople and France in Syria ... in the event Britain found itself at war with either or both of these powers, "our communications with India by the Suez Canal might be placed in considerable jeopardy."[12]

Kitchener's position was in direct opposition to that taken by Asquith and Grey, both of whom had agreed to Russian Foreign Minister Sergei Sazonov's request on 7 March 1915 for formal recognition of Russia's claims to specific Ottoman territory. This included "Constantinople, the Asiatic shore of the Bosphorus, the islands in the Sea of Marmora, and the islands of Imbros and Tenedos." A week later, on 14 March Asquith and Grey also bowed to French claims to "annex Syria together with the region of the Gulf of Alexandretta and Cilicia up to the Taurus range."[13] For Kitchener, agreeing to these demands in the interest of preserving the Entente had put British interests in the region at risk.

Further discussion of Britain's war aims in the Middle East continued several days later at the 19 March meeting of the War Council, at which strong differences of opinion were displayed. The Admiralty and War Office wanted to establish a strategic base at Alexandretta and incorporate Mesopotamia into the Empire, while the India Office preferred Haifa or Acre, and was undecided about Mesopotamia. On the other hand, the political department of the India Office thought Mesopotamia an ideal place for dumping the surplus Indian population, while "the military department wanted to preserve the military integrity of the Ottoman Empire so as to provide a barrier against a potential Russian threat." Given this lack of consensus and an urgent need for the matter to be resolved, on 8 April Asquith appointed an Interdepartmental Committee on Asiatic Turkey under Sir Maurice de Bunsen, assistant under secretary of state at the Foreign Office. The De Bunsen Committee, as it came to be known, was charged with deciding on the aims and objectives of British Middle East policy.[14]

Members of the Committee included representatives from the Admiralty, the War Office, the Colonial and India offices and the Board of Trade. To represent the War Office, Kitchener selected Maj. Gen. Sir Charles Callwell, War Office director of military operations and intelligence, and appointed Sir Mark Sykes as his personal representative on the Committee. As the recognised authority on Ottoman affairs, Sykes quickly came to dominate its proceedings.[15]

Sykes later described his unique position on the De Bunsen Committee as Kitchener's personal representative. Because Kitchener wanted to be kept informed daily of the details of the meetings, Sykes would report to Fitzgerald each night at York House on the various problems that had come up for discussion, and "received instructions as to the points Lord Kitchener desired should be considered". He continued, "this I did as best as I could by explaining the views which he approved of or suggested."

On the rare occasions when Sykes's opinions differed from Kitchener's, he would "argue the case out with Fitzgerald, and prepare memoranda, etc. until we could agree."[16]

One matter over which the two disagreed was Alexandretta. On 10 March Kitchener had insisted in the War Council meeting that Alexandretta must be included in any postwar division of Ottoman lands given to Britain; if Britain were to maintain its control over Egypt. Palestine and its port of Haifa, in his view, "would be of no value to us whatsoever" if Alexandretta was "in other hands."[17] Kitchener was adamant on this point in War Cabinet meetings, as he "wanted Alexandretta to be a British port with a through connection to the Euphrates." However, apparently Sykes was able to convince him otherwise and he "was ultimately reconciled to Haifa."[18] Thus, Sykes proposed Haifa rather than Alexandretta on Kitchener's behalf to the De Bunsen Committee at its meeting of 17 April 1915.[19]

Otherwise, Sykes's says of his arrangement with Kitchener in the year he worked for the War Office: "I acted, Fitzgerald spoke, he inspired." Yet, despite this, Sykes was more than Kitchener's puppet, being far too full of his own ideas and opinions.[20] However, the two agreed on most things so he stayed close to the line set by the war secretary. Like Kitchener, Sykes advocated moving the Caliphate south, out of the reach of Russian influence and French financial control. As for the latter, both "assumed that [postwar] Ottoman finances would be largely controlled by the French in view of the large French investment in the Ottoman public debt."[21] Sykes's desire for a buffer between British and French interests would later be realised in the combination of the Anglo–French Accord of May 1916, afterwards known as the Sykes–Picot agreement and the Balfour Declaration of November 1917, in both of which he was to be intimately involved.

Sykes's position on the Committee was unique. Unlike the other members, he was not a senior official but merely a personal representative of Kitchener's. At the age of 36 he was also the Committee's youngest member. Initially, his youth and lack of seniority had caused some of the members to have doubts about Sykes, but his first-hand experience and extensive knowledge of Ottoman Asia quickly gained their respect. No one else could discuss the area's terrain or its strategic importance like Sykes.[22] Throughout the course of the Committee's deliberations, he proved to be "the most active and concerned of all the participants ... submitting memoranda, refuting others, [and] providing detailed maps and interpretative material on little-known subjects, such as the Kurds and Caliphate." Furthermore, his strong personality and reputation as the government's Middle East expert led to his becoming its "most outspoken member."[23]

Thus, through Sykes, Kitchener was able to dominate the Committee and Sykes was to become the single most important man in Whitehall responsible for Middle East affairs during much of the war. Kitchener needed

> a young politician who knew the Middle East, and young Sir Mark Sykes was one of the handful of Members of Parliament who knew the area. As a Tory, he shared many of Kitchener's sentiments and prejudices. In every sense they were members of the same club.

Despite his appointment as the war minister's personal representative on the Committee, Sykes barely knew Kitchener and in the year or so he worked for him he was not to get to know Kitchener much better. Nevertheless, the other committee members assumed that he spoke with the war minister's full authority and, along with his first-hand knowledge of the region and his persuasive personality, the "relatively inexperienced M.P. controlled the interdepartmental committee."[24]

When he was not involved with De Bunsen Committee meetings or reports, Sykes was in Parliament, or at the War Office, where he worked under Maj. Gen. Sir Charles Callwell, the Director of Military Operations and Intelligence, preparing "information booklets to be used by the Mediterranean forces." It was tedious work for a man of action and one who was used to being in charge and it was thus not much to Sykes's liking.[25] At other times, he was in Newcastle preparing his battalion to go to France.

During the De Bunsen Committee's deliberations, Sykes inserted his own views and proposed a postwar partition of Ottoman Asia that would confine "Turkish sovereignty to a Turkish kingdom in Anatolia and [partition] the remainder of the empire among the various European Powers."[26] He further argued that in the partition scheme:

> [U]nder it or under a scheme of spheres "we stand square with our Allies, with instruments we can adhere to, boundaries we can see, and interests we can respect, and consequently shall be able to unite in a cooperative policy with permanent purpose and unanimity."[27]

In May, back from visiting his troops in France, the ever-creative Sykes presented another scheme for the Committee to consider: devolution. He explained it in another memorandum and in several meetings that followed. Devolution entailed dividing "the Ottoman Empire into five historical and ethnological provinces: Anatolia, Armenia, Syria, Palestine and Iraq-Jazirah.[28] Sykes argued that through decentralisation these provinces would be freed "from the vampire-hold of the metropolis" so that they would have the "chance to foster and develop their own resources." In his view, "devolution would counter 'the evils of Turkish rule' from Constantinople, which had 'enabled a small party of individuals to engross the whole power of the Empire in their hands.'" Moreover, "if the five proposed Ottoman provinces employed foreign advisors without reference to Constantinople, Great Britain would gain influence in all the Asian provinces without interference from the Turks, the Germans, the Russians, or the French."[29]

For three months the Committee studied every aspect of proposed postwar divisions of the Ottoman Empire and the ramifications of each. Everything was done with Russian and French territorial claims in mind to avoid upsetting the Entente. Although Maurice Hankey, secretary of the Committee of Imperial Defence and secretary of the War Council of the Cabinet, controlled the agenda of the meetings, "it was Sykes who outlined the alternatives that were available to Britain." He reviewed each proposed territorial settlement for the Committee, outlining the pros and cons of each. These included the

> annexation of the Ottoman territories by the Allied Powers; dividing the territories into spheres of influence instead of annexing them outright; leaving the Ottoman Empire in place, but rendering its government submissive; or decentralizing the administration of the empire into semi-autonomous units.[30]

In its final report, along with the proposed zones of interest and Ottoman independence, Sykes's plan for partition was dismissed by the Committee as the "least likely to reconcile British interests with the continued existence of Turkey in some form, while still deterring aggressive designs of the present allies." The consensus was in favour of devolution, a "scheme based upon a decentralization of authority." This decision was reached despite the fact that partition was the most popular plan among the Committee members, who had spent most of their time considering it, and even listing six advantages in its favour.[31] The sticking point with Sykes's partition scheme had been the proposed construction of a 1,000-mile railway to connect Haifa with the Euphrates in Mesopotamia. In response, Sykes prepared a memorandum, "The Question of a Railway connecting Haifa with the Euphrates," in which "he played up the economic angle," rather than its strategic advantages. He showed

> the part the railway would have in the development of southern Mesopotamia — "It is a railway that, sooner or later, would be built for

business purposes." Such a rationale appealed to the India Office planners and to the Admiralty's preoccupation with Persian Gulf oil.

His concluding remarks, in which he emphasised the importance of Haifa as a Mediterranean base for Britain, rather than Alexandretta, also appealed to Whitehall. It would avoid problems in dealing with the French *concessionaires* in the Levant.[32]

Past experience of dealing with the French in Egypt and the Russians in Persia had developed in Whitehall officials an "official distaste for any but British administration ... [and] made [them] only too happy to avoid any outside interference."[33]

In late May 1915 the Committee reviewed all four proposals, placing them "in order of difficulty of attainment from the 'largest, most difficult' − partition and spheres of interest − to the easiest − devolution ... and chose the latter," because "it required no additional British efforts. Moreover," they decided, "if devolution failed, then they could always turn to the more ambitious schemes." The government never officially approved the Report, however "many of its strategic, political and religious assumptions subsequently guided Sykes and other makers of war policy." Later, "the maps Sykes had prepared by the War Office for the De Bunsen Committee were to be used again and again during the war." Despite its unpopularity with the Committee, the idea of a rail link between Palestine and Mesopotamia persisted. So did "the assumption that Great Britain was better off involving itself with the Arabs and the religious questions in Palestine, than letting the old Ottoman Empire be taken over by other Great Powers."[34]

Although it was not the proposal advocated by the De Bunsen Committee, partition would end up being the British government's objective the following year. At that time, Sykes would present partition − without the rail link − as Britain's goal in his secret negotiations with France's François Georges-Picot, in what would come to be known as the Sykes−Picot agreement of 1916.

By the time the De Bunsen Committee published its final report on 30 June 1915, Sykes was out of the country. As he wrote his wife on 6 April, Kitchener had "decided that I ought to go right round the Middle East and report on the various situations."[35] So, in early June, after the Committee's deliberations had been completed and its report written, he left London on a special mission for Kitchener. Officially, he was "to check reactions of British officers in the East to the De Bunsen Committee report and to make observations on the military and political situations in the Near East and India." In doing so, Sykes was given access to "the highest British circles in the Balkans, [the] Eastern Mediterranean, Egypt, Aden, India, the Persian

Gulf and Mesopotamia, [and] . . . was to report to General Callwell on what he heard and saw." Sgt. Walter Wilson, his private secretary from Sledmere and the Territorials, accompanied Sykes and in Marseilles the two men were joined by Jerusalem-born Antoine Albina, who acted as their interpreter. From there, the three men headed east.[36]

Their trip was to take them to Athens, Gallipoli, Crete, Sofia, Cairo, Aden, Karachi, Simla and Basra. At each place Sykes made extensive observations and commentaries. He interviewed royalty on three continents, together with government officials, soldiers of all ranks, religious officials, prisoners of war, local notables, journalists and others to achieve what he hoped would be as wide-ranging a view of the war in the East as he could possibly get.

In Athens, Sykes met exiled Ottoman Prince Sabah-ed-Din, a nephew of the deposed sultan Abdul Hamid II, whose mother was the sultan's sister. An early member of the Young Turk movement, the prince split with the Young Turks at the 1902 Paris Congress of the Committee of Union and Progress and, along with a group of Ottoman liberals, founded the League of Decentralization and Private Initiative in 1905. After the 1908 Young Turk Revolution, his organisation opposed the Young Turks and supported the Sultan's unsuccessful counter-revolution. As a result, it was banned and the prince went into exile along with several others.[37]

In his 12 June 1915 report, Sykes told Maj. Gen. Callwell of his meeting with the exiled prince and listed Sabah-ed-Din's terms and conditions for supporting Allied efforts to overthrow the Young Turk's Committee of Union and Progress (CUP) government:

> [T]he [ceding] of certain small territories to the Allies ... the neutralization of the Bosphorus and the Dardanelles and the dismantling of the forts ... the retention of Constantinople by Turkey ... the adoption of a wide scheme of reforms, including decentralization and the establishment of local autonomy in all parts of the Empire, under a scheme of regions compatible with national aspirations [and] the transfer of the Caliphate by the Sultan to a member of the Koraysh.[38]

Sabah-ed-Din also told Sykes his party wanted a separation between religious and state affairs. This "would enable the Ottoman Government to legislate on a basis of religious equality in Turkey more easily when the Sultan was no longer a religious emblem." One thing the prince was most concerned about was Russia gaining possession of Constantinople.[39]

Sykes made several more stops on his journey and held interviews at each; including one with Sir Valentine Chirol, the former director of

The Times foreign department, who was in Crete at the time. Having retired from *The Times* in December 1911, Chirol was knighted in 1912 for his distinguished service as a foreign affairs advisor and rejoined the Foreign Office (he had served previously from 1872–1876) as a diplomat as soon as World War I broke out. He was sent to the Balkans. Travelling through Greece, Macedonia, Bulgaria, Serbia and Romania, he met foreign officials and heads of state to help convince them to join the Allied side. He was sent to Cyprus in the early stages of the war by the Foreign Office "to coordinate the information of officials already on the spot, as well as to give the hesitant countries [to join the war] 'a more authoritative account of [Britain's] position and ... intentions.'"[40]

In his interview with Sykes, Chirol severely criticised the De Bunsen report. While generally agreeing with the Committee's decision to aim for devolution, he posed questions on matters he believed had not been addressed by the Committee; in particular, the continuing construction by Germany of the Berlin to Baghdad railway.[41]

Sykes reported that Chirol believed that "it has proved so powerful an instrument of political expansionism that the first condition of any peaceful or stable settlement ... must ... be the elimination of this form of German influence." Chirol was also concerned that under devolution the Ottoman sultan would retain the position of caliph, believing it should be transferred to another. Otherwise:

> that the survival of even a marginal Turkish Sultanate over so large an area, as is contemplated ... will leave the door open to a continuance of the pan-islamic propaganda initiated by Abdul Hamid and prosecuted of late by the Turkish [CUP, or Committee of Union and Progress].

To this he added, "I doubt whether the menacing character of that propaganda has yet been sufficiently appreciated."[42]

Sykes arrived in Cairo in July, where he met High Commissioner Sir Henry McMahon and General Sir John Maxwell, who was commanding the British troops in Egypt, Oriental Secretary Ronald Storrs, Brig. Gen. Gilbert Clayton, director of military intelligence and many others. He soon settled in at Shepheard's Hotel to spend most of the next six months of his mission based there.

After meeting the major British officials, Sykes interviewed prominent Cairenes to discuss the Arab desire for independence from Ottoman domination. He began with the Egyptian sultan Husayn Kamel (r. 1914–1917), who had recently become ruler of Egypt. The British had deposed his nephew Khedive Abbas II on 19 December 1914 while he was on an extended visit in Constantinople, because they feared Abbas would side with the

Ottomans. The British then put his uncle Husayn Kamel in his place and declared Egypt a British protectorate. Husayn Kamel was then given the more regal title of sultan instead of the traditional title khedive, which had been given by the Ottoman porte to Egyptian rulers since 1867. This made him "the first Egyptian ruler of his line not to be appointed by Imperial decree of the Ottoman Sultan."[43]

His new title made Husayn Kamel's position weak, according to Sykes, "because ... as Sultan [he] has no moral sanction of any kind in the eyes of Moslems." The title of sultan was one that "the Caliph alone can confer, yet he has received his office at the hands of a Christian power." Sykes sensed the new sultan "feels this very keenly." He told Sykes it was his hope that in any division of the Ottoman Arab lands, Syria

> should be included in the Government of Egypt ... [and] in the event of the Caliphate being assumed by the Sherif of Mecca ... that the Sherif would ... confirm the Sultan of Egypt in his present title.[44]

This raised the subject that, according to Sykes, was on everyone's mind and had to be addressed: the question of Syria in any postwar settlement. He reported that the prevailing belief was that both Britain and France wanted Syria for themselves. So in his report Sykes recommended caution when asking for opinions about Syria or the "solution of the Turkish question."[45]

He also interviewed a number of prominent members of the Cairo community, who represented a cross-section of the various ethnic and religious groups and were produced for him by General Clayton, director of military intelligence.[46] Not surprisingly, in his meetings with the first two people he interviewed, both of whom were of Syrian origin, the inevitable question of Syria was raised.

Said Pasha Shucair,[47] a Muslim notable, told Sykes that the annexation of Syria by France was unacceptable to both Muslim and Christian Syrians. He said this was due "to the petty methods of French administration, the probability of navigation laws and tariffs in favour of French interests, the ruthless methods of exploitation practiced by French concessionaires and the influence which financiers have with the French Government." Furthermore, as a Muslim he was concerned that "the French would develop religious antagonism by supporting Christians against Moslems ... which would prove contrary to the general good." As for the devolution scheme Sykes described to him, Shucair believed it would work only "if the various regions were under some form of European control, the Imperial army and fleet completely abolished, and the imperial taxes subjected to international control."[48]

Dr Faris Nimr, a wealthy Syrian Christian originally from Beirut, was the founder and editor of the popular Cairo newspaper *Al Mokattam*.[49] Although he was pro-British like Said Pasha Shucair and both held similar views, he differed in his opinions of the future of the Arab Ottoman territories. "He strongly objected to the division of Syria and Palestine, either as annexed territory or under devolution." However, speaking from a sectarian perspective, Nimr

> thought if Damascus could be included in Palestine under British protection, and the Lebanon retained in Syria, there would be a considerable advantage to the Syrian Christians by the elimination of a large body of Moslems and a consequent increase in the size of the Christian minority in Syria.

As for devolution, he maintained that "the retention of even the shadow of Ottoman power under any devolutionary scheme was dangerous to Syrian interests owing to the ingrained subservience of Arabs to Turks in moments of crisis." Sykes noted in his report that "[n]either Dr. Nimar nor Said Pasha Shucair had the slightest hope of an independent Syria holding together for a day."[50]

Late July found Sykes in Aden. At the tip of the Arabian Peninsula in southern Yemen near the entrance to the Red Sea, the port of Aden was an important outpost of the British Empire. It had been in British hands since 1839, when the British East India Company landed Royal Marines there to occupy the area to stop pirate attacks on British shipping to India. Eventually, the British Indian government extended their control over Aden and its immediate surroundings (an area of about seventy-five square miles). Soon it became an important transit port and coaling station for trade between Europe, British India and the Far East. In 1869, the opening of the Suez Canal greatly increased its commercial and strategic importance, and since then the port of Aden had been one of the busiest trading ports in the world.

In a lengthy report to Maj. Gen. Callwell, Sykes described the military situation in Yemen and what he learned from interviews with recently captured Ottoman Arab prisoners and deserters. His interview with Maj. Rauf Bey from Baghdad of the Arab Ottoman Army proved quite revealing. Bey told Sykes, "if Constantinople fell, the people of Baghdad would probably declare themselves independent, under a Caliph who would probably be a member of the Koraish."[51] Rauf Bey told Sykes he doubted "the Arabs should be able to rule Irak without European assistance particularly in the domain of finance." Otherwise, he believed that anarchy would become widespread "before a settled government could be established from within or without."

And, although some organisation "allied with the Pan-Arab movement is at work among the Arab officers in the Ottoman Army ... it was incoherent in plan and undecided in policy."[52]

Sykes also interviewed Lt. Col. Harold F. Jacob, the British acting resident in Aden, who had been stationed in Yemen in one capacity or other almost continuously since 1904.[53] Jacob held quite the opposite view from British officials in Cairo on the caliphate. He believed that transferring the caliphate from the Ottoman dynasty to the Sharifian family would not be a good idea. Such a move, in his view:

> [W]ould result in a general turmoil in Arabia, which might end in a religious revival such as produced the Wahhabi movement resulting in the Caliphate falling finally into the hands of those who would imbue it with a renewed vitality.

Given the proximity of Arabia to India, this "might prove the focus of unrest and intrigue [adding] to the danger." Furthermore, he believed that a "moribund Caliphate in an atrophied Turkey ... would have fewer potentialities of danger than a Caliphate situated in Arabia where the vital spark of Islam survives."[54] Sykes made no comment on Jacob's contrarian views, which were to prove startlingly prophetic.

On his return to Cairo, Sykes interviewed prominent Egyptian politician Sa'ad Zaghloul. A distinguished lawyer of independent means, Zaghloul was a member of the Egyptian upper class and a politician. He was married to the daughter of former Egyptian Prime Minister Mustafa Fahmi Pasha and, although he was active in nationalist politics, he was a moderate and generally acceptable to the British. Appointed a judge, minister of education (1906–1908) and minister of justice (1910–1912), in 1913 Zaghloul became vice president of the Legislative Assembly.

Zaghloul told Sykes, in what was to soon become his major *cause célèbre* as leader of the Egyptian nationalists after the war, the "declaration of a Protectorate was a disheartening blow, inasmuch as it put an end to the theory that the occupation [begun in 1882] was not a permanent institution." However, in words he would later disavow, he added that "the idea of an absolutely independent Egypt was not one which could be entertained, and that for purposes of defence, finance, and foreign relations Egypt must always depend on some other Power." Instead, he suggested that public opinion towards the sultan would be greatly improved if he gave Parliament "local autonomy in matters which concerned native affairs alone." Otherwise, Sykes noted that Zaghloul seemed to regard the Caisse de la Dette,[55] Suez Canal, mixed courts [separate European and Egyptian courts], capitulations [exempting Europeans from local prosecution, taxation, conscription, and search of their households] Anglo-Sudanese administration, the army, and foreign relations as beyond the [Egyptian] Chamber's powers as far as legislation or executive control were concerned.

To this, Sykes added prophetically, "It must be admitted that the existence of a Chamber with purely consultative powers must always give rise to hopes and ambitions that it will eventually become something more."[56]

By this time, Sykes's interviews in the area had brought him to a number of conclusions. Two were uppermost in his mind. He advised Maj. Gen. Callwell that in order to implement "a sound Anglo-Arabian policy," political and military control over Aden should be transferred to Egypt. This would remove it from the currently existing cumbersome dual-control system under Bombay and the Government of India. To this he added that anything to do with the Arabs should be done through Cairo, not India.[57] In another letter to Callwell a few days later commenting on centralising control, Sykes pointed out that the Turkish forces opposing both Aden and Egypt operated under a single centre of command in Damascus.[58] This was not to be his last remark on the subject.

CHAPTER 3

ISLAM, INDIA, IRAQ AND THE ARAB BUREAU

Although only "a few with inner knowledge" could divulge its "decisive" role, in general {the Arab Bureau} had served as "no mere collecting ... agency for general intelligence" but had "advised with authority upon the highest and most delicate questions affecting British policy and diplomacy".[1]

Sykes interviewed others in Cairo, including those who were not pro-British. The most prominent of these was the controversial Syrian pan-Islamic journalist and reformer Rashid Rida. Together with other Arab intellectuals, in 1912 Rida founded an organisation in Cairo named the Ottoman Decentralization Party (*Hisb al-Lamarkaziya al-Idariya al-Uthmani*), which sought to "impress upon the rulers of Turkey the need for decentralizing the administration of the empire ... [and] to mobilise Arab opinion in support of decentralization."[2] He was also the cofounder and publisher of *Al-Manar* with his mentor Muhammad Abduh, the late grand mufti of Egypt and founder of the modern Salafi movement in Islam. *Al-Manar* was dedicated to Islamic commentary, which "covered the range of reformist concerns – Quranic exegesis, articles on theological, legal, and educational reform, [and] *fatwas* on contemporary issues .... [and called] for a reinterpretation of Islam."[3]

Sykes described Rida as

a leader of Pan-Arab and Pan-Islamic thought ... a hard uncompromising fanatical Moslem, the mainspring of whose ideas is the desire to eliminate Christian influence and to make Islam a political power in as wide as field as possible.

To Sykes, "His mental arrogance is, I think, attributable chiefly to the idea that Great Britain is afraid of Islam, and that British policy first and foremost is planned to soothe Moslem opinion and to concilitate [sic] Moslem prejudices." Rida told Sykes that "the fall of Constantinople would mean the end of Turkish military power, and therefore, it was necessary to set up another Mohammedan state to maintain Mohammedan prestige." He also said:

> Egyptian Moslems would never be reconciled to British tutelage, and that Indian Moslem discontent would increase as time went on, [and] that when Turkey fell Islam would require the setting up of an absolutely independent Arabia including Syria and Mesopotamia, under the Sherif [of Mecca].[4]

Rida made no attempt to moderate his words, or hide his feelings. "He never pretended for a moment that if Great Britain assisted in this scheme [of overthrowing the Ottomans] there would not be the slightest diminution of discontent either in India or Egypt." He noted that "the liberties and consideration given to Moslems by Great Britain in the past did not seem to him to be causes for gratitude." If this was not bad enough, noted an apparently surprised Sykes:

> [H]e is so obsessed with the imaginary idea that Islam actually is an independent world state, and that Moslems can dictate British policy almost in the tone of conquerors, that he cannot bring himself to make the slightest concession or hold out any hope of actual friendship and loyalty on the part of Moslems.

In what Sykes viewed as a veiled threat, Rida told him

> that if Great Britain did not fall in with Moslem views she would lay herself open to the great danger of a permanent alliance between Moslem opinion and the great power of Germany which was the leading force in material and political science, which would survive the war no matter who won.[5]

Rida was equally blunt about what he wanted to see in the postwar Ottoman Arab lands. He wanted the Sherif of Mecca to

> rule over Arabia and all the country south of the line [of] Ma'arash, Diarbekir, Zakhu, Rowanduz [roughly corresponding with the Taurus Mountain range, where it divides modern Turkish Anatolia from Lebanon, Syria and Iraq], and that the Arabian chiefs should each rule in

his own district, and that Syria and Irak should be under constitutional governments.

Rida was adamant, Sykes wrote,

> resolutely [refusing] to entertain any idea of control or advisers with executive authority of any kind. He held that Arabs were more intelligent than Turks and that they could easily manage their own affairs; no argument would move him on this point: the suggestion of partition or annexation he countered by the statement that there were already German officers who had become Moslems, that more would do so, and that England would hardly dare annoy her numerous Moslem subjects in India and elsewhere.[6]

Reacting to Rida's comments, Sykes pointed out that he had "no great following but that his ideas coincide with those of a considerable number of the Arab Ulema [religious scholars]." Ever the imperialist, he added: "It will be seen that it is quite impossible to come to any understanding with people who hold such views, and it may be suggested that against such a party force is the only argument that they can understand."

Little did Sykes know at the time that Rashid Rida was to become one of the most influential and controversial scholars of his generation. He was deeply influenced by the early Salafi movement and the movement for Islamic modernism, an ultra-conservative reform branch or movement in Sunni Islam founded by his mentor, Shaykh Muhammad Abduh. Rida's ideas would later influence twentieth-century Islamist thinkers into developing a political philosophy of an "Islamic state"[7] of the modern Jihadist movement. One thing, however, was clear to Sykes from his interviews and the persistent rumor-mongering. As he told Maj. Gen. Callwell, it was "that Great Britain and France should come to some understanding as soon as possible with regard to Syria."[8]

The beginning of August found Sykes still in Cairo at Shepheard's Hotel writing a detailed report analysing what he termed as "certain intellectual forces at present developing in the Islamic world of the Near East, *i.e.*, Egypt, Arabia and the Ottoman Empire." It was written, he stated, in order to "give the outlines of a coherent general policy with regard to Islam in the Near East." He added that even though he was "unacquainted with the stage of Indian Mohammedanism . . . I think the suggestions included in this dispatch might be of interest to the India Office."[9]

As a study of the intellectual and political makeup of the Muslim populations in these regions, it was remarkable for its breadth and scope as well as its authorship. An untrained, non-academic observer, Sykes described in detail the various regional differences in populations so often overlooked in most Western appraisals of Sunni Islam, which was usually viewed as monolithic. Focusing solely on the majority Sunni Muslims, he divided them into two major groups, Ancients and Moderns, further subdividing them into classes I, II and III. These groups and classes were then applied to each area and region, noting the local and regional differences.

The Ancient group, Class I, he described simply as orthodox:

> This type of mind is soaked in Islamic learning and prejudice: is hard, unyielding, bigoted and fanatical; desires no change, and is wedded to a close observance of formulæ and nice distinctions of cleanliness of person and propriety of conduct. The advance of Europe has embittered these thinkers against Christians, till they are even more violent and sour in their sentiments against Christendom than their forebears.

Those in Class II are "meticulous and scrupulous in matters of form, belong [ing] to a type that Islam has always known since the days of Abu Bekr, the first Orthodox Caliph." He continued, "That is the type of mind which is generous, kindly, tolerant and hopeful, with a strict sense of duty which is tempered by a profound sense of justice and devotion in religion which is divorced from political ambition." Those in Class III are fewer in number than the other two Ancient classes, noted Sykes. They are "the body of educated Moslems who, while devoted to the culture and habits of the past, are moving along the path of unorthodox mysticism, which knows no immutable formulæ nor restriction in thought."[10] These were the Sufis of Islam.

He went on to explain in detail how these groups in their various stages and areas functioned and interacted, and what might be expected from them:

> In the problems which His Majesty's Government is now facing in the Levant, Egypt, Arabia and Irak, the above factors, racial and spiritual will doubtless play a considerable part, and in deciding on a policy it might be profitable to examine the way in which the various influences work.

In like manner, he examined in detail the existing situations in the Western desert (Libya), Syria, Anatolia and Constantinople, Upper Mesopotamia and Iraq ("the country lying between the triangle Mosul-Aleppo-Basra, Baghdad") and Arabia as a whole.[11]

At this point, Sykes noted that "if the above analysis is correct (and I feel some confidence that it is) our policy towards Islam may fall into a definite line." He went on to explain what British policy should be towards each group and its various classes in order to succeed in dealing with them. As far as the Ancients were concerned, Sykes believed that Class II were Britain's greatest hope. "If by wise and tactful methods we can increase its power and obtain its active support, much will be done to ensure the peace not only of our own borders but of mankind as a whole." He noted that

> The Azhar and the Government of Mecca, which at present both lean in this direction, with sympathisers among the Ulema in Syria, Anatolia, and Irak, could practically sway the whole tendency of ancient Islam[ic] thought and curb the uninformed masses of peasant, and urban, and nomadic population.[12]

While his report may not have stirred up much interest at the time in the political departments of either the War Office or India Office, it certainly added to Sykes's growing reputation as the government's Middle East expert. In retrospect, it was an insightful analysis of the political and religious situation existing at the time in the areas he visited and shows the seeds of future conflicts soon to afflict the region.

After sending his report to London, Sykes headed for India, where for some time his visit had been viewed with growing trepidation by the Government of India and Viceroy Lord Hardinge.[13] On learning of his plans to go to India in his meeting with Sykes in July, Valentine Chirol had written to Hardinge and described Sykes as being of "great ability & knowledge of the East, but he is wayward & eccentric, and rather lacks ballast & knowledge."[14] No doubt this was not encouraging news for the viceroy. In fact, in June he had written to the Secretary of State for India Austen Chamberlain protesting that he feared that "a 'grave risk of friction' and 'duplication of work'" would result from Sykes's visit. With its own government and army, India "was not ready to start taking orders from a War Office emissary."[15] Responding from the War Office, General Callwell sought to assuage the viceroy's concerns by assuring him that "Sykes would only be studying questions arising from the De Bunsen Committee and that he was merely a visitor without executive authority."[16]

Once in India, Sykes returned to his mission of surveying attitudes and opinions about the De Bunsen Report and related matters. The reserved and taciturn Viceroy Lord Hardinge, already unhappy about Sykes's visit, restricted his responses to Mesopotamia. He kept his thoughts to himself when Sykes presented "the entire Eastern picture," including "Cairo's sherifian and Arab schemes." Writing afterwards to Valentine Chirol on 20 August, Hardinge remarked that "Sykes takes himself very seriously. He knows a good

deal, but seems unduly impressed with the importance of the Syrian Arabs."
In a letter to Austen Chamberlain a week later, Hardinge

> expressed his disapproval of the devolutionary proposal for Asiatic
> Turkey on the grounds that "Sykes did not seem to be able to grasp the
> fact that there are parts of Turkey unfit for representative institutions."
> No record survives of what Sykes thought of the viceroy.

Hardinge was anxious over London's decision to support the Sherif and Sykes's
visit because of his concern over the admiration of Turkey of Indian Muslim
and non-Muslim nationalists. He was strongly against "encouraging the
Sherif's revolt because [of the possibility of its] ... dividing Islam, or stirring
up trouble for the Caliph, or exposing the Pilgrimage to hazards." He
believed "the Sherif would be regarded as a rebel both in India and
Afghanistan and that the risk of attaching blame to Britain for embroiling the
Moslem Holy Places, or the Hejaz, in the war" was one that should not be
taken. Hardinge's words would fall on deaf ears, because "London, Cairo and
Khartoum recognized the risk involved but were in favour of taking it."[17]
Such was the influence of Kitchener at the War Office.

Sykes's visit to India prompted in him an unanticipated negative response
that would be subsequently manifested in a negative attitude toward the
subcontinent, its people and its colonial government. India was not as he
remembered it from his visit as a fourteen-year-old with his father twenty
years previously. "India lived up neither to his boyhood memories nor to his
personal views of what British rule should be like." He

> had an aversion to the Hindu religion, their shrines giving him "the
> horrors – so red and greasy and mysterious." He wrote to Edith, 'the
> sacred bulls which [roam] around the streets' might amuse her more
> than they did him as they took food from stalls that people could have
> eaten.

And, "like many British in the East, Sykes preferred the Moslems because
he thought they made better soldiers. What he called 'the Hindu Moslem,' on
the other hand, was in his opinion, 'an absolute degenerate.'" Despite this,
"he could not understand why the British in India let the Moslems get away
with as much as they did in wartime," like letting them go on the Hajj.
He saw this as especially bad practice, because "thousands of Moslem fanatics
are to be given an extra dose of fanaticism. Moreover, "the resentment of the
Indian Moslems towards the British nagged at Sykes as he contrasted them
with friendly Arabs he had met." He told Edith it seemed to him "everyone

feared upsetting 'religious susceptibilities,' a phrase which is beginning to get on my nerves."[18]

He visited some upper-class Indians while in India, including a maharaja, but did not spend much time interviewing local leaders, journalists or businessmen. He did meet a number of Anglo-Indians who worked in the government offices and had little good to say about them.[19] Their "conventions struck him as a hideous cultural transplant of Victorian England. All the stuffy pretensions of the middle class had been grafted on to India." He noted that "'the Anglo-Indian of low degree is accustomed to travel with loads of servants, bedding, etc, and expects all men to don evening dress on all occasions." The way Anglo-Indians kept to themselves disgusted him. Their "hours of labour," much of which was spent doing anything but labour, "were not convenient for anyone in a hurry," as he was. So, after his time in India he was more than ready to return to the Middle East.[20] He was given a detailed historical and demographical report in a "Memorandum on Indian Moslems," prepared for him as a reference by the Indian Government's Political and Foreign Department in Simla, to take back with him.[21]

One section in the Memorandum in particular that caught Sykes's attention and prompted his concern was titled "The Intellectuals and Old School." To him, this was evidence of the "attitude of the 'Intellectuals' towards education, and the so-called revolt against the 'old school' teaching, [was] identical with the attitude of the Young Turks towards the Ulema." The difference between the two, he noted, was that "the Turkish Ulema are a learned and cultivated body of well-trained clergy with considerable prestige, whereas the Indian old school Moslems are disorganized, atrophied, and feeble so far as learning is concerned." In contrast, the Indian "intellectuals" were like the Turkish Committee of Union and Progress and threatened the established order in India. In his view, they were

> trying the same game as the Committee of Union and Progress: that is to engross all political power in the hands of a clique of journalists, pleaders, and functionaries; to oust the clerical element, but to retain its power to excite an ignorant mob to massacre or rebellion when necessary.

Sykes believed the Indian authorities and policy makers

> should face facts ... An old school Moslem may be fanatical, but his fanaticism has a logical basis, and may be tempered or assuaged, or even reasoned with; there is in his mind an element of righteousness which

can be appealed to, and an element of love of justice which makes him sooner or later amenable.

This was not so with the intellectual, he maintained:

[W]ith an imitation European training, with envy of the European surging in his heart, who is an agnostic and has no belief whatever in religion, but sees in Islam a political engine whereby immense masses of men can be moved to riot and disorder, is far more dangerous.[22]

He came away from his visit to India with negative feelings and attitudes upon which he would repeatedly expound afterwards, with far-ranging consequences.

The next stop on Sykes's tour was Iraq. In Basra he interviewed enemy prisoners, including Turks, Syrians, Iraqis, Kurds, Greeks and Armenians. He was hosted there by Lt. Arnold T. Wilson, the resident political officer. Fresh from India and not one to keep his opinions to himself, Sykes immediately made it quite clear to Wilson what he thought of India and its government. He told him that he believed Iraq was an imperial matter and should be under the control of Cairo and London, not India. If this did not upset Wilson, as an officer serving in the Indian Colonial Service in Iraq, it did not endear him to the visiting MP. Sykes's criticism of what he saw as the lack of effort being made to win over the Iraqi Arabs to support Britain did not help either. Why were "greater efforts [not been made] to win over local sheikhs, raise guerrilla bands to attack the Turkish flanks and so on?" As a relatively junior officer, Wilson "must have felt constrained to suffer this tactless onslaught from his aristocratic and distinguished official visitor."[23]

Wilson later recalled Sykes's visit to Mesopotamia in his book, *Loyalties Mesopotamia 1914–1917*. In addition to being "unimpressed with the efficiency of the British administration in India," he wrote that Sykes

was too short a time in Mesopotamia to gather more than fragmentary impressions. He had come with his mind made up and he set himself to discover the facts in favour of his preconceived notions, rather than to survey the local situation with an impartial eye.

Wilson added that Sykes had an "impetuous energy, [a] ... genius for happy, but not always accurate generalizations, and [an] ... intense interest in everything he saw." In particular, he believed London's Middle East expert was overly concerned to "do justice to Arab ambitions and satisfy France!"[24]

Wilson read Sykes well, and Kitchener's agent at large would soon be put to the test in an attempt to reconcile these two contradictory goals.

After leaving Iraq, Sykes put together a fifteen-page, three-part memorandum of his observations in India and Basra. He began by stating that there were two important military and strategic considerations in Iraq: "reinforcements of the Army, and the occupation of Baghdad." To this he added a third: the Iraqis' suspicions of British intentions in Mesopotamia worked against the British. His interviews of local notables and others in Basra made it clear to Sykes that there was "fear in their minds that we should retire or let the Turks in again." He recommended this issue should be resolved to provide them with "some certainty to their future."[25]

Under the section called "Future Policy," Sykes expanded on the points he had made earlier about the differences between India and Mesopotamia and mixing the two. Their cultures were polar opposites and to introduce "Indian methods and Indian personnel should be merely temporary and should form no part of our future scheme." He added:

[The] Indian administration has grown up in the course of years [in a foreign land] and is based on traditions and social customs which have no counterpart in Iraq. The introduction of Anglo-Indian and native Indian officials directed from India will mean inevitably that Irak will develop on Indian lines.

Therefore, he believed it would be a great mistake "to impose artificially an alien and lower grade of civilization upon a people who have a natural tendency to a higher and more progressive social state." Mesopotamia should be connected instead with Egypt and Sudan, with people related to them by race, language, religion, history and culture.[26]

Sykes continued to expand over several pages on the Indian connection with Mesopotamia, providing numerous reasons why it should not be prolonged and why control should be given to Egypt and the Foreign Office. He was particularly concerned that with an Indian administration would come its Muslims, and with them "Indian seditionism," which could unleash "from the very outset . . . powerful forces which will begin spreading unrest and disaffection at the first available opportunity" in an otherwise peaceful Iraqi population, whose "best elements look to us for good and firm government."

While stating that it was difficult to generalise about India's Muslims, he did so anyway. In a section entitled "Indian Moslems and the War," he compared what he saw as a great gulf between the Islam of the Arabs, Turks and North Africans, and Islam in India. In India, he claimed, there existed an "extreme ignorance of Indian Mohammedans as a whole of Islamic theological doctrine as taught in the schools and universities of the Al Azhar, Damascus, Constantinople, and Kairawan." This made it difficult, he believed, "to impress on Indian Moslems the absolutely hypocritical attitude of the Committee of Union and Progress, the enmity of the Turkish Clerical party for the Committee, and the sordid motives which impelled the Turks to war." Thus, in Sykes's view, to employ them in Iraq would put Britain at "a considerable disadvantage to us as rulers and to open an avenue for our enemies." This was because "the bulk of Indian Moslems sympathised with the Turks more or less, the educated because they have been affected by Young Turk propaganda, the uneducated because they have no learned theological body to keep them straight."[27]

Sykes was convinced all Sunni Muslims in India (the vast majority of Muslims), educated or not, looked to the Ottoman caliph as the leader of Islam. Therefore, for them to see the British occupiers of their country at war with the caliph might lead Indian Muslims into conflicting loyalties; and hence, pose a potential threat to their British rulers, or become a liability or even a fifth column.[28] While such broad generalisations may seem far-fetched, it probably would have been true to say those who were opposed to British rule in India might see this as one more reason to oppose their colonial masters. For most, however, including those who worked for the British Indian government and served in its armed forces, and the remaining Indian Muslims, while a war with the caliph could potentially place them in a position of divided loyalties, as Sykes suggested, it was less likely to do so. They had a vested interest in the British Raj.

It was clear that Sykes was convinced that putting Indian Muslims in Iraq was a risk not worth taking and one that could seriously affect the outcome of the war in the Middle East. Because of this, he spared no efforts in his Memorandum attempting to convince Whitehall just how disastrous such a decision could be. He did have a point, for when the policy makers in London declared war on the Ottoman Empire they had not taken into serious consideration the possible repercussions among the Muslims in the British Empire. However, when the Ottoman Empire allied itself with Germany and declared war on Britain, the British government had little choice but to reciprocate.

Sykes was not alone in his concern. It was certainly a major cause for alarm in Cairo. Anticipating the problem and with Kitchener's approval, officials there had acted quickly, seeking to replace the Ottoman caliph as leader of the world's Muslims with the most obvious candidate: the Sharif of Mecca. From their perspective, replacing the Ottoman caliph with the Sharif was Britain's weakest link in the East; they knew it, and so did the Germans.

Soon after war was declared on the Ottoman Empire, the Shaykh al-Islam, the highest Ottoman religious official, declared *jihad* (generally meaning a holy war against infidels) in Constantinople against the Entente Powers on 7 November 1914:

> [This] declared it a sacred duty on all Moslems in the world, including those living under the rule of Great Britain, France or Russia, to unite against those three enemies of Islam; to take up arms against them and their allies; and to refuse in all circumstances, even when threatened with the death penalty, to assist the governments of the Entente in their attacks on the Ottoman Empire and its German and Austro-Hungarian defenders.[29]

As Tilman Lüdke explained in his book, *Jihad Made in Germany* this was a call to all Muslims to wage war against their colonial masters, which was just what Sykes had feared. Moreover, it was instigated by the Germans. It was a tactic to distract Britain, the world's foremost "Muslim Power," as well as France and Russia, both of which had large Muslim populations fighting in the European war. If successful, Germany hoped "they would be confronted with rebellions in their colonies of a scale apt to terminate their colonial reign." As was pointed out to the German high command, "no imperial army could reasonably hope to subdue India's 60 million Muslims, who might possibly be joined by their Hindu compatriots for the sake of freedom of their country." Despite this, the proclamation of jihad caused little stir in Great Britain or Europe except among the orientalists. Understandably, it was viewed with much concern in India, North Africa and the Middle East, where past experience had shown that "in almost all resistance movements against the French or British colonization ... Islam had played an important role for rallying the local population against the *infidels.*"[30]

Certainly, "even a partially successful *jihad* might have proved a serious threat to the allies, which neither England ... nor France, ... nor Russia could afford to disregard."[31] Sykes further noted that "the general political situation arising out of the war between the Entente Powers and the Ottoman Empire is really evolved out of the efforts of Germany and Turkey to mobilize Islam against Great Britain and Russia."[32] To counter this, he listed the positive factors the Entente Powers had going for them, including discontent with CUP rule in Turkey, the dislike of the "clergy for the Young Turks," Arab dislike of the Turks and "disunion between Sunni and Shias." In order to take advantage of these, he suggested working in concert with Britain's allies:

1) to back the Arabic-speaking peoples against the Turkish Government on one consistent and logical plan ... 2) (a) to support the anti-Committee Turkish parties, and (b) the influence of the Sunni Mohammedan clergy wherever it is antagonistic to the Committee ... 3) to propagandise Islam in a definite and offensive manner, not making an apology for our acts, but attacking the enemy on the score of injustice, crime, unorthodoxy, and hypocrisy, in our own Press in the native Press, and by means of leaflets.[33]

Over the next few pages Sykes elaborated on how he believed these things might be accomplished. The remainder of his report dealt with the military situation of the Middle East, in which he discussed strategy, troop strengths, and the history of the Ottoman army and suggested maneuvers, among other things.[34]

In another report, "Policy in the Middle East III, The Arab Movement," also written during his trip, Sykes gave a detailed description of the Arabs in the Middle East in "an analysis of the various human, religious, and political factors of which it is composed." He classified the Arabs into four groups: "1) The Arabs of Arabia proper ... 2) the Arabs of Mesopotamia ... 3) the Syrians ... [and] 4) the Arabs of Northern Irak and Jazirah." Under each classification he listed religious affiliations, that is, Sunni, Shi'a, Wahhabi, Boudis (Ibadis of Oman), Christians and Jews, and gave an abbreviated history of each region and group, as well as whether they were pro- or anti-Turkish, pro- or anti-Ally or neutral in the war. Some of his statements on potential alliances between Arabs and even their potential as a so-called Arab movement, undefined but loosely outlined as anti-Turkish and possibly pro-Ally, were more wishful thinking than fact; and possibly to some extent even disingenuous.[35]

All in all, this report was more of a general demographic study of the Arabs under Ottoman suzerainty and an analysis of existing anti-Turk sentiment for support of a British-promoted Arab revolt than anything else. Despite its title, nowhere in his report did Sykes present any significant evidence that the Arabs he interviewed were actually planning to revolt against the Turks. Nevertheless, he concluded his report by saying, "On the above groundwork we have an Arab movement, which is natural, spontaneous, but unorganized." His argument for the existence of an Arab movement was weak, based primarily on race, that is, on being Arab, and a natural dislike of the Turks.[36]

In an attempt to bolster his argument and give it substance, Sykes referred in passing to an unnamed organisation without giving any numbers of its membership, saying it included "military officers drawn from all parts, the notables of Syria, the clergy, the Christians of Syria," as if it were evidence of widespread anti-Turkish support. (In fact, at that time there were two Arab secret societies to which he could have been referring. These were *Al-Fatat*, founded in Paris in 1911 for civilians, that is, the notables and the educated elite, and *Al 'Ahd*, founded in 1914 for Arab officers serving in the Ottoman army, which had only recently united in common cause against the Turks.)[37] He mentioned that these groups supported the Sharif and added they had "spread the idea among the Arabian chieftains," but noted little else. This was all the evidence he found – without giving any details – to convince him to state as fact that the "groundwork of the movement is real, but nevertheless the movement is incapable of action except with strong active support."[38] From the scant evidence he gave, if there were to be any Arab revolt at all it needed not only British material support and guidance but also motivation. Otherwise, there would probably be no revolt at all.

Sykes was unable to end his report without once again making a comment about Indian Muslims. In the "Final Note" appended to the report, he wrote:

Indian Moslems are politically and racially against the Arabs. The Arabs regard the Indians with contempt on account of their poverty, physique, and ignorance of religion. The Indians being pro-Turkish are anti-Arab. This is an immense benefit to us in the event of the Arab movement succeeding.[39]

What the War Office made of all this is unknown.

Another problem that became immediately apparent to Sykes during his trip east was the pressing need for an organisation to oversee and control all British activity in the Middle East. This was something Brig. Gen. Gilbert Clayton, director of intelligence and chief of the Arab Bureau in Cairo knew all too well and Sykes quickly learned from his six-month fact-finding mission. As he noted in a report, the areas of authority between the political and military administrations in the region were the ill-defined; "so subdivided as to create much departmental duplication, inefficiency, and internecine rivalry. The resulting confusion bred ignorance, intrigue, and practical paralysis." He described the mass of telegrams that resulted from these multiple authorities, as "a perfect babel of conflicting suggestions and views, which interweave and intertwine from man to man and place to place in an almost inexplicable tangle."[40]

In what has been referred to as a Gordian knot, eighteen people had been authorised "to advise on the content and direction of British policy in the Middle East ... Worse, these officials were scattered among numerous and often rival agencies, departments and bureaus." While most of them were in Egypt, they were also in the Sudan, the Red Sea, Mesopotamia and Aden. These included General Sir Archibald Murray, commander of the Egyptian Expeditionary Forces based in Ismailia, the Mediterranean Expeditionary Force under General Sir Ian Hamilton based in Alexandria and the British Army in Egypt under General Sir John Maxwell based in Cairo. It also included Sir Reginald Wingate, Sirdar, commander-in-chief of the Egyptian Army and commander of military operations in the Hijaz, who was based at Khartoum in the Sudan, and Vice Admiral Sir Rosslyn Wemyss, who commanded the East India Squadron and the Red Sea Patrol in the Red Sea. General Murray's intelligence network was headed by Brig. Gen. Clayton until 1916, when he was replaced by Thomas Holdich, director of military intelligence for General Murray, when Clayton was assigned to political intelligence and control of the Arab Bureau and Sudan Agency in Cairo. Military command in Mesopotamia, along with political supervision of eastern and southern Arabia was under the control of the government of India. However, in Mesopotamia Lt. Gen. Sir Percy Lake was the British military commander along the Tigris, while political matters were under Sir Percy Cox, Lake's chief political officer.[41] Lt. Col. Harold F. Jacob, the British resident in Aden, who handled southern Arabia including Asir, the Yemen, and the Hadhramaut reported directly to his superiors in Bombay. Meanwhile, in Cairo the high commissioner was responsible for Egypt and reported directly to the Foreign Office.[42]

If this was not bad enough, besides departmental conflict of departmental authority and overlaps, personality clashes and jealousies added to the confusion. "In both Mesopotamia and Egypt, Cox did not get on well with Lake, nor Clayton and Wingate with Murray. McMahon distrusted both Wingate and Murray – and they were all suspicious of Lake and the Indian viceroy, Lord Hardinge." The situation was no different in London, where

> the Admiralty, the War Office, the India Office, and the Foreign Office all sparred for influence over the conduct of affairs in the Middle East. Each department held its own views on how best to incorporate Arab policy into various military and political contexts.

In short, Britain's "Middle East policy was reactive rather than anticipatory, creating an environment in which indecision and disagreement flourished and leaving a vacuum into which officials on the spot in Cairo, Jeddah, or Basra, were forced to step." In addition, the "diffusion of authority was made worse by the vague and often curious definition of boundaries. This was especially true in Arabia, where the Indian view of the extent of political supervision differed radically from the one prevailing at Whitehall" and, as a result, there was an intense "contest between Cairo and Delhi for influence in London."[43]

Sykes constantly came up against these conflicting areas of authority and responsibility throughout his six-month fact-finding trip and became convinced of the need for a single organisation centrally situated to coordinate everything in the east that affected the war effort. This was a difficult situation at best, but it had been made worse by war. So, with the help of Clayton in Cairo, Sykes provided London with the facts and an apparent solution with his idea of an Arabian Bureau. He maintained that it was critical to establish such an organisation if Britain was to formulate a coherent policy in the Middle East that would succeed. Something had to be done to consolidate and coordinate all Middle East operations.

In earlier attempts in London to get official interest in creating such a Near Eastern Bureau, Clayton had "lobbied vigorously against centering its activities anywhere other than Cairo." In a memo to the War Office, he wrote: "I am of [the] very strong opinion that this section should be formed here, and attached to the Intelligence Department." He continued:

> [I]t is by working on the spot, where they would have the full benefit of all the detailed information which we get, and have the opportunity of seeing all the various people who gravitate to Cairo, and they could be able to get a really clear grasp of the whole situation.[44]

Apparently, the idea of an Arabian Bureau, as Sykes referred to it, or Near Eastern Bureau, as Clayton called it, had been discussed at length by the two men while Sykes was in Cairo. These discussions prompted Sykes to write Lord Robert Cecil at the Foreign Office about the idea and ask his opinion. A personal friend and fellow Unionist MP, Cecil was Parliamentary Under Secretary of State at the Foreign Office and son of the former three-time Prime Minister, the 3rd Marquis of Salisbury. "Dear Bob," Sykes wrote on 4 October:

> I think that I have got so far that the best thing I can do is to write you a letter and put you in possession of the situation as it strikes me. This letter is for yourself, George Clerk,[45] Fitzgerald, and if you think fit Austen Chamberlain."[46]

Sykes told Cecil one thing he had observed on his trip that bothered him everywhere he travelled was the lack of coordination throughout "the whole of our organization between the Balkans and Basra." This contrasted markedly with "the German scheme of things which is highly co'ordinated, though evidently well decentralized." In "Afghanistan, Persia, Mesopotamia, Southern Arabia, the Balkans and Egypt you find ... different parties [are] putting up a local offensive or defensive on almost independent lines, and quite oblivious to what the others are doing." While not blaming anyone in particular, Sykes attributed this to

> our traditional way of letting various offices run their own shows, which was alright in the past when each sector dealt with varying problems which were not related, but its bad now that each sector is dealing in reality with the common enemy.[47]

He then listed the areas he believed were most lacking in coordination, which should be in close connection with each other:

> The Dardanelles expedition is not in close touch with Egypt. Egypt runs the Red Sea down to Jeddah and only now and again hears what is toward in Yemen and Aden. Egypt is not in touch with the Ibn Saud question which is run from Mesopotamia. The Indian government regards the Mesopotamia expedition from a purely Indian point of view and not as an Imperial question. The Persian littoral of the Persian Gulf is divided between the Indian political and the Foreign Office.

Sykes believed that to succeed against the organised enemy in the East operations must be coordinated. To do this, "a new department under a secretary or Under-Secretary of State should ... be started, this would be the department of the Near East and would be responsible for policy and administration of Egypt, Arabia, and Mesopotamia." Personnel could be drawn "from the Gulf officers of the Indian Service, the Egyptian service, and the Levant consular service." Its London staff could come from the Foreign and India offices, thereby forming "a liaison between the two, and would in days to come when the war is over be of great value in linking together our ideas and peoples between the Mediterranean and the Persian Gulf." Since the area was occupied mostly by Arabs who spoke the same language, combining operations under a single department would give the government a single organisation "to deal with the Arab situation [in] both national, strategic, and economic [matters]," as well as having personnel there who were well acquainted with the area. This would allow the government to pursue a single consistent policy throughout the region.[48]

Sykes then turned his attention to "our attitude towards Islam, in this matter all that I have said comes in with double force." From having been there, he could attest to the fact that "Egypt, [the] Dardanelles, and India are all working either on different lines or at least out of touch with one another." Those reading his papers and dispatches would be familiar with the situation to which he referred, which was of great concern to Sykes. This was that in India people are quite often totally ignorant of the methods of the Committee of Union and Progress, the attitude of the Turkish Ulema, the aspirations of the Arabs [etc.], similarly Egypt is entirely in the dark as to the foundations of Indian Mohammedan trouble.

In contrast with this, Germany through its "agents [are] in close touch with the whole question, and this matter will not end with the war." Herein lay the problem, according to Sykes, and it would continue long after the war. So, something should be done to present Islam "in its true perspective."[49]

Sykes believed it was necessary to deal with this problem right away and he was the person to do it:

I believe that the best way to deal with it immediately would be to give me two advisers, . . . one from the Foreign Office, and the other from the India Office or Indian Government, a man who is in close touch with the Indian Moslem problem as a whole.

With these advisers he would "establish an Islamic Information Bureau in Egypt, which would be in direct correspondence for the receipt of information with India, Zanzibar, Athens, Teheran, Sofia, [and] Mesopotamia." However, he "would only issue information to the Director of Military Operations, War Office or some other centre for circularization, thus information would only be issued after it had been considered and approved of in London." Thus organised and situated, Sykes was convinced that

[I]t would be possible to do a very useful and necessary work. The establishment of such an office would be a short cut to co'operation, and those with whom I have discussed the matter in those parts believe that it would be of assistance, for instance Sir Percy Cox.

He then asked Cecil to get back to him as soon as possible with his opinion on the establishment of such an organisation.[50]

After reading Sykes's letter to Cecil, the Secretary of State for India Austen Chamberlain telegraphed Viceroy Lord Hardinge in India on 10 October as follows:

To combat German and Turkish propaganda proposed to establish Bureau at Cairo under general orders of DMO, but under control of Sir Mark Sykes, assisted by Philip Graves and Hennessy. Function to communicate information to departments and persons concerned in

London, India, Mesopotamia and Mediterranean, and prepare propaganda material for Indian, British and French press.

Given the Viceroy's existing hostility towards Sykes, his negative response was not surprising. Some time later, Hardinge telegraphed Chamberlain to say he was "[e]ntirely opposed to Bureau carrying out any kind of propaganda activity in India. Although Bureau would be under control of the Director of Military Operations." He also added that he doubted "whether personnel named ... possess necessary military knowledge."[51]

After learning the Viceroy's response, Sykes wrote to Clayton that the Indian government's concern was that the creation of a new bureau would mean the surrender of control over the relations between its forces in Mesopotamia and the local Arab population. Moreover, "Indian officials were also fearful of the reaction that anti-Turkish propaganda might provoke among Indian Muslims."[52] Despite Hardinge's disapproval, by the end of the year after Sykes's meeting with the War Cabinet in December, the general outline and organisation of a Bureau would be approved.[53]

CHAPTER 4

THE HUSAYN–MCMAHON CORRESPONDENCE, THE ARAB REVOLT AND ADVISING THE WAR CABINET

These amateur diplomatists are to my mind most dangerous people and Mark Sykes in particular owing to his lack of ballast. Still they are all the vogue at the present time and I am not sure that we may not see the civilians yet occupying high military and naval posts because they are amateurs.

Lord Hardinge[1]

While he was still in Cairo in November 1915 Sykes learned that the protracted negotiations between High Commissioner Sir Henry McMahon and Sharif Husayn that had begun in July 1915 had reached a critical point. At issue was Amir Abdullah's request for British support of his father, who sought independence from Constantinople, and the British desire for the Sharif to lead an Arab revolt against the Ottomans in support of the Allied war effort. Knowing Sykes was on his way back to London, officials in Cairo wanted him to learn everything about these negotiations so he could provide specific details on them to the officials in Whitehall. As part of his instruction, he was introduced to Lt. Muhammad Sharif al-Faruqi, a Syrian Ottoman army deserter who "was an influential member of [al] Ahd, the secret society of Arab nationalists."[2]

Sykes was told that al-Faruqi had been sent to Sharif Husayn to explore a possible arrangement between the British and Sharif Husayn.[3] Then, before meeting al-Faruqi, Sykes was shown "correspondence relating to the Arab movement and the Sherif" by Sir Milne Cheetham, Counselor at the Cairo Embassy. On 19 November his telegraph to Maj. Gen. Callwell described what he had learned about the proposed Arab revolt. "Two difficulties strike me," his telegram reads, "(a) Arab want of confidence in our might (b) Difficulty of making arrangements with Arabs inoffensive to French susceptibilities, based on financial interests and historic sentiment."[4]

After making suggestions on military strategy in the region, Sykes addressed the second of the two issues, which he believed was the most important and how best to handle it. First, the

> Entente to agree with the Arabs to recognize, respect, and protect Arab provisional governments of Beyrout, Aleppo, Damascus, Jerusalem, and Hedjaz ... during the war, and to guarantee above areas as minimum of independent Arab territory after the war in excess of Arabia proper.

Next, in an apparent contradiction of what he had just written on the independence of Arab governments in the same areas, he wrote:

> Great Britain, Russia, and Italy to engage not to obtain concessions in vilayets of Aleppo, Beyrout, Damascus ... without approval of French Government, and to recognize the spirit of previous agreements between French Government and Ottoman Government with regard to educational establishment in same areas in suggested Arab independent State or States.[5]

Moving from the Levant to Mesopotamia, Sykes then advised Maj. Gen. Callwell that, in his opinion, "the vilayets [administrative areas, or provinces] of Baghdad and Basra are incapable of self-government, and a new and weak state could not administer them owing to Shia and Sunni dissension." He continued:

> We might agree with Arabs to administer these provinces on their behalf, allocating certain revenues to their exchequer or exchequers, this corresponding to their demand for subsidy, and further agreeing that, in the event of the population of the vilayets of Baghdad and Basra not producing a sufficient number of administrative personnel under British supervision, [the] deficit will be made good from Arab state or states.[6]

Sykes ended his telegram with the warning that he made these suggestions because he believed "the situation is critical." Should Britain limit itself to defending the Suez Canal and let the Turks and Germans "re-establish their prestige, and so work a real Jehad with Arab support," it would result in "strong repercussion in North Africa ... Persia, [the] Caucasus, and Afghanistan" and would eventually require more Entente troops "under less favourable circumstances than at the present." Thus, Russia, France and Italy would all be directly affected: "At the present the Arabs are anti-Turkish, and the Taurus and Armenian snow hampers our enemies' movements." Thus, Sykes believed that "now is our only chance of foiling the German plan of involving all the Entente Powers in defensive operations against Islam."[7]

On 21 November Sykes sent another telegram to Maj. Gen. Callwell, in response to a letter from the latter over difficulties with France. Sykes told him that he had spoken to al-Faruqi on the matter and obtained an impressive list of concessions from him. In retrospect, it is hard to believe that Sykes and the Cairo officials believed that this man, who was supposedly a member of a secret society of Turkish army officers and who otherwise had no official capacity, could speak on behalf of the Arabs. However, this was probably because al-Faruqi told the officials in Cairo and an eager Sykes just what they wanted to hear. So they enthusiastically believed what he told them.

Al-Faruqi had been "ADC to the commander of the 12th army corps stationed at Mosul which was transferred to Syria at the outbreak of the war." Here he joined a secret society of Arab army officers, whose members "engaged in subversive activities in Syria and tried to encourage mutinies and desertions in the Ottoman 4th army, which had been assembled to launch an attack on the Suez Canal." Their activities soon came to the attention of the army's commander, Jamal Pasha, who ordered an investigation and made arrests. Some of the officers, including al-Faruqi, were imprisoned, but when no evidence was found against them they were released. Still under suspicion, they were posted away from Syria and al-Faruqi found himself in Gallipoli, where he deserted to the ANZAC forces fighting there and was sent to Cairo.[8]

In Cairo al-Faruqi was interviewed by Brig. Gen. Clayton, who was impressed by him. His claim to being "a prominent member of the Young Arab party (military) and his contention ... his family [was] one of some eminence among the Arabs"[9] was confirmed by Aziz Bey Ali El Masri, a former Ottoman officer now working with Sharif Husayn, who was also a member of the same secret society. Al-Faruqi's description of the officers' secret society and its activities were "impressive and grandiose, if somewhat hazy," according to Clayton, "with a Central Office at Damascus 'in continual communication with Headquarter Office,' branches 'in every important town of station,' a cypher, and a treasury amounting to £100,000 accumulated from members' subscriptions." Al-Faruqi told Clayton it was

> so powerful that "neither Turks [n]or Germans had dared to attempt to suppress it, though fully aware of the fact that its attitude has been, at least passively hostile, and in the cases of many of its members actively sympathetic towards the Allies, more especially Great Britain."[10]

Furthermore, according to al-Faruqi:

> [T]he members of the society had sent an officer to Mecca who, on their behalf, had paid allegiance to the Sharif. They had also taken "a solemn oath on the Koran that they will enforce their object and establish an Arab Caliphate in Arabia, Syria and Mesopotamia at all costs and under any circumstances, sacrificing for this object all their efforts and property and, if needs be, their lives."[11]

Given his credentials and his having been cleared by Clayton, Sykes believed that in dealing with al-Faruqi he was talking to the right person and would get from him all the information and assurances he needed. On 21 November he telegraphed Maj. Gen. Callwell at the War Office in London: "The following is the best I could get ... Arabs would [grant] ... France ... a monopoly of all concessionary enterprise in Syria and Palestine being defined as bounded by the Euphrates as far south as Deir Zor and from thence to Deraa [in southern Syria], and along Hejaz railway to Ma'an [in today's southern Jordan]." Second, The "Hejaz railway as far south as Amman could be sold to French concessionaires" with agreement "to employ none but Frenchmen as advisers" and, third, that "Arabs would agree to all French educational establishments having special recognition in this area."[12]

As for British interests, Sykes told Callwell:

> Arabs agree to an identical convention with Great Britain as regards the remainder of greater Arabia, viz., Irak and Jazirah, and North Mesopotamia. Further, Arabs would agree to any territory north of the greater Arabian frontier being French possessions under the French flag.

Also, "Arabs would agree to Basra town and all cultivated lands to the south being British territory." In addition, the

> Arabs would be prepared to make a treaty with the Entente Powers: — (1) Undertaking on their part to have no diplomatic relations with Turkey-Germany or Austria for a period of 15 years (2) On part of Entente Powers to guarantee to protect independence of Arabs. Further, a treaty of alliance with Entente Powers giving them freedom of movement in Greater Arabia, and use of railways in Arabian area for duration of war ... Entente troops to evacuate territories on cessation of hostilities.[13]

Al-Faruqi insisted that these terms and conditions, including a landing near Alexandretta and other nearby ports, had to be agreed to; otherwise the Sharif would not lead the revolt. It is not clear if Sykes told al-Faruqi that was impossible, but in his notes Sykes wrote that such an agreement was out of the question. He ended the telegraph to Callwell with the remark: "I am convinced of necessity of efficient action at earliest possible moment to enable Arabs to move."[14]

Al-Faruqi's proposal of an Arab revolt radically changed the direction of British policy that had been recommended by the De Bunsen Committee. Not surprisingly, it was the Committee's leading light, Sir Mark Sykes, who would be the uncritical bearer of his message. In this, "the British miscalculated badly. In fact, there was no general Arab revolt ... [and], at the time, no real support for the idea of Arab nationalism" other than among a small group of Western-educated elite. Reading between the lines of Sykes's memoranda and telegrams there appears some evidence to support this. He was determined to support the Arab revolt proposed by his friends in Cairo, encouraged and supported by Lord Kitchener in London. It never occurred to Sykes, Storrs, Clayton, anyone else in Cairo, or even Kitchener in London, that the Arab revolt to overthrow the Turks they were being encouraged to support by al-Faruqi was based on false information. They wanted to believe it, because it fitted in with their postwar plans and those of France for the region. So they needed little encouragement to do so.[15]

In his reports to Maj. Gen. Callwell, Sykes, therefore, enthusiastically supported Sir Henry McMahon's correspondence with Sharif Husayn and the promises made to the Sherifian leader. Based on

> al-Faruqi's claims, Cairo jumped to the conclusion that only [the] prompt satisfaction of the Sharif's demands would keep the Arabs from [an] alliance with the enemy. These dubious arguments were vigorously endorsed by Kitchener in London and Grey – against his better judgment – gave way to it.[16]

With Kitchener's assent, Grey cabled Cairo, "telling the high commissioner to be as vague as possible in his next letter to the sharif when discussing the north-western – Syrian – corner of the territory [Husayn] claimed." Thus, Grey gave McMahon virtually sole responsibility for reaching an agreement with Husayn.[17] However, it would be the deliberate vagueness and ambiguities in McMahon's second letter to Husayn of 24 October 1915, in which he sought to limit the Sharif's territorial claims so as not to conflict with French claims in Syria and British interests in southern Iraq, that was later to haunt Britain in its dealing with the Arabs.

Sykes told Callwell that he "felt that the Arabs would side with the Ottoman Turks 'in the event of our letting this opportunity go.'"[18] While there is no evidence Sykes had anything to do with McMahon's second letter to Sharif Husayn, it is not unreasonable to think that he may have. He was in Cairo as Kitchener's representative at the time and was always ready to inject himself into anything he believed required his "Eastern" expertise. So, it is entirely possible he may have had some input. In his three-volume biography of Kitchener published in 1920, Kitchener's personal secretary Sir George Arthur gives a possible hint of this:

> After consultation with the Sirdar [Wingate] Kitchener sent Sir Mark Sykes, who thoroughly understood his part, to the East, and there resulted in October an agreement with the Sherif formulating the promise of the previous November [Kitchener's 1914 letter]. We undertook, if the Arabs shook of Turkish supremacy, to support them with cash, comestibles [food], and cartridges, and to recognise Arab independence south of latitude 37° [roughly today's Syria and Turkey border], except in the provinces of Basra and Baghdad, where British interests require peculiar measures of administration, and any locality where England was not free to act without prejudice to France. This agreement rendered the so-called "Arab movement" practicable, and brought about the final revolt of the Arabs against the Turks.[19]

Not coincidently, it would be Sykes who later played a key role in obscuring the truth of Britain's duplicity in his dealings with Sharif Husayn to ensure its agreement with France under Sykes–Picot while also prolonging the Arab revolt.

Soon after he arrived back in London, the prime minister summoned Sykes to a meeting at Downing Street. Apparently, the detailed and extensive reports of his trip sent to the War Office in November had caught the attention of the prime minister and the War Council. On the morning of 16 December 1915, with a map and notes on what he planned to say, Sykes hurried to 10 Downing Street. He had been instructed

> to advise [the prime minister] and the cabinet on how they might resolve a row about the future of the Ottoman Empire that looked like it might tear Britain's fragile alliance with France apart. 'By extraordinary luck,' Sykes put it afterwards, 'I was allowed to make a statement to the war council.' What he said was to shape the modern Middle East.[20]

The 11:30 am meeting to which he hurried that day would bring Sykes into the inner circle of what Sir Maurice Hankey, Cabinet secretary of the War Cabinet, would later refer to in the title of his two-volume work as the "Supreme Command"[21] of Britain's war effort. In addition to Hankey, who took the meeting's minutes, those present who questioned Sykes extensively about his trip East included Prime Minister Herbert Asquith, War Secretary Lord Kitchener, Minister of Munitions David Lloyd George, and the First Lord of the Admiralty, Arthur Balfour. Hankey captioned the notes of the meeting as, "Evidence of Lieutenant-Colonel Sir Mark Sykes, Bart., M.P., on the Arab Question."[22]

Sykes was asked to brief the Committee on his trip and comment on the details and observations made in his 28 October memorandum. In doing so, he found himself at the centre of a discussion on the potential of the Arab revolt. Prompted by Asquith, Sykes briefly described for the Committee where he had been on his travels. In referring to his 28 October memorandum, he specifically addressed the Arab Question, its areas of strength and weakness geographically:

> The fire, the spiritual fire, lies in Arabia proper, the intellect and the organising power lie in Syria and Palestine, centred particularly in Beirut. I should like to mention that the intellectual movement, which is behind the Arab movement, is not revolutionary like the Young Turk, because education in Syria, unlike modern education in India and in Turkey, has been confined in Syria to the property-owning classes, and consequently you have [g]ot a lot of very poor men who have got a little education and greater ambition.

Elsewhere, he continued,

> in the Mosul district the movement is influenced by the Kurds, but east of the Tigris the Kurds are pro-Arab. [In] the region of Diarbekir [south-eastern Anatolia], and ... that north of Aleppo, the Arab movement is spoiled to a great extent by the Armenian question and by Turkish influence ... [Here] there is not so much chance of co-operation between Christian and Moslem.

As for Mesopotamia:

> The Arabs round Kerbala and to the south of Baghdad are very much cut off from the rest of the Arab movement by Shiism – by the Shia religion. They have a certain sense of race and breed, but they do not fall in with the other people.[23]

Following this introduction, the committee members questioned Sykes on a variety of issues pertaining to the Arabs, the region, and his report. In response, Sykes made two points. The first was that "nearly all [the Kurdish officers are] at one with the Arab officers" and, the second was that

> the ideal, running through nearly all the military members of this organisation, is nationalism and religious equality. All the officers I have spoken to want to bring in the Arab-speaking Christians and to give them religious equality. That is a strong feeling.

This latter comment seemed to surprise Balfour. "Equality with the Mohammedans? Yes," responded Sykes,

> the Arab army officers want to establish an Arab state in which the Christians shall be recognised as Arabs first, and not to go on religious lines. The second force is that of the Syrian Christians, like Dr. Faris Nimr and others and certain Syrian Moslem intellectuals, and a few religions leaders who ... have the same ideas as the army officers.[24]

Sykes continued, there is a

> third force [of] ... uneducated notables at places like Homs, Homa, Baghdad and Nablus, who are bigoted and who want to establish an Arab State which shall be a Mohammedan State, and a good number of the Ulema and religious leaders are on their side.

As for the Arabs of the Hijaz, Sykes described them as "the last force. Wherever there are nomadic Arabs, there is a sense of breed, and they are not fanatical, and they would fall in with the Sherif, as would also a large number of the Kurds."[25] The inference here was that they would follow the Sharif whichever side he chose.

Curiously, given the broad scope of his presentation about the Arabs of the region, Sykes made no mention of Ibn Sa'ud, with whom the British government was negotiating at the time. While there is no indication he was aware of this, given the time he spent in Mesopotamia with Sir Percy Cox, the British resident in the Gulf who was in charge of negotiations with Ibn Sa'ud, he undoubtedly knew about it. However, since this matter was being dealt with by the Indian colonial government and was in their area of responsibility, he may have felt it was not a matter for his report. So, when Kitchener and Lloyd George asked about the Wahhabis of Ibn Sa'ud, Sykes described them merely as "a Sunni sect," quickly dismissing them as "a dying force" and said nothing more about them.

Among the four groups of Arabs he did list, Sykes noted

> two common factors to the whole of the four of those schools of thought. One is that they *must* ask for theoretical independence, otherwise, if they ask for an obvious European tutelage, the Committee of Union and Progress will take the reactionary party over on their side.

As for the second, he said that "practically all the Arabs are pro-English and not anti-French, but [they are] frightened of the French":

> [T]hey have obtained a great deal of their culture from the French, but they are frightened of financial exploitation, and ... of French colonial methods: that is, bringing a lot of French, Italians, Portuguese, and other people to colonise; and the Christians are as afraid of them at the bottom of their hearts as the Moslems. They like the French ... [and] French culture, but they are frightened of French methods.[26]

When Kitchener questioned whether this attitude toward the French was found only in Syria, Sykes responded: "That feeling, Sir, is prevalent everywhere, because there has been so much propaganda by Syrians, and they know what the French do, so that feeling runs pretty well all through." He added that "the chief difficulty seems to be the French difficulty, and at the root of that ... lies in Franco-Levantine finance," and he gave examples.[27]

Sykes's mention of the French Nationalist Party's "sentimental" interest in Syria prompted the prime minister to question him on this. Sykes explained it went back to the Crusades, during which time French nobles and knights fought and lived in the region; whereas, "the financial group," had more practical interests. They worked "on the fears of the French colonial party of an Arab Khalifate, which will have a common language with the Arabs in Tunis, Algeria, and Morocco ... They are afraid, I think, of a Khalifate, or an independent State, speaking the same language as their Mohammedans."[28]

Sykes expanded further on French designs on Syria: "I think that the financiers have three objects: ... if the *Entente* wins they want to have Syria, Palestine, and North Mesopotamia." However, "if the *Entente* fights only to a draw, [their ambition] is to maintain Turkey intact, and work the 1914 concessions that were got on that loan [prewar French loan to the Ottoman government], and the Syrian railways, for all they are worth." If they lose, Sykes told the ministers that the French believed they could "square ... with the Germans later on."[29]

He told the ministers he believed that it was important to counter the designs of the financial group in France on Syria and elsewhere in the Middle East. To do this required a diplomatic approach

[which showed] great sympathy with the clerical feeling in France, and to point out that if matters are allowed to drift they will lose their real anchorage in Syria, owing to the anticipated massacre of the Syrian Christians in the same way that the Armenians were massacred.

To this Sykes added that it should also

be pointed out to the French colonial party that, if the Arabs come under the influence of the Committee of Union and Progress, they will be much more formidable to them than they will under their own Sherif, when they will be quarrelling amongst themselves, as they always do.[30]

All this was a great concern of the French clerical party, he said, as well as the colonial party. Under capitulations granted France by the Ottomans during the reigns from Francis I (r. 1515–1547) to Louis XIV (r. 1643–1715), France was granted the right "to protect French Christians and foreign clerics in the Turkish Empire," which was eventually extended "to include the right to protect all Christians regardless of nationality, including Turkish subjects." Over time, the impact of the French *mission civilisatrice*, or "civilising mission", was brought about through French Catholic missionaries and their schools, French traders and merchants, and the use of the French language in schools, trade and commerce throughout the region. "The Ottoman Empire, therefore, was 'not merely a field of economic activity for France. It is also, and above all, a territory for the radiation of her intellect and the expansion of her culture.'"[31]

Sykes then took the opportunity to raise the issue of the Indian Moslems, describing what he believed was a mutual dislike between the Indians and Arabs, much of which he attributed to CUP propaganda. As a result of this there was strong support among Indian Muslims for the Caliph and Turkey. It was his belief that the way to handle the problem was "not to mix ourselves up with religious squabbles which have to do with the Khalifate."[32]

Despite saying this, Sykes then told the startled ministers he believed that if Britain did not show its support for Sharif Husayn, the Sharif would be killed "and a Committee of Union and Progress nominee put in his place." Not only that, but the consequences of the lack of British support for the Arab movement would be very damaging on several fronts. It would give the Turks and Germans Mecca. It would also give the notables and mobs free rein to exterminate the Syrian Christians without any fear of retaliation, because there would be no British troops there to defend them. In support of this, Sykes referred to the recent Armenian massacre of April that year, where the Turks had killed up to 1.5 million Armenian Christians. However, he failed to mention that it was Turkish troops, not notables or mobs, that killed the Armenians. Moreover, the brutal slaughter of Armenians was a reaction to a planned rebellion by Armenians who were seeking independence and resistance to conscription and forced labour by the Turks. The Turks believed all this was inspired by the Dardanelles campaign, prompting fears the Armenians might become a fifth column in the Allies' war against Turkey. It was not specifically because they were Christian.[33]

These deliberate omissions and the blatant exploitation of this recent horrific event did not seem to bother Sykes. He knew his audience. While Balfour was not religious, both Asquith and Lloyd George came from strong Christian backgrounds. This was not the first time Sykes would stretch the facts to sway an audience to his way of thinking, nor would it be the last.

With this, Sykes continued his monologue uninterrupted by any further questions. He described to the ministers further possible atrocities that could happen if the CUP and its supporters continued unrestrained. "The anti-Committee elements will be destroyed among the Arabs, the intellectual Arabs will be hanged and shot, and the officers that matter will be exterminated too." If this were not bad enough, he predicted:

> The Arab machine will be captured and the ignorant notables and the fanatical people will then be left alone, and they will become subservient to the Committee of Union and Progress, and the Germans will then oblige the Turks to combine terrorism with concessions to the ideal of Arab nationality.[34]

As this had already happened under Jamal Pasha in Syria, doubtless his point was well taken.

Sykes then raised the issue of the growing German influence in the region in support of the dark picture he was painting. He mentioned that

> Mr. Koch, of the German Consulate at Aleppo, is beginning to talk of Turco-Arabia, and I understand that Baron Oppenheim is on the same path. I think that we shall live to see Islam pretty solid; then we shall be confronted with the danger of a real Jehad.

He added that there was a real possibility that "Mesopotamia may become a scene of major operations if we intend to hold there, and a stream of people going uninterruptedly to Persia, to Afghanistan, and considerable unrest in India and the Soudan." Sykes took this opportunity to remind the ministers that if Sharif Husayn was not supported, "next year, if the war is in progress, the Indian pilgrims will go to a Committee Sherif, and not to a Sherif [Husayn] who is known to be well disposed to us. That is, the pilgrims will be going to Mecca."[35]

Mention of Oppenheim's name with German plans for the Middle East and the possible ramifications of their success no doubt also got the attention of his audience. Baron Max von Oppenheim was well known in Cairo, where he had lived from 1896 to 1914. He was an attaché at the German Agency from 1896 to 1909, as well as an archeologist, and he had gained fame and a reputation for his excavation of Tell-Halaf. According to Oriental Secretary Ronald Storrs, from 1905 on Oppenheim was the German Agency's Oriental Secretary and "known to us all as 'the Kaiser's Spy.'" He curried favour with Ottoman officials and nationalists alike, and lost

> no opportunity of reminding the Extremist Press ... that Islam was threatened with extinction by Europe, that England and France were at the head of the anti-Islamic movement, that the Sultan was the last hope of the Faithful and that Germany was the friend of the Sultan and therefore the only Moslem-minded European Power.

At the time Oppenheim was not taken very seriously.[36]

On his return to Berlin at the beginning of the war, the German Foreign Office asked Oppenheim to come up with an idea to defeat the Allies in the Middle East. He submitted his memorandum, *Denkschrift betreffend die Revolutionierung der islamischen Gebiete unserer Fiende*, "How to revolutionise the Islamic territories of our enemies," to the Foreign Office in October 1914. The memorandum proposed that the Sultan "proclaim a 'qualified' jihad immediately; qualified in the sense that it was to be fought [only] against Britain, France and Russia, not against all *kafirs* (infidels)," that is, not against Germany and the other Central Powers. Additionally, it "was to be accompanied by propaganda to be carried out from a central institution in Istanbul modeled on the propaganda institution Oppenheim proposed to set up in Berlin, the 'Nachrichtenstelle für den Orient' (Intelligence Office for the East, hereafter IOfE)." To supplement its work in the field, "some 'tactful and qualified' Germans, who were to keep up the pretense to be only friendly advisers to the Turks, should act as supervisors of this Ottoman propaganda institution." Oppenheim headed the IOfE, which he had founded with his own money in September 1914, which was to develop the necessary propaganda in Berlin.[37] Following von Oppenheim's recommendation, jihad "was proclaimed by *Shaykhülislam* Ürgüplü Hayri Bey on November 14."[38]

While little came of Germany's efforts to stir up Britain's Muslim population in the end, at the time it was a matter of grave concern at Whitehall and the War Cabinet. "The British had faced and crushed local uprisings inspired by Islam in India and the Sudan in the years before the war. They took the sultan's threat, which was of a different magnitude, extremely seriously."[39] In fact, because of Germany's effort to influence the pan-Islamic movement "large numbers of British troops [were diverted] to the Middle East that could not be used in the major war theatres in Europe." Therefore, it can be said Oppenheim "drew considerably more attention from Britain than perhaps he warranted ... and that consequently he even contributed, although in a limited way, to influencing British policy." Certainly, mention of his name by Sykes helped get his message across about the need for Britain to take action in the Middle East.

Sykes then turned the ministers' attention from the German threat in the Middle East to several other areas he believed needed more immediate attention. As far as Syria was concerned, he told them, we "ought to settle with France as soon as possible, and get a definite understanding about Syria." It was also necessary "to organize a powerful army in Egypt which is capable of taking the offensive; and, thirdly, to co-ordinate our Eastern operations." This would bring into being "one machine, and one definite problem: link up Aden, Mesopotamia – the whole of that as one definite problem for the duration of the war." With all this in place, Sykes felt it would then be "worth backing the Arabs, no matter what ground we may have lost to the north of Haifa." It would also be

> worth backing the Sherif party of the Arabs in Damascus and in Lebanon – and those who escape massacre – and also, I think, it is worth considering where we are going to be on the defensive and where on the offensive.

As far as Sykes was concerned, he told the Committee, "Egypt should be the base of offensive operations, because the climate is good, and it is a good place to keep and train troops."[40]

Balfour asked Sykes, "What sort of arrangement would you like to have with the French? What would you say to them?" Sykes replied:

> I should like to retain for ourselves such country south of Haifa as was not in the Jerusalem enclave ... I think it is most important that we

should have a belt of English-controlled country between the Sherif of Mecca and the French.

The prime minister asked, "You mean the whole way from the Egyptian frontier to Haifa, except the enclave?" Sykes replied, "I think it could be argued to the French that they were not giving up very much." Balfour then asked, "What do you leave the French in Syria?" It would leave the French "from Acre up to as far as they like to go round the Gulf of Alexandretta," Sykes replied.

To Kitchener's request for further details involving possible French claims in that area, he told the war secretary: "They are only giving up what lies between Acre and the beginning of the Jerusalem enclave, which will be about 20 miles." This prompted Balfour to point out:

[W]e have always regarded this 90 or 100 miles of desert upon her eastern side as a stronghold of Egypt; now you propose still further east of that to give us a bit of inhabited and cultivated country for which we should be responsible. At first sight it looks as if that would weaken and not strengthen our position in Egypt.

At this, Kitchener interjected:

I think that what Sir Mark Sykes means is that the line will commence at the sea-coast at Haifa. These Arabs will then come under our control, whereas if we are off the line we lose control over the south.

Balfour then asked, "What do you mean to give exactly?" Sykes replied, "I should like to draw a line from the 'e' in Acre to the last 'k' in Kerkuk."[41]

At this point, Lloyd George changed the line of questioning by asking Sykes, "Do you propose that this should be the first step before you take any military action?" Sykes was brief. "I think it is essential that we should know where we are." This prompted Lloyd George to ask, "Before you begin?" Sykes responded, "I think we should begin to prepare for military action." Lloyd George pressed Sykes further. "Is it your idea that there should be a great offensive in Egypt, which will sweep up into Syria?" A suddenly modest Sykes responded, "Well, Sir, I do not like to dictate."[42]

Apparently not wishing to leave things hanging, Lloyd George continued his questioning of Sykes:

With regard to the proposition you put before us, that we should be on the defensive in Mesopotamia, but that we should be on the offensive in

Egypt, what does the offensive mean there because you said something about supporting the Arabs in Damascus and Lebanon? You could not support them without troops?

Sykes agreed troops needed to be sent. Lloyd George continued, "It means a great offensive from Egypt sweeping up through Syria?" Balfour interrupted, "Unless you land in Syria?" Sykes agreed. Lloyd George ignored Balfour and asked Sykes, "I take it you mean Egypt?" Sykes explained he meant that "Egypt would be a storehouse for the troops." Lloyd George continued his line of questioning. "You might attack from Alexandretta?" Sykes agreed, but added "or across the Sinai Peninsula, as had been done on several occasions." Lloyd George asked whether it would be a small force, to which Sykes replied, "Yes." When asked whether there had been any large forces attempting such manoeuvre, Sykes said there had been very large armies in the past. Kitchener interjected, "Ibrahim Pasha's force, for instance." Sykes agreed.[43]

Returning to the subject of the French, Kitchener again raised the question of the need for an agreement with the French: "If you cannot come to an arrangement with France, may you not be straining your relations with France very gravely if you assume you have come to an agreement with them and take action in Syria?" Answering his own question, Kitchener said: "My opinion is that before it can become a military problem we must know what the French actually demand – not what they demand, but what they insist on having. They demand the whole of Syria." After stressing this point, the war secretary continued:

> Their demands are very much indeed as Sir Mark Sykes has told us, but how much will they give way on that? If they give nothing, all these operations will be taking place and be a source of the gravest anxiety to us.

To this, the prime minister said, "That will not do." To which, Lloyd George added, "We are not quite as simple as that."[44]

At this point, Asquith took over the meeting. "We must have a political deal. We must come to terms with the French, which means we must come to terms diplomatically." Sykes agreed. The prime minister then posed a question. "I wonder," he asked, "at the present moment if they are inclined to allow us to get good terms out of them?" Sykes said he thought so and suggested two approaches to use. The first was to convince the French colonial party of the serious implications of just "what a Committee of Union and Progress Sherif means, and point out what they have done in India and what they might do elsewhere." As for the clerical party, he believed that pointing out the danger this would mean to a French Syria could convince them. Then, lacking the support of these two parties, the financial group would not be in a strong enough position to refuse an agreement that would protect their interests. Sykes told the ministers he had discussed this with French officials in Cairo and they both saw his point and changed their opinion "that France should have everything ... [admitting] that giving up to Acre is giving up very little – only that small strip." To this he added: "it is not taking anything away from them if we sacrifice even Hauran to them."[45]

After hearing Sykes's evidence, the War Cabinet

> agreed that Lord Crewe, who was acting temporarily for Sir Edward Grey at the Foreign Office, should contact Lord Bertie,[46] the British ambassador in Paris, as to the advisability of Sykes making an official visit to test the reaction of the French.

Bertie quickly responded opposing making any such overtures at the time. He told Crewe that

> even if Sykes pressed his views as his own, the French would regard them as official. To talk to anyone in the French Government, Bertie pointed out, Sykes would have to be "armed with introductions," which would naturally lead Paris to assume that he was not a Don Quixote.[47]

Thwarted for the moment from seeking an agreement with the French but not giving up, Sykes turned to concentrate on his Arab plans. Meanwhile, his reports, with their observations and suggestions, would bring about major changes and a shift in strategy with an increased focus on the war in the Middle East at Whitehall. Full of his success after advising the War Cabinet, Sykes was certain "once he had gained cabinet approval for launching an Egyptian military offensive and a propaganda campaign," he would be sent to Cairo "to run the new Arab show."[48] So sure was he that he had already begun working with Brig. Gen. Clayton in Cairo to find people to work under him in at a "Near Eastern Bureau" in Cairo.

In a letter dated 13 December 1915, Clayton wrote Sykes of his efforts to stir up official interest in the organisation, as well as find staff to fill its positions with the help of D.G. Hogarth, from the Naval Intelligence Division:

> I have already started the nucleus of a Near Eastern office, but so far I am confining myself to making it deal with political suspects of all kinds, and pan-Islamic propaganda ... I am only waiting to hear from you, and

to get another man or two, to expand it and to take on the study of the higher political questions and also the initiation of propaganda on our own account.

Clayton ended with the remark: "I hope to see you here again before very long."[49]

After his meeting with the War Cabinet, Sykes outlined his ideas for an Arabian Bureau in a "Memorandum on the Constitution and Function of the Arabian Bureau," which he wrote over Christmas and Boxing Day and presented on 28 December.[50] The Bureau's purpose, he wrote, was "to harmonize British political activity in the Near East" and keep the Foreign Office, India Office, Admiralty, War Office and Government of India "simultaneously informed of the general tendencies of German and Turkish policy." The Bureau's second function "would be to co-ordinate pro-British and pro-Allied propaganda relating to the war in the East not only for Arabs, but for India, England and neutral countries." And, as he had already told Cecil, he offered to head the Bureau, assisted by staff in Cairo and London.[51]

Besides Cecil, Clayton and Sir Percy Cox, Sykes had already approached a number of influential people about his idea of an Arab Bureau. However, by 28 December he wrote to Clayton about the mixed reception his proposal had been received at Whitehall. While his friends at the War Office, Foreign and India offices were generally supportive, the same could not be said for the Admiralty. There, Sykes told Clayton, the prevailing opinion was that Cairo needed intelligence rather than propaganda for the Arabs.[52] Since 1914, the Admiralty's Red Sea Patrol "had been involved in intelligence activities along the Hijaz coast," and its patrols "were also assigned to prevent Turkish mine-laying and enemy infiltration into Egypt and Sudan." So, it was understandable that they would have a strong opinion on the matter and when the idea of an Arab Bureau became "a high priority item in London and be championed at the highest levels," the Admiralty would seek to control it. Moreover, "the political influence that the new bureau might command was not lost on anyone."[53]

Anticipating this, in the early autumn of 1915 Clayton had begun a correspondence with the Admiralty on the subject of an Arab Bureau and by late December

Naval Intelligence even dispatched a set of its own guidelines to him explaining primary functions for the proposed bureau. This scheme was ambitious, envisioning an agency that would closely coordinate Cairo's propaganda and intelligence activities with those of the government, refine and homogenize British political activity in the Near East, and harmonize Allied propaganda with India's special concerns.[54]

The Admiralty organisation, largely based on Sykes's plan, was divided into three departments, each headed by a secretary reporting to the director. "The first secretary would handle the press and propaganda requirements of London, the Persian Gulf, Mesopotamia, and the Arab world." The second secretary was responsible for advising "the director and the first secretary on activities most likely to conflict with the government of India in addition to supplying both the Anglo-Indian and vernacular presses in India." It was the responsibility of the third secretary to organise and collect "native agents in Egypt, Syria, and the Hijaz for purposes of intelligence and propaganda." Mesopotamia had no department or secretary, but a roving bureau officer "would tour Mesopotamia, India, and the Persian Gulf, reporting back to Cairo via the established British authorities in those regions." Not one to think small, Sykes also envisioned liaison offices for the bureau "in London (to the Committee on Imperial Defence), Basra (to Cox, the chief political officer), and New York (to remain in touch with the Arabs in America and disseminate useful propaganda there). Under this arrangement, Allied attachés were to keep abreast of the type of propaganda contemplated for distribution, in case deletions or revisions should prove necessary)."[55] While Sykes personally admired Capt. Reginald "Blinker" Hall, in charge of Naval Intelligence at the Admiralty, he "viewed the Admiralty as a staid, discredited department owing to its recent debacle in the Dardanelles, whose voice would carry little weight in political questions."[56]

However, Clayton had his own plans for an agency based in Cairo. As noted in his 13 December letter to Sykes, he had already taken steps to gather around him candidates from various departments with considerable background and expertise in the Middle East:

> Desiring a special status for this new bureau, he took preemptive steps to save the unit from the grasping talons of other covetous departments, whether in London, or in Cairo itself. He envisioned considerably more freedom of action for the proposed unit than Sykes had in mind.

Clayton admitted to Capt. Hall in a note that his ideas for the bureau were "somewhat different from those of Sykes in that what I want to start is a Bureau here which will be a center to which all information on the various questions connected with the Near East will gravitate."[57]

On 6 January 1916, Sykes's friend Sir Maurice Hankey, secretary of the Committee of Imperial Defence, arranged for a conference to meet and discuss the matter at the Committee of Imperial Defence Secretariat. Among those attending the meeting was Lt. Gen. Sir George Macdonogh, Director of military intelligence from the War Office, who chaired the meeting, Lancelot Oliphant from the Office of Political Affairs at the Foreign Office, Arthur Hirtzel, secretary of the Political Office of the India Office, and Capt. Sir Reginald "Blinker" Hall, representing the Admiralty.

All were friends or supporters of Sykes's idea of creating a single entity to coordinate British operations in the Middle East, and had been given full authorisation by their departments to come to an agreement to establish the Arab Bureau. The meeting ended in a series of interdepartmental compromises, in which the Arab Bureau was placed under the Foreign Office and funded by the War Office, with representation of the Government of India on its staff. In order to mollify the concerns of the Admiralty, Capt. Hall named his own candidate to head the Arab Bureau – D.G. Hogarth, then working in Naval Intelligence in Cairo.[58] With the Arab Bureau an established fact and headquartered in Cairo as he wanted, Sykes was not disappointed at losing control of his creation, because by this time "he was already involved in secret diplomacy with the French – negotiations leading up to the Inter-Allied Agreement of 1916, which would be known as the Sykes–Picot agreement."[59]

Soon, under Brig. Gen. Clayton's wise and skillful direction the Sykesian creation of the Arab Bureau was considerably modified from the original idea to become an independent and highly effective tool in the Middle East war. Despite frequent opposition to it from the highest levels, under the patronage of Lord Kitchener, the secretary of war and the Foreign Office, the Bureau's junior officers, including T.E. Lawrence and others

were able to wander freely around the Middle East with inexhaustible supplies of money, in defiance of generals and in open contempt of official policies and campaign strategies, at a time when the Allied cause and the lives of millions of men hung in the balance.[60]

CHAPTER 5

THE SYKES–PICOT AGREEMENT

Sykes–Picot drew lines in the Middle East sands that blood is washing away.[1]

Less than a month after Sykes met the War Committee on 16 December 1915 and Lord Bertie preemptively dismissed the suggestion that Sykes should approach the French government about a Middle East agreement, the prime minister decided it was time to discuss the matter with the French. French aspirations in Syria and the Middle East after the Fashoda crisis of 1898 and the sensitivity of Anglo-French relations that had resulted in the Anglo-French Entente Cordiale in 1904 were sources of constant concern to the Asquith. Meanwhile, events in French political circles had brought the French government to a position that made such a discussion timely, if not necessary, if Britain were to realise its objectives in the Middle East.

The change in Paris had been gradual and long in coming, and it was resisted along the way by a French government more distracted by the immediate concerns of a war on its doorstep than any colonial acquisitions. However, political battles went on between the Comité de l'Afrique Française and the Comité de l'Asie Française over taking advantage of the war to acquire new territories in Africa, that is, the German colonies, and in the Middle East, anticipating a postwar breakup of the Ottoman Empire. Thus, the reopening of discussions between Britain and France about the disposition of Middle East territories was not so much an about-face as a victory for the Comité de l'Asie Française, championed by the indomitable will and personality of French diplomat and staunch member of this Comité, François Georges-Picot. Having recently returned from Beirut, where he had been the French consul, his efforts at the Quai d'Orsay were given a boost by a casual remark made by Sir Mark Sykes about his mission on his way east while on a stopover in Paris. Simultaneously, news was received at the Quai d'Orsay of secret overtures being made to Sharif Husayn of the Hijaz by the British of a deal that would affect the postwar distribution of Ottoman territory in the Middle East.

In 1912 the French government had asked the "British government to confirm it had no designs on Syria. The British, who then thought Syria an empty, worthless place, were happy to oblige."[2] Later, they would come to regret it. In a letter from Paris to Lord Crewe[3] dated 21 December 1915, Lord Bertie wrote: "When we disclaimed any political aspirations for ourselves in Syria the French took it that they might have them."[4]

In early February, when he learned that plans were being considered for a landing at Alexandretta as part of the British Dardanelles campaign, Albert Defrance, the French minister in Cairo, became alarmed and notified Paris:

> Suspecting that the British were reneging on their 1912 commitment about Syria ... On 8 February 1915 the French foreign minister, Théophile Delcassé,[5] reminded Grey of their two-year-old agreement and forcibly asked him to stop his officials plotting.

Commenting soon after on the matter to Churchill, Grey told the First Lord of the Admiralty: "I think it is important to let the French have what they want. It will be fatal to cordial cooperation on the Mediterranean and perhaps everywhere if we arouse their suspicions as to anything in the region of Syria." So Kitchener's plan for a landing at Alexandretta was quickly dropped and six weeks later on 25 April a compromise landing was made at Gallipoli in the Dardanelles.

Meanwhile, in France the Comité de l'Asie Française:

> a small but thick-skinned group of imperialists ... began to put pressure on Delcassé to lay claim to Syria and Palestine. Many of Comité's supporters were diplomats working at the French foreign ministry at the Quai d'Orsay and were concerned that the French government had not announced any formal "war aims."[6]

Its claim to Syria also appealed to French Anglophobia by maintaining that "imperial expansion ... denied the rapacious British gains at French expense." Central to French claims on the region was what the Comité de l'Asie's leader, Robert de Caix,[7] argued was France's

> hereditary' right to Syria and Palestine because it was "the land of the Crusades" ... [He brushed] off the "latent discord of race and religion" that his forebears had left behind and insisted that three centuries of

sporadic bloodshed had in fact established "a very special bond of union between the Franks of France and the world of Islam."[8]

In the French Senate, Pierre-Étienne Flandin, a member of the Comité de l'Asie Français and wartime co-leader "of what became known as the 'French Syrian Party' (the pressure group within the colonialist movement)" applied further pressure on the government.[9] He

> needled the French foreign minister for the "regrettable lethargy" of his diplomats in failing to defend French interests in Syria, and called on his government to "save from death millions of fellow humans being hunted down by the Red Sultan's mercenaries."

Despite all this pressure, Delcassé was unmoved.[10]

However, in late July 1915 Delcassé received news from Cairo that caused him some concern. In a meeting with French diplomats in Cairo on 28 July before he left for India, Sir Mark Sykes revealed a British scheme "to build a railway between Basra and the Suez Canal once the war was over, and [they] wanted control of the territory over which it ran." In order to do this, since

> the most direct route was not viable, because drifting sand would clog the track, he explained that the railway would have to trace a huge arc through the stony desert further north, through Damascus, to reach the sea at Haifa.

While he professed no interest in Syria and not realising how his remarks would be taken, Sykes magnanimously "offered the French Alexandretta and the nearby port of Adana as well as a share of Palmyra" as a consolation. As he probably saw it, he was merely discussing a scheme that was in vogue in Cairo at the time but was not official. However, given his official position, it was not taken that way. He did not have authority, in any case, to offer any deals to the French. An amateur in the diplomatic world and full of his own sense of self-importance, the naive Sykes was prone to indiscretion about what he said and to whom. He did not shy away from making grandiose statements or gestures, however speculative they may have been at the time. Not surprisingly, his loose talk – labelled "a full account of 'English designs'" – was wired to the Quai d'Orsay the same day.[11]

On learning of Sykes's "plans," the Comité de l'Asie Française applied further pressure on the beleaguered foreign minister. Delcassé advised Jules Cambon, the French Ambassador in London, that Sykes ambitions could not be realised "without the risk of one day posing a problem to Anglo-French relations." On 31 August, "Cambon duly informed the British ... that his government 'would not tolerate any infringement' of its 'rights' in Syria." His warning, however, was drafted for him by an attaché from Paris who had newly arrived in London to join the French mission, François Georges-Picot. Recently returned to France from his position as French consul-general in Beirut and an ardent member of Comité de l'Asie Française, Picot "bore a long-standing grudge against the British ... [and] his appearance in the British capital heralded a much more hard-nosed French approach."[12]

On 15 October 1915, anticipating problems once the French became aware of British negotiations with Sharif Husayn, Grey suggested to Cambon that "France appoint a representative to discuss the future borders of Syria because Britain wanted to back the creation of an independent Arab state, but did not tell Cambon just how far discussions with [Husayn] had already gone."[13] By this time the failure of the Dardanelles campaign was a foregone conclusion and the decision to evacuate Gallipoli was expected shortly, so Grey turned his attention further east. On 21 October 1915, he proposed formal talks with France to settle the postwar spheres of influence in the Middle East. By this time, the foreign secretary was "dogged by failing eyesight, 'inhumanely busy and tired'" by affairs of state and willing "to defer to the judgment of other ministers on matters affecting the conduct of the war.[14] On Asiatic Turkey the views to which he chiefly deferred were those of War Minister Lord Kitchener, the leader of what was in effect an 'Egyptian party,' as influential in Britain as the Comité de l'Asie Française was in France."[15]

Meanwhile, "on 28 October (the very day that McMahon warned London that he had 'reason to suspect that' [Albert] Defrance had 'wind of the recent interchange of messages between Mecca and ourselves')," the French minister in Cairo had "just confirmed the British were communicating with Sharif Husayn by asking al-Faruqi about the correspondence, [and] reported to Paris what he had learned."[16] Apparently, al-Faruqi was not above sharing his intelligence, and an alarmed Defrance immediately notified Paris. Now the pressure began in earnest for Britain and France to meet and discuss what both hoped to achieve from the war in the Middle East.

When consulted by Paris, France's ambassador to Great Britain, Paul Cambon, suggested that his new adviser at the embassy in London, François Georges-Picot, was the best person for the job. He assured Premier René Viviani that the recently returned French consul from Beirut "understands Syrian questions better than anyone." Viviani already knew Picot, as he and his family were well-known activists in the French colonialist movement. His father, Georges Picot, was founder of the Comité de l'Afrique Française and his elder brother Charles was its treasurer. The younger Picot acted as an advocate of the colonialist party in the Quai d'Orsay and was a dedicated proponent of a French Syria, which included the whole of Syria and Palestine except Jerusalem.[17] In 1915 he had inspired a parliamentary campaign in Paris against the ministers who were prepared to give way to British interests in the Middle East. The Lyon and Marseille Chambers of Commerce had also sent resolutions to the Quai d'Orsay in support of a French Syria. Shortly afterwards, proponents of a French Syria took control of the Committee on Foreign Affairs of the Chamber of Deputies.[18]

With such credentials, Picot's appointment was quickly approved, and he returned to Paris where "he drafted the official telegram to Cambon announcing his own appointment as French delegate. Picot then also drafted his own instructions. 'This,' he wrote ironically to his friend Defrance, 'is a good way of ensuring that my instructions are satisfactory.'"[19] From this point on Picot "became the ventriloquist for the French government's policy in the Middle East."[20]

The role Picot thus acquired for himself was quite unique in French diplomatic history. Merely an acting embassy secretary who had been only been in London for six months, he had replaced his superior, the French ambassador, in negotiating the future of France in the Middle East and been given a relatively free hand to do so. If this were not unusual enough, Ambassador Paul Cambon "was both the unofficial doyen of the French diplomatic corps and a man with a singular sense of his own importance. Picot's extraordinary influence was due in part to his skill in winning the ambassador's confidence." However:

[I]t was due even more to the nature of the subjects he negotiated. Cambon would never have willingly allowed a subordinate to take the leading role in European negotiations. But despite his previous experience as French resident in Tunisia and ambassador at Constantinople, he regarded the partition of the Arab Middle East . . . as involving the sort of technical detail in which he was not required to immerse himself.[21]

So he was more than happy to have someone as knowledgeable as Picot on his staff to whom he could turn; someone who also represented the powerful Syrian lobby both inside and outside the Quai d'Orsay. Nevertheless, Picot would work closely with Cambon, often seeking his advice and support throughout the negotiations.

The instructions Picot wrote for himself included the French demand for "the whole of Syria 'broadly defined.'" Realising compromises would need to be made once negotiations began, he further instructed himself: "Our task is to make our demands and to abandon ground only foot by foot if compelled to do so; that way we shall always have some ground left." Earlier, Cambon and the leaders at the Quai d'Orsay had expressed their reluctance to claiming

Palestine to the south and territory as far east as Mosul, but eventually Picot prevailed. "You are to insist, he [instructed] himself, that our possessions stop only at the Egyptian frontier." Anticipating objections to French control over the Christian holy places in Palestine, he further instructed himself: "But it will be easy for you to reply that the objection which would be made to France would apply equally to any other owner of southern Syria." Moreover, "the existing rights of other powers in the Holy Places could be adequately protected by diplomatic protocols." As for defending the inclusion of Mosul in the French claim:

> [H]e would merely have to stress Russia's opposition to a British (rather than a French) presence so close to her frontier: Faced with an argument of this nature, they [the British] are not likely to show themselves intransigent on this point.

No doubt the Comité de l'Asie Française were pleased that Picot had included the colonialist agenda in his instructions. The new French President, Aristide Briand,[22] who also served as his own foreign minister, was too preoccupied with the war at home to engage in the matter and he quickly approved Picot's instructions. Later he was to encourage rather than tone down these instructions.[23]

When a meeting was finally convened a month later between representatives of both countries at Whitehall to discuss the Syrian borders on 23 October it was not the British or French cabinet policy that held sway but that of the two imperialist pressure groups — Kitchener's Egyptian party and the Comité de l'Asie Française. Sir Arthur Nicolson,[24] permanent under secretary at the Foreign Office, chaired the meeting and led the six-member British delegation — all senior officials, two each from the Foreign Office, India Office and War Office — to face the single French delegate, François Georges-Picot, a middle-ranking diplomat as the first secretary of the French Embassy in London. As was noted later, "Seldom in the history of major diplomatic negotiations between major powers has one side been so comprehensively outnumbered and outranked by the other." However, Picot was not intimidated. Despite his momentary initial surprise because he expected to meet Lord Kitchener alone, in his opening remarks Picot made his position clear. He told the British officials "that 'no French government would stand for a day which made any surrender of French claims in Syria.' And the Syria which he claimed was as largely defined as in the instructions which he had drafted for himself."[25]

Nicolson suggested Picot "moderate his demands . . . [stressing] the danger that the Arabs would join forces with the Turks. Such a union, he insisted, would present 'a grave and immediate danger' not merely to Egypt and India but also to French North Africa." But Picot ignored Nicolson's advice, stating how France had no problems in its colonies. "In Tunisia, in Algeria and in Morocco," he noted:

the natives had replied in large numbers to our appeal ... and had demonstrated their loyalty by their heroic defence of our territory. In Egypt, on the other hand, the unanimous feeling of the public was unhappily quite different.

He condescendingly then offered French aid:

Although we were not personally threatened, we would willingly take account of the different situation of our Allies and should be happy to come to their assistance. Still, it would be necessary to know what sacrifices would be asked of us.

Ignoring the challenge in Picot's remarks, Nicolson shared with him the details of British negotiations with Sharif Husayn, adding that "Arab participation in the war against the Turks would depend on the promise of a large Arab state," which could affect French claims to Syria. If such were the case, he "suggested that the Syrian interior could still become a French sphere of influence even under Arab sovereignty." To this, Nicolson added that there could be a "possibility of direct French control over Lebanon and the Mediterranean coast further north." Picot was unimpressed. He replied, "'nothing short of a French annexation of Syria would be admitted by the French public,' and merely agreed to submit the British proposals to his government." With this, the first round of negotiations ended in a stalemate.[26]

All appearances to the contrary, Picot was actually pleased with "Nicolson's admission that the Mediterranean coast should not form part of the future Arab state. His insistence on France's right to direct control over the Syrian interior was merely designed to strengthen his bargaining position." Moreover, "after 'sufficient opposition' to impress the British, he was quite prepared to restrict his claim for direct rule to the coastal region and Cilicia [the south coastal region of Anatolia]." Picot's instructions for the second meeting "authorized him to accept indirect control over the Syrian interior provided that, in return for this 'attenuation of our sovereignty' the French zone of influence stretched far enough east to include Mosul."[27]

Talks resumed in London on 21 December, with Picot announcing, in a ploy to get what he wanted, that

> out of loyalty to her ally, France would accept a zone of influence in the Syrian interior, rather than the direct control which was hers by right. This painful renunciation of French rights, however, clearly required major concessions from the British.

While he was pleased that "Britain was prepared to accept direct French rule over a five hundred mile band of territory from Cilicia in the west to the Persian frontier in the east," Picot was less so

> with the British proposals for the Mediterranean coast. Though Alexandretta and Latakia were to come under direct French rule, Lebanon (together with Beirut and Tripoli) was to form "a nominal part" of the new Arab state under the administration of a French governor. The questions of Palestine and Mosul were to be left of further discussion.[28]

An indignant Picot rejected the British proposal for Lebanon. For both Picot and Cambon, with whom he consulted, "the British insistence that Lebanon as well as the Syrian interior form part of the Arab state represented a retreat from the solution which Nicolson had been prepared to accept during the first round of negotiations." While neither felt a need to tell the British so, "they believed that direct French military control over a postwar Greater Lebanon would be necessary to prevent religious war between the Christians and the Muslims who would be included within its frontiers." They also believed that unless such controls were in place once the Ottoman authority was removed, existing communal rivalries would erupt into violence and internal strife in Lebanon. The frustrated duo concluded that

> the main reason for this hardening of the British line was the addition of Sir Mark Sykes to the British delegation. On a number of points, the French believed, Sykes had adopted "the extreme positions" of the Arabs and the English colonialists and had at least partially converted Sir Arthur Nicolson to them.[29]

Sykes's growing influence in Whitehall had not gone unnoticed, and with good reason. As a result of his travels in the Middle East and elsewhere over the previous year on behalf of the War Office, he had greatly impressed the Foreign Office and others with his voluminous and detailed reports, meticulous maps and insightful comments, in particular on the Arabs and Middle East, most recently at the 16 December meeting of the War Cabinet. Sir Henry McMahon noted in a letter to Sir Edward Grey on 15 December that Sykes was "a very capable fellow with plenty of ideas but at the same time painstaking and careful."[30] So it was not surprising to find him added to the British delegation at the 21 December talks and quickly emerge as the main British negotiator.[31] On his return to Paris after the meeting, a disgruntled Picot described the reception he had received in London as decidedly chilly. Such was the state of the Anglo-French discussions in late December 1915.[32]

In his appearance before the War Cabinet a few days earlier, Sykes stressed that Britain must come to an agreement with France on what they wanted in a postwar settlement in the Middle East, particularly in Syria. He agreed with Lord Kitchener that this had to be decided diplomatically before any military action could take place in the region, otherwise it could seriously harm Britain's relations with its French ally. Sykes was also concerned about French ambitions in Mesopotamia, in particular their interest in Mosul. So, to counter the strong pressure exerted on the French government by those with interests in the Middle East, and successfully promote British interests in the region as well as secure French cooperation, he suggested that British diplomacy should address the concerns of the major factions behind French aspirations in the region; namely, the French clerical and colonial parties. These concerns were over the possibility of a CUP-controlled Arab Caliphate in Istanbul and Sharif in Mecca. So negotiators might work on French fears of how this would affect the Arabs in their colonies of Tunisia, Algeria and Morocco.[33]

While this approach was not taken in the official negotiations on 21 December, it can be imagined that Sykes would have used it in later discussions with Picot when the two worked together as the sole representatives of their countries. While there is no record whether he did or not, Picot represented the French colonial party and related interests and no doubt would have found these compelling arguments. This would have been all the more reason for Sykes to use them.

Immediately after the 21 December meeting, the Foreign Office asked Sykes to take over the talks and work with Picot in private discussions. After this, "almost daily" talks were held between the two men at the French embassy. While Sykes would have preferred the French to be "out of Syria altogether in return for compensation in Africa," which would allow Syria, Palestine and Mesopotamia to be "under the Sultan of Egypt and the spiritual dominion of the Sherif of Mecca, both guided by British advisers," this was not to be. France would not be denied what it viewed its rightful claim to Syria. However, since Sykes "considered it essential for Britain to secure

Palestine and an outlet on the Mediterranean for the Mesopotamian railway,"
he was willing to give up territory "in the north-east of the proposed French
sphere so as to provide a buffer between British and Russian territory."[34]
This was possible because his instructions from Nicolson had been relatively
general, giving Sykes much leeway in the negotiations. He was told merely
"to outline 'the requirements of the various parties' – of the Arabs, of France,
and Great Britain, and of the religions concerned with the Holy Places."[35]

Despite their initial mutual suspicions, the two men soon established a
good working relationship and became close friends. Both shared a strong
Catholic faith and were ardent imperialists. It also helped that Sykes liked
France and its culture, and spoke French. However, although both men
were aware of the British negotiations with Sharif Husayn and the
commitments made to him, including the sharif, his representatives or any
Arabs in their discussions was never considered. Nor would it be, for as ardent
imperialists both were working on behalf of their respective countries to
secure the best arrangement they could in the region. To involve its
inhabitants would only frustrate and complicate matters, if not negate their
ambitious plans.

In Cairo, the officials responsible for the Husayn–McMahon correspondence (begun earlier and conducted simultaneously with the Sykes and Picot
discussions) were well aware of British and French ideas of a postwar
settlement "and of the commitments and obligations by which both felt
themselves bound." Despite this, they continued the part they were assigned
to play and set into motion by Kitchener. Interestingly enough, although he
gave a different impression in his correspondence, McMahon actually favoured
giving the sharif "a 'spiritual' caliphate ... [not one with the] Caliph actually
governing extensive territories in the Levant or Mesopotamia. His province
was, as McMahon put it, 'independent Arabia': as the context indicates, this
territory began south of 'Palestine.'"[36] Consequently, the Arab state to be
carved out of postwar Ottoman Arab lands Sharif Husayn was encouraged to
hope for by the British was actually far less than what he expected. In London,
the British and French negotiators, Sykes and Picot, made sure of this as they
divided the region up between Britain, France and their Russian ally.
The result was nothing less than a shameless imperialist land grab; a scramble
for new territory at the expense of the Arabs that at the end of the war US
President Wilson would do all he could to prevent. Worst of all, in this case,
"the left hand *did* know what the right hand was doing," and did it anyway.

Under Sykes and Picot, both of whom had full authority from their
governments to negotiate the terms and conditions of any agreement reached,
negotiations went quickly and relatively smoothly. "Sykes quickly dropped
the British demand to make Lebanon and Mediterranean Syria nominally part

of the proposed Arab state" that had earlier so upset the French. "He also readily agreed to include Mosul in the French zone as part of a buffer region between the British and Russian Empires." A more reluctant Picot "agreed to concede Haifa and Acre as the Mediterranean terminus for a British trans-Asian railway. He was encouraged to make this concession by the belief that Sykes was under strong pressure from the Admiralty to demand Alexandretta instead." However, they reached an impasse over Palestine. While agreeing that special arrangements should be made for the holy places they both demanded the whole of Palestine and "each flatly refused to concede the other's claims." In a note to Cambon on 3 January 1916, Picot frankly admitted: "[I]t seemed to me that prolonging the controversy indefinitely was pointless, and risked embittering the issue without any likely result." For his part, Sykes concurred, noting that "Palestine was a subject on which the French were 'hardly normal': Any reference seems to excite memories of all grievances from Joan of Arc to Fashoda."[37]

So on 3 January Sykes and Picot drafted a preliminary agreement. In it Palestine "was to form a so-called 'Brown Area' [in yellow on the map] under an international administration whose precise form was to be decided after consultation with Russia and later with Italy and the representatives of Islam." Excluded from this area were Haifa and Acre, which by previous agreement would go to Britain for its Mediterranean terminus. In a Blue Zone "France was allowed to 'establish such direct or indirect administration or control' as she desired," in an area "comprising a wide band of territory from Cilicia [in southern Anatolia] to the Persian border and the Mediterranean coast almost as far south as Acre." The Red Zone in southern Mesopotamia gave Britain administrative rights comparable to those France had in the Blue Zone:

> The area between the Red and Blue Zones was to form "an independent Arab state or confederation of Arab states," divided into "A" and "B" zones within which France and Britain respectively would have exclusive rights to provide advisers and economic assistance.[38]

This was not what Sharif Husayn expected, or what he was told or had bargained for, and it would deliberately be kept hidden from him for the next two years.

Both Sykes and Picot considered the Palestine International and the Brown Zone in the agreement as a temporary solution, not a final one. So they included an accompanying memorandum with the document in which their positions on various possible postwar scenarios were noted. In it, Picot "restated France's 'direct interest' in the Ottoman Empire 'since the earliest times' and explained that, 'on the hypothesis that there were no other circumstances to be considered, the French Government might be expected to desire commercial and political predominance up to the Egyptian border.'" To this Sykes added, "that the British 'ideal solution' also included Palestine."[39]

While no documents have survived from their three days of negotiating the memorandum for Sir Arthur Nicolson in January 1916,

> it seems likely that Sykes believed he had the upper hand. He had Picot's acceptance of the Arab "confederation," under the titular leadership of the Sherif of Mecca, that denied Syria to France except for the coast, as well as acceptance of international protection of the Holy Places and a British strategic link in Palestine with Mesopotamia.

Certainly, "given France's earlier demands for all of Syria and Palestine, Sykes was pleased with what he took to be Picot's moderation."[40]

The two sides contrasted sharply in how they came to decisions on their war aims in the Middle East and in drafting the Sykes–Picot memorandum. In March 1915, the inter-departmental De Bunsen Committee had been formed "to advise on Britain's 'territorial desiderata.' British policy making was thereafter bedeviled by what Sykes denounced as 'the political confusion arising out of a multiplicity of councils and counselors and overlapping functionaries.'" Despite his being put in charge of the negotiations and given great latitude in his talks with Picot, Sykes's "talks with Picot were closely monitored by four different government departments." This was in stark contrast with the French side of the negotiations, where "only the Syrian party inside and outside the Quai d'Orsay exercised a significant influence on the negotiations." As its representative, Picot

> was not exposed to the conflicting views of rival departments. The French cabinet as a whole would merely rubber-stamp an agreement in whose negotiation it had played no part. Even the President, constitutionally responsible for treaties with foreign powers, was bypassed.[41]

Moreover, Picot and Cambon put pressure on their government "to ratify the agreement as soon as possible." They were concerned that if there were delays and the Arab revolt did not materialise, the British might change their mind and "decide not to settle spheres of influence with France after all." Prime Minister Aristide Briand agreed, and even suggested some additional concessions that might be wrung from the British. None of these, however, "were to be pressed if they threatened to compromise the success of the negotiations as a whole" and none were.[42]

In Whitehall things were not so easy, nor was everyone pleased with the agreement. Criticisms were straightforward and blunt. Lt. Gen. George Macdonough, director of military intelligence at the War Office, wrote: "It seems to me we are rather in the position of the hunters who divided up the skin of the bear before they had killed it," adding, "Such proposals are 'academic,' for the British ought to be getting on with their Egyptian offensive, bringing 'the Arabs in on our side as early as possible' instead of worrying about the French." The strongest criticism, however, came from the powerful director of Naval Intelligence, Capt. W. Reginald "Blinker" Hall. Appointed director of naval intelligence (DNI) in November 1914, Hall soon transformed British intelligence gathering. Nicknamed "Blinker," "because of his unmistakable facial twitch, Hall was charismatic, devious, innovative and entirely ruthless in achieving his aims." Also, as

> an empire builder, the new DNI had no qualms about concerning himself with matters well outside the strictly naval. Hall wanted a finger in every pie and it is largely because of his remarkable drive and insistence on a totally free hand that naval intelligence became by far the most powerful of the main British intelligence gathering organisations during the First World War.[43]

So it was no surprise he had something to say about the agreement.

In Hall's opinion, in the Sykes–Picot agreement, "France was the only party to gain advantage, Great Britain having relinquished a naval base in Alexandretta. This and more had been sacrificed for French co-operation with the Arabs, whose military usefulness he doubted." However, although Hall agreed recent British reverses in the East

> had brought "matters in Arab countries very near a crisis," he thought that instead of negotiating with the French and making promises to the Arabs, the British ought to make a landing in Palestine. In a word, "Force is the best Arab Propaganda."

He "also pointed out that the Sykes–Picot memorandum did not even give the British control of all the railways within the 'Brown' area of Palestine despite the fact that British railways communications with Mesopotamia was a 'strategical necessity.'"[44] Hall added that "France had no justification for claims to Palestine and that it was an error to ascribe to the Arabs 'a general desire for unity ... they will never be united.'" He reiterated once again that, "England should secure for itself, in order to assure its position in Egypt, at least exclusive control of all the railways in southern Palestine."[45]

Hall also noted that others were interested in Palestine:

> The Jews throughout the world had not only "a conscientious and sentimental interest" (as mentioned in the memorandum) but "a strong *material* and a very strong *political* interest" in the future of Palestine, and were likely to oppose "Arab preponderance in the southern Near East"; and he suggested that "In the Brown area the question of Zionism ... be considered."[46]

This last comment, coming as it did from Hall, has long been a matter of controversy. While no records have been found tying Hall to any Zionist groups or organisations, or showing any Zionist sympathies, he knew Home Secretary Herbert Samuel and, apparently, had read his memorandum on Palestine. He also knew Zionist leader Dr Chaim Weizmann, a reader in chemistry at the University of Manchester, who was working in government laboratories at the time to develop synthetic acetone for the war effort. It is possible Hall saw the possible strategic importance to Britain in southern Palestine that would ensure a future British railway terminus there to connect Mesopotamia with the Mediterranean. Hall was extremely knowledgeable and well connected. His word was not taken lightly and his opinions carried weight with his superiors and others in key positions. So, for Sykes to ignore his comments would not be prudent. It is possible, therefore, that Hall's comments were responsible for changing Sykes from being anti-Semitic and openly antagonistic towards Jews to seriously considering Zionism as a possible aid to the war effort.[47]

Although this was apparently not discussed during Sykes's negotiations with Picot in January 1916, as Hall anticipated the Jewish Zionist interest in Palestine soon became a matter of major importance. It was to occupy much of Sykes's time and effort over the next two years and provide him with the solution he sought for Britain to wrest control of Palestine from the French. However, in 1900 a much younger Lt. Mark Sykes held different views about Jews. When he was a soldier in South Africa he accused them of being to blame for the Boer War. He wrote his fiancée Edith Gorst, "I would extort the last farthing from the most jingo loyal Jew in the British Empire before I'd fine a traitorous gentile." Soon, his mother's accumulated debts, at the time amounting to some £120,000, necessitated his return to London from fighting in South Africa to settle her debts again, including "£10,000 claimed by 'an accursed Jew' named Sanguenetti, [and] £4,000 in interest on

Sanguenetti's claims." This did nothing to endear him to Jews, nor did the "'Jews of the most repulsive type' who jabber about the mines all day long" whom he met on the boat on his return to his regiment in South Africa afterwards. "It was for 'these beasts' that he was supposed to be fighting," he angrily grumbled in his letter to Edith.[48] Sixteen years later, however, Sykes would look favourably at Jewish Zionism.

In late January Sir Arthur Nicolson met his committee and sought to allay their concerns. With Sykes absent from the proceedings, he addressed the comments and criticisms of the Sykes–Picot memorandum. He agreed with Hall that Britain should have control of its railways in Palestine and

that the "Brown" area would require further attention. None the less, the Permanent Under Secretary at the Foreign Office insisted that giving in to some French demands was the only way of winning their approval for a British offensive from Egypt.

To this, he added that "the Arab 'confederation' provided Great Britain with an escape clause: 'If the Arab scheme fails the whole scheme will also fail and the French and British Governments would then be free to make any new claims.'"[49]

Sykes believed his secret negotiations with Picot were necessary before there could be any military offensive by the Egyptian Expeditionary Force against the Turks. This inspired his negotiations with Picot to reach an agreement as quickly as possible with France in order "to launch a forward Arab political policy early in 1916." However, the military motivation "evaporated when Asquith supported [Lt. Gen. William] Robertson's contention that no such offensive ought to be undertaken, that Egypt could be defended from its own borders."[50] Robertson, who served as chief of the Imperial General Staff from 1916 to 1918, was committed to a Western Front strategy focusing on Germany and was against what he saw as peripheral operations in other fronts.[51]

The prime minister's decision infuriated Sykes. Thus, following the king's speech at the opening of the new Parliamentary session on 17 February 1916, he took the opportunity to speak. Sykes chided the government's appallingly haphazard and wholly inefficient approach to the war in the East and the lack of any effort to do anything about it. With the recent catastrophes of the Gallipoli campaign in the Dardanelles and Maj. Gen. Sir Charles Townsend's surrender at Kut in Mesopotamia fresh in everyone's minds, his criticism of the war effort in the East received attention across party lines. He explained how

when Aden required directions, it had to consult with Bombay and Delhi, which in turn had to inform the India Office, which had to meet jointly with the War Office and Foreign Office, which had to report to the War Committee, which usually sought the opinion of the High Commissioner in Egypt, who had then to report back to the Foreign Office and War Committee.

Not only was this convoluted process time-consuming but it was both contrary to the efficient prosecution of the war and costly. Sykes estimated the process "cost £250 [over £26,250 in 2018][52] and took sixteen days." He then told the members of the House: "If we muddle, if we go on muddling, and if we are content to allow muddling, it will not be a question of a draw, but the War will be lost."[53]

Without mentioning the prime minister, Sykes continued, telling the House that Britain needed a strong leader who could make up his mind. One solution that might help he said, was establishing "a small War Committee to determine policy" to help that leader. Leopold Amery immediately followed Sykes. A Conservative backbencher from Birmingham South, Amery took a similar approach and urged that a small new committee should run the war. Years later, in his autobiography, *My Political Life*, Amery recalled that day and the events that led to Sykes speaking out. He noted that the prime minister did not seem at all pleased with what they had to say. With an undisguised glee undimmed by the years, Amery recalled that "Asquith's occasional interjections, and still more the contemptuous expression on his face, showed what he thought of my argument, and, naturally, even less of my undisguised attack upon his own unfitness to lead."[54]

A Conservative politician and journalist, Amery was born in India in 1873 to an English father serving in the Indian Forestry Commission at the time and an English mother of Hungarian Jewish ancestry. He was a contemporary of Churchill at Harrow but, unlike Churchill, Amery was both a brilliant scholar and athlete, winning many academic prizes and scholarships in addition to representing Harrow in gymnastics. At Oxford he gained a First in Classical Moderations in 1894 and another in Classics in 1896, and was a runner-up to the Craven scholar in 1894 and the Ouseley scholar in Turkish in 1896. He also won a half blue in cross-country running. Elected a fellow of All Souls College, Amery was a polyglot who had learned Hindi in childhood and later French, German, Italian, Bulgarian, Turkish, Serbian and Hungarian. He shared an Eastern outlook with Sykes and the two parliamentary colleagues soon found themselves working together in the War Cabinet Secretariat, where some of their suggestions to coordinate the war effort and improve its efficiency would be put into place.

It was not Amery's speech but Sykes's less partisan one, however, that proved an immediate sensation. Reporting it the following day, the press was full of praise. *The Times* noted, "By his tolerant outlook, his information, and the pains he is at to make his speeches real contributions ... [Sykes] has the respectful attention of all parties." The *Manchester Guardian* "praised 'the intrepid member for Central Hull' and the *Daily Telegraph* regarded his speech as 'admirable contribution.'"[55] His remarks caught the attention not only of the press but of other influential members of parliament as well. The day after his speech Sykes was invited to meet Lloyd George in his room at the Commons. While there is no record of what was said, the two men had much in common, including a "shared impatience over Asquith, and frustrations about dealing with the Turks." Sykes's remarks also found approval with Frederick Scott Oliver, a prominent British political writer, businessman and Unionist Party (UK) politician, who advocated tariff reform and imperial union for the British Empire. Oliver wrote to Sykes, "You seem to have nearly succeeded last night in jabbing the dragon in the pot of his belly." He then invited Sykes to dine with himself, Lord Milner, Leopold Amery and Geoffrey Robinson, Editor of *The Times*.[56] "Given the big role these men were to play in the coalition Lloyd George formed [later that year], this recordless rendezvous anticipated more wide-spread attacks on Asquith's leadership in the course of 1916,"[57] in particular, that of Lord Milner, the former High Commissioner in South Africa, who had played an influential leadership role in the formulation of British foreign and domestic policy from the mid-1890s and was to play a major role in the future Lloyd George War Cabinet.

On 4 February, after much discussion, the War Cabinet had finally agreed on the draft of the Sykes–Picot memorandum, "though not the preamble in which Picot had sought to state France's claim in principle to the whole of Palestine. Four days later the French cabinet also approved the agreement." However, Britain insisted that it would only come into effect "after the proclamation of the Arab Revolt" and with the consent of Russia, their wartime ally.[58] The need for both countries to consult Russia before making any agreement derived from the Treaty of London signed by Britain, France and Russia in September 1914, in which all parties had "agreed to consult each other on peace terms, which meant that Russia had to approve the recent secret Anglo-French negotiations on the Ottoman Empire." While Picot was appointed by the French government to go to Russia, Sykes volunteered. Sir Arthur Nicolson agreed to let him go, knowing "the two men got along well and that Sykes might possibly be of use to the British ambassador, who would be in charge."[59] So once arrangements were made for the trip in early March, Sykes had to leave his newfound popularity in certain quarters of the press and Parliament and join Picot in Petrograd on a mission for the Foreign Office.

On his arrival in Petrograd Sykes reported to the British Ambassador, Sir George Buchanan,[60] who told him that Picot had arrived before him.

Moreover, the French ambassador and the Russian Foreign Minister Sergei Sazonov[61] were already at odds over the division of Ottoman territory as outlined in the memorandum. While British designs were on territory that lay far to the south, France's claims bordered those of Russia and Russia was unwilling to give in to French demands. This left Sykes with little to do while the French and Russians argued. So, he took the opportunity to take in the sights of Petrograd, attending the ballet and opera.[62]

The French and Russians eventually came to an agreement over their designs on Ottoman territory to be incorporated in the agreement, and Sykes was free to return to London. Before leaving St Petersburg, he had an audience with Tsar Nicholas II and, apparently, he made a good impression on the autocrat, who invited him to dine with him at the palace that evening. This favourable impression was reconfirmed later, when Sykes received word after his visit that he had been made a Commander of the Order of St Stanislaus.[63]

Meanwhile, back in London after a flurry of last-minute discussions between France and Britain and consultation with the Russians, the Asia-Minor Agreement, with some minor changes, "was formally ratified by the two governments in an exchange of letters between the British Foreign Secretary, Sir Edward Grey, and the French Ambassador, Paul Cambon," on 16 May 1916.[64] Keeping the agreement secret was "a reflection of the fact that, even by the standards at the time, it was a shamelessly self-interested pact, reached well after the point when a growing number of people had started to blame empire-building for the present war."[65]

With the signing of the agreement, Sykes's connection with the Foreign Office ended and with it his office at Whitehall. Not one to wait for things to happen, he turned to his friend, Sir Maurice Hankey, secretary of the War Committee and the man in charge of the Committee of Imperial Defence Secretariat. Their shared Eastern views had led to the two men earlier becoming friends. Hankey admired Sykes and suggested that Sykes should come to work for him at the Secretariat. This appealed to Sykes, who would be able to continue working on Eastern matters and avoid "the routine of any one department in Whitehall." After his role as a member of the Secretariat was agreed, Hankey wrote to "the heads of the Admiralty, and of the Foreign, India and War Offices [specifying] the work that Sykes would undertake." Together with their approval, Hankey received the approval of the prime minister.[66] In his new position Sykes would have more freedom in his activities and direct involvement with the War Cabinet.

Meanwhile, the Sykes–Picot agreement would soon become problematic for Sykes and the British. As if he anticipated the Bolshevik Revolution that was to occur three months later, Sykes "warned London policy-makers . . . the risk the old agreement involved − 'if at any moment the Russian extremists

[Bolsheviks] got hold of a copy,' then 'they could make much capital against the whole entente [sic]!'" Therefore, it was crucial that "the idea of annexation really must be dismissed."[67]

Meanwhile, continuing criticisms of the Sykes–Picot agreement were not lost on Sykes. By now, few saw the agreement as one cast in stone and it was felt that its use as a vehicle to guarantee French cooperation in the Middle East was long past. Many believed that Britain had gone too far in making British concessions to their ally. "On more than one occasion after its signature Sykes disclaimed responsibility for its provisions, and expressed disagreement with them." In response to the criticism, Sykes claimed that he acted "not as a principal but as an agent, bound by the views and decisions of ministers."[68]

Finally conceding revisions needed to be made to the agreement, Sykes would prepare a "Memorandum to the Asia-Minor Agreement" in August 1917 in which he outlined what he believed should be done. The key to these changes, he insisted, was for France and Britain to close any gaps in the agreement that left open any possibility for annexation, which would lend credence to claims raised by the non annexationists. Britain and France must also agree

> not to annex but to administer the country in consonance with the ascertained wishes of the people and to include the blue and red areas in the areas A and B [Then] we shall be on much firmer ground at a Conference.

He ended his three-page memorandum with the note: "I want to see a permanent Anglo-French Entente allied to the Jews, Arabs and Armenians which will render pan-Islamism innocuous and protect India and Africa from the Turco-German combine, which I believe may well survive the Hohenzollerns."[69]

As subsequent events in Russia would soon prove, Sykes was right. On 23 November 1917, after overthrowing the Kerensky government the Bolsheviks published the full text of the original and unchanged Sykes–Picot agreement in *Izvestia* and *Pravda*, to the great embarrassment of the Allies. Coming as it would three weeks after the publication of the Balfour Declaration, this would add to the distrust growing between the British and the Arabs. But all of this was in the future.

CHAPTER 6

WAR CABINET SECRETARIAT

Mark Sykes could have made a reputation in at least half a dozen careers. He was one of those few for whom the House of Commons fills . . . . {He could have been} a first-class music hall comedian; holding a chance gathering spellbound by swift and complete changes of character . . . or a tragic actor.

Sir Ronald Storrs[1]

Now he had a desk in Whitehall Gardens, Sykes was able to visit relevant departments on daily rounds and "read dispatches from officers in the East." He published the pertinent material he found in a weekly digest, which he called the *Arabian Report*, which was to be the equivalent of the weekly *Arab Bulletin* published in Cairo by the Arab Bureau.[2]

Shortly after moving into his new office at the Committee of Imperial Defence Secretariat on 5 June 1916, two events occurred almost simultaneously that would greatly affect Sykes's activities. The first shots of the Arab revolt were fired on the Turkish garrison in Medina and the British cruiser HMS *Hampshire* carrying Lord Kitchener on a diplomatic mission to Russia struck a mine in the North Sea off the Orkney Islands. Kitchener, his staff – including Lt. Col. Oswald Fitzgerald – and most of the crew of 655 were either drowned or died of exposure. While the revolt would take time to develop, Kitchener's death brought immediate changes at home. Lloyd George replaced him as secretary of state for war, which ended Sykes's privileged position in government circles and in the Middle East as Kitchener's Man. Although Lloyd George was an Easterner who supported Sykes in War Committee meetings, it was not the same as having the legendary K to promote and back him.

For much of the rest of the year, Sykes found himself frustrated in Whitehall without his powerful patron. He remained a key advisor to the government despite not having an influential sponsor in the halls of power.

He found it difficult to arouse interest in the war in the Middle East when he was confronted by politicians and generals who were more concerned with a difficult war closer to home. There was some hope, however, "knowing that Lloyd George as head of the War Office wanted to challenge the General Staff's opposition to Eastern campaigns." This encouraged Sykes to use the declaration of the Arab revolt by Sharif Husayn as an excuse to champion their cause and do all he could to arouse interest in war in the Middle East.[3]

In a memorandum dated 20 June 1916 and provocatively titled "The Problem of the Near East," Sykes painted "a gloomy picture of the current situation in the East and [called] Asiatic Turkey 'the last German colony.'" Unless things changed, he predicted that

> If Germany can retain her position in Turkey [at the end of the war] she keeps the Baghdad railway, the Turkish army, her hold on the Caliphate, and has a white man's country in the uplands of Anatolia to colonise. [Thus a] post-war Germanised Turkey gives Germany her military bases of attack on Egypt and India, her political bases for the fostering of internal trouble wherever we have Mohammedan subjects, an international pawn in Palestine, which gives her a hold at once over the Zionists, the Papacy and the Orthodox, a strangle hold on Russian in the Bosphorus, and a monopoly of certain oilfields essential to maritime, aerial, and industrial power.[4]

With such possibilities in the offing, he saw the Sharif of Mecca's revolt against the Turks as the only hope of countering such a disaster. Sykes urged the War Committee to adopt "a definite policy with regard to the Arabs, and consideration of our contingent intentions in the Middle East." He reminded them that "having launched the Sherif on his rebellion it is imperative that we should see that he keeps his head above water." Furthermore, should the Sharif lose to a combined Ottoman and German force "with guns, discipline and unfettered control of the air [it] will put the Sherif in a bad position to defend himself." It will also look bad for us, "who lured him on to rebellion by gifts of money." Then he painted a stark and gloomy picture should the Sharif's revolt fail:

> Soon all the Muslim world will know what has occurred; if our protégé is successful all will be well, but if he is driven out of the holy places and they are retaken by the Turks a terrible ferment will be set up among Mohammedans, we shall have played with fire and probably set our house in a blaze. The Turks will be in Mecca for good and will use it as a

fountain of sedition, excitement, and anarchy. This will be [as] serious for France as ourselves and almost as serious for Russia.[5]

To prevent this, Sykes advised that a change of attitude toward the Arabs as a whole was necessary at home and abroad. A Middle East policy should be developed with Egypt at the centre of war activities for the region and "a reconsideration of the political status of the Persian Gulf and Mesopotamia" was needed.[6]

In an apparent response to Sykes's memo, the General Staff at the War Office issued a memorandum on 1 July 1916 entitled "The Sherif of Mecca and the Arab Movement." It summarised in detail all the events involving British strategy and planning in the Middle East, specifically on Sharif Husayn and the Arab Movement that had developed since early 1915. A single appendix was attached to the memorandum entitled, "Points Relating to the Agreement Between the Sherif and Great Britain," which spelled out exactly what had and what had not been agreed to in negotiations with the Sharif.[7]

A few days later on 6 July Sykes was called to a meeting of the War Committee to discuss the situation in the Middle East. Brig. Gen. Gilbert Clayton, director of intelligence and head of the Arab Bureau in Cairo, happened to be in London at the time and attended the meeting with Sykes. Prime Minister Asquith began by asking Sykes for his views "on the political side of the Arab situation." While downplaying his own military opinion as being "not of much value," Sykes was nonetheless blunt in his remarks and, as usual, free with his advice.[8]

Opening with his frequent complaint about the lack of coordination of policy in the East between Cairo and India, Sykes told the Committee: "I do not think we shall get our value out of the military decisions if we do not have a co-ordinated political policy." He then turned his attention to what he viewed was the bad situation currently existing in Egypt. While ascribing no blame to Sir Henry McMahon and saying that the high commissioner "had done extraordinarily well," he said that under McMahon it had been "Business as Usual," which had led "to a "bad atmosphere in Egypt." Sykes told the ministers: "You want to wake up Egypt – civil and military – something very stiff, something to vitalise the civilians, then you get the morale of the army immensely improved. I am confident about that."[9]

He noted it was generally understood that McMahon was a temporary replacement for Kitchener, who was expected to return to Egypt after the war and resume his position there once his duties at the War Office were over. In his absence a complete lack of discipline had been allowed to develop in Egypt. Without Kitchener returning, Sykes suggested, not too subtly, that

McMahon be replaced. He told the War Committee: "I strongly advise that you wake up Egypt to the utmost, and get into the Government of Egypt some ruthlessness – that if people do not wake up they are made to do so without consideration, even sometimes unjustly."[10]

He then turned his focus to the Government of India. While praising its agents in the Gulf as "the best political staff that you could possibly have," he stated that in his opinion "the Government of India is not capable of running a pro-Arab policy satisfactorily." Noting that the Arabs were naturally divided (by tribes and region, urban and nomads, Sunni and Shi'a), it was virtually impossible to pursue a single coherent Arab policy in the war against the Turks by having "two centres of political control." They knew "Sir Percy Cox's instructions come from India, that our people in Egypt are running the Sherif, and that India will be against the Sherif. They know there are two seats of control and the Arabs play upon it." He had spoken to Lord Kitchener – through Fitzgerald – about his belief that "co-ordinated political control ought to be under one man." In response, Fitzgerald told him Kitchener believed "the military control was going to the War Office and the political control would naturally follow." However, this was not done while Kitchener was alive.[11]

If this were not bad enough, Sykes told the ministers that he believed "that the mass of people, the Indian Moslems, are anti-Arab," which would naturally influence the British in India and give them an anti-Arab attitude. And, to make matters worse, it was "the sort of friction the Arabs try to play upon." Another area that would cause problems between Arabs and Indians was "that the Indian Government is followed by minor officials in the service," as well as "moneylenders, and all the parasites who follow the train of the Indian Army." Adding to this there is tension between the Indians and the Arabs over the Caliphate, with the Arabs in revolt against the Ottoman Caliph and the Indians supporting him.[12]

Sykes pointed out these problems did not apply "to the higher type of officials in the Army, but merely to the collection of clerical and small shop people who come behind the Indian Army," which would only serve to increase the animosity between the Indians and Arabs. He also noted that the Indian Government "naturally works on Indian lines ... and it imposes law in a way that Arabs do not quite understand, which produces another bad influence." He added: "The last reason is one that seems very serious, and that is that there is an old long-standing feud between India and Egypt."[13]

Sykes then told the War Committee that if they wished to win the war in the East, they must develop a single and consistent Arab policy to be followed by everyone. He told them what he believed that policy should be:

Towards all Arabs, whatever their condition, whether independent allies as Ibn Saud and the Sherif, inhabitants of protectorates, spheres of influence, vassal states, we should show ourselves as pro-Arab, and that wherever we are on Arab soil that we are going to back Arab language and Arab race, and that we shall protect or support Arabs against external oppression by force, as much as we are able, and from alien exploitation. That were we to govern we shall employ Arabs in the administration where we do not employ Englishmen, and that where we employ Arabs we intend to give them the executive authority of their rank, and that we do not recognise any social distinction between an Englishman and an Arab of the same rank, and further that we do not intend to introduce the idea of a ruling race.[14]

Sykes further expanded on the race issue as another serious problem in involving the government of India in the Middle East: "You cannot run Arabs by white men who are their equal," which is how the Arabs have been dealt with by the white man for centuries. This is in stark contrast to India, where "you have all the old traditions of black and white," and such an attitude by English administrators from India will cause problems with the Arabs:

For that reason, I think, one thing is essential: I suggest the existing staff be kept intact and that Sir Percy Cox should be made High Commissioner of the whole of the Arab part that is now being run by India, putting him under the direction of the Foreign Office.[15]

Sykes urged

that this be done, because we are a month late. The Sherif rose on the 6th, we got the news on the 10th, and now it is the 6th of July. We ought to have done it six months ago, but one thing and another has happened and it has not been done.

He ended by saying:

I beg we get the Arabs under one Command now, so that we can deal with all these discordant influences from one centre, and directly under the War Committee, so that there is an immediate response. In that way you can get the moral and the material additions.[16]

Apparently, Hankey's talk with Sykes on 21 June in which he asked him "to moderate his attitude toward the Indian Gov^t in the Arab question"[17] had little effect and Sykes's criticism of the Indian government would continue to cause friction with the Indian government.

The day after the meeting Sykes prepared a memorandum for the War Committee to send to Cairo along the lines he suggested in the 6 July meeting, which was attached to the minutes of the meeting as Appendix 1.[18] The same day he also drafted a letter to Sir Austen Chamberlain, secretary of state for India. Apparently, the two men had a conversation the day before in which Sykes again pleaded his case for the Indian government to relinquish its control of Mesopotamia, Arabia and Aden to a single authority. That way, he told Chamberlain, there would be a unified central command to fight the war in the Middle East.[19]

After the meeting Sir Edward Grey "agreed to send a telegram to Cairo, but this satisfied neither Sykes nor Lloyd George." In their view, more had to be done to correct the situation. At the War Committee's next meeting, to which Sykes was not invited, "the head of the India Office protested against giving up Mesopotamia to the Foreign Office." At this point "Asquith dropped the subject with the observation, 'None of the Committee was impressed by Sykes's proposals.'"[20]

The rest of the summer found a frustrated Sykes behind his desk in London where, despite his lack of success with the War Committee, he persevered in his attempts to gain support for the Arab revolt. Each day he painstakingly made his rounds of Westminster, the Foreign Office, the India Office and the War Office, before returning to his office at the Committee of Imperial Defence Secretariat in Whitehall Gardens. There, he would sit at his desk and closely scan the daily newspapers – both local and foreign – "ready to pounce on any writer who smacked of sympathy for the enemy, particularly Turcophiles like Marmaduke Pickthall."[21]

Some years later, Sykes's son Christopher Hugh Sykes wrote that his father had once been a lover of Turkey but by this time had forgotten what "had once moved [him] to admiration ... [of] the virtues of the simple people of Turkey." As for Pickthall, Herbert and others of the Turkish Party in England, Sir Mark Sykes had remarked in the Commons in 1914:

[T]hey represented the Disraeli school of Tory foreign policy and their central belief was that "the disappearance of the Ottoman Empire must be the first step towards the disappearance of our own." Unless Great Britain had the goodwill of the Caliph in Constantinople, they argued, her place in Asiatic affairs would be fatally weakened.

Moreover, "they esteemed the Turkish character and abhorred Russia."[22] None of this reflected the current state of affairs and Britain's relationships with the two countries, which made them particularly abhorrent to Sykes. So he was furious when Pickthall and Aubrey Herbert began to "appear on London platforms to urge making a separate peace with Turkey."[23]

Sykes, Herbert and Pickthall had come to know each other earlier during their travels in the Middle East and through a shared interest in Eastern affairs. However, despite their differences over Turkey, Sykes remained a close friend of Herbert, but not of Pickthall. Unlike Herbert, who served in the military, fought and was wounded at Mons, served as an interpreter at Gallipoli and Kut and worked at the Arab Bureau in Cairo, Pickthall was deemed to be "a security risk. His talents as a linguist and as an authority on Syria, Palestine and Egypt could have been used but his reputation as 'a rabid Turcophile' prevented him from being offered a job with the Arab Bureau."[24]

In early 1916 Pickthall attempted to broker a peace arrangement with the Ottoman government through Dr Felix Valyi, whom he had met in London. Valyi was pro-Ottoman and the editor of the *La Revue Politique Internationale* in neutral Switzerland. A Hungarian Orientalist, Valyi had contacts in Switzerland with the Turkish government. Pickthall sought permission to go to Switzerland and through Valyi "build bridges between the British and Turkish governments. Not surprisingly, his application for a passport was refused." Pickthall quite openly sent copies of his letters to Sykes, seeking his support with the Foreign Office. He did not receive much sympathetic support. After correspondence between the two beginning on 25 May in which Sykes tried to explain to Pickthall the futility of his scheme, Sykes finally sent him a brusque letter on 10 July 1916, in which he wrote bluntly, "I do not consider that it is proper that you should assume absolute friendship to an enemy State . . . and further speak in a distinctly hostile tone of your own Government." Pickthall persisted for another month, until the matter was finally dropped.[25]

In addition to his daily routine in Whitehall, Sykes made a number of public speeches during the summer of 1916. In June, at a meeting sponsored by the lord mayor of London, he joined Lord Bryce,[26] recently retired ambassador to the USA, to discuss the 1913 Turkish massacre of Armenians. Later, in October Lord Bryce – with the assistance of historian Arnold J. Toynbee – was to publish *The Treatment of Armenians in the Ottoman Empire, 1915–1916: Documents Presented to Viscount Grey of Fallodon*, also known as the *Blue Book*, which was presented to both Houses of Parliament by order of His Majesty King George V.[27]

In August Sykes spoke at the Conservative and Unionist 1900 Club, where he gave a talk entitled, "After the War," in which he stated that Germany's historic yearning to expand into eastern Europe, known as *drang nach osten*,[28] has taken on a new meaning and has within it "the perils of a German take-over in the East." In September

> his speech on German War Aims, complete with a huge map illustrating the *Drang nach Osten*, given to "hundreds of wounded soldiers from the overseas dominions at Claxton Hall .... was so dramatic that he had many requests for copies and had 600 reprints made of it and his map.[29]

The event was even reported on 22 September in an article in the New Zealand newspaper, *Feilding Star.* It described the speaker, "Sir Mark Sykes, [as] one of the most independent and far-seeing members of the House of Commons, on the world-aspect of the war and what we really are fighting for." Referring to his map, the article noted that Sykes "defined clearly what was meant by the German '*drang nach osten.*' It was essential that we should break the German line to the East," adding, somewhat prophetically, "as otherwise it would only give the Germans another 30 years to make their preparations for a new war on a greater scale."[30] The article reported that Sykes ended his speech with the following dramatic and jingoistic words:

> The Allies were fighting for the law and toleration of the Roman Empire, the civilisation and chivalry of the Middle Ages, and the true democracy which came from the French Revolution. If we fail, humanity would fail. The maintenance of peace is a thing worth living and worth dying for, and which, innate in you, brought you here to fight for the right cause, and which means, in the last realisation, to bring peace on earth to men and goodwill.[31]

In all his public speeches Sykes made a point of mentioning the war in the Middle East and emphasising its importance. In doing so, he helped to popularise the phrase Middle East, which first appeared in an article by US naval officer Alfred Thayer Mahan in the September 1902 issue of London's *National Review*, an influential Conservative monthly. This referred to "the territory that lay between Constantinople and India, and ran from Egypt to Russia." It would take some time before the term caught on in Whitehall, "but Sykes's map showing the *Drang nach Osten* caught on in Fleet Street like wildfire."[32]

Seeking to build on the success of his map, Sykes contacted the Admiralty, where he was put in touch with Professor H. N. Dickson, the director of I.D. 32, the Geographical Section in Naval Intelligence, an organisation established at Dickson's suggestion by Capt. Reginald

"Blinker" Hall in 1915.[33] Together, the two men spent many hours at the Royal Geographical Society working "on a more ambitious project – an atlas of Western Europe and the Middle East illustrating geography, history, linguistic and religious distinctions, including information on everything from rainfall to the Jewish Diaspora." The map developed by Sykes and Dickson was made up of "different coloured and marked translucent sheets [that] had to be placed over the basic map and then illuminated from behind to be seen," on a "cumbersome glass-topped table with an electric light underneath for this purpose." After failing to get their invention patented, Sykes had fifty numbered folios produced in English and French as "The Statesman's Atlas."[34]

Meanwhile, Sykes continued to be frustrated by his inability to raise interest in the Arab revolt among top government leaders. His efforts were hampered by news from the Hijaz of repeated delays in the start of the revolt. This was due primarily to the lack of guns, ammunition and military support faced by the unsophisticated and untrained Sharifian Bedouin forces who were to fight against a modern Turkish army. Despite this, the revolt had begun on 5 June when the joint forces of Sharif Husayn's sons Faysal and his older brother Ali began operations against the largest Ottoman garrison in the Hijaz at Medina.

Faysal had just returned from Damascus, where he had pleaded in vain with the Turkish governor, Jamal Pasha, for the lives of twenty-one Arab nationalists who were condemned to be hanged the month before and to find the Arab Movement crushed. So he returned home to join his father and brothers in launching the revolt themselves. It was clear there would be no broader revolt that would join them in an uprising against the Turks in Syria. The most they could do was to challenge the Ottomans in the Hijaz, where their "success depended on their ability to mobilize the notoriously undisciplined Bedouin to their cause."[35]

With only 1,500 tribal volunteers, the princes faced over 11,000 Turkish troops garrisoned at the terminus of the Damascus to Medina railroad. While Faysal's forces "engaged the [Turkish] Hejaz Expeditionary Force (HEF) outside the perimeter of Medina. Ali's men ... ripped up 150 km of railway track to block the progress of Turkish reinforcements from Medain Salih (175 miles) to the north." Meanwhile, word of the princes' revolt spread to Mecca and Jeddah, where surprised Turkish officers panicked on hearing "that 3,000–4,000 armed Bedouin had already gathered to charge the garrison in the name of the Sharif." Not only were the Turks surprised: the "speed of events took Britain by surprise as well." A few days later, Oriental Secretary Ronald Storrs, D.G. Hogarth, and Kinahan Cornwallis[36] of the Arab Bureau arrived in Jeddah intending to make an assessment of the military situation and give Abdullah £10,000 as encouragement for action.[37] Instead, they found they were too late, as the action had already begun.

In the early hours of 10 June, after his pre-dawn morning prayers Sharif Husayn leaned out of the window of his palace in Mecca "and fired his rifle at the nearby Turkish barracks, officially beginning the Arab revolt. His followers attacked the Turkish units in the city and cut the water supply." After recovering from their shock "the Turks responded, [and] one party fired towards the Great Mosque and struck part of the holy Kaaba, an accident that would be held against the Ottoman government for the remainder of the war."[38] Elsewhere, "sustained rifle-fire was opened on the barracks in Mecca and on the Hamidiya building in which [were] housed the offices of the Government; and a siege was laid on all the Turkish troops in their several strongholds." It was to last for almost a month. When the barracks were finally taken on 9 July, Ottoman control over Mecca was brought to an end and the Sharif's forces quickly took over "all the garrison posts, guard-rooms and Government offices, as well as all the official quarters and buildings in the city."[39]

The Turkish garrison in the port city of Jeddah was attacked the same day as Mecca. Several thousand tribesmen loyal to the Sharif attempted to take the city but were unable to because of their lack of "modern weapons, guns and artillery ... [and] were held off by a Turkish garrison of 1,500 men." However, unlike the siege of the Holy City of Mecca, on this occasion British warships at the port were able to provide support. They "shelled the external Turkish positions, and seaplanes dropped bombs outside the perimeter of the walled city." Without reinforcements from Mecca, the Jeddah garrison had no choice but to surrender on 16 June. In the north, Rabigh and Yanbu were captured, while in the south Ta'if was put under siege. Abdullah bided his time in the siege of Ta'if, until "the garrison surrendered unconditionally on the 21st of September, with the Governor-General of the Hejaz, Ghalib Pasha, as the chief prize."[40] With this, there was no turning back. The only city left in Ottoman hands was Medina, which, although it was besieged at the beginning of the revolt by sharifian forces, remained in Turkish hands for the duration of the war. However, the taking of Jeddah the Hijaz's "chief port for supplying food and materials to the Holy Places was a military and political triumph."[41]

In London reports of the revolt and its repercussions continued to pour in from Cairo, Bushire, Aden and elsewhere in the Middle East, and found their way to Sykes's desk in Whitehall. In Turkey, however, news of the revolt was kept from the public, even to the point of denying it had occurred at all. In an announcement on 29 June the official paper in Damascus, *Al-Sharq*, reported that there had been tribal attacks on a few posts in the Medina area, but failed to mention the capture of Mecca and Jeddah by sharifian forces. Its earliest mention of the Sharif was on 2 July, when an imperial *firman* announced Husayn's dismissal, without giving a reason, and the appointment of his cousin Sharif Ali Haidar to replace him as Amir of Mecca. It was a month later, on 26 July, before "a distorted and belittling version of the facts was allowed to appear in *Tanin* (Constantinople)." For several months

> the [Turkish] Press continued to describe the Sharif Husain's movement as an act of personal insubordination, provoked by British intrigue, and one which was in the process of being crushed with the help of the population and tribesmen of the Hijaz who had remained loyal to "the caliphate and the Prophet's injunctions regarding the sacred duty of *jihad*."[42]

On learning of the revolt, the head of the Fourth Army and military governor of Syria Jamal Pasha in Damascus

> issued orders for wholesale arrests [of Arab leaders]; and his military police ... laid hands, in Damascus alone, on some forty of the principal residents there, threw them into prison and subjected them to various forms of atrocious torture.

Many were put in solitary confinement, flogged, beaten and starved, and another "120 other Arab notables from all over Syria [were] arrested and deported to Anatolia." The repression of rights proclaimed by Jamal Pasha under martial law in Syria intensified and in October the autonomy granted to Lebanon in 1864 was revoked.[43]

Back in London, Sykes read the reports of the revolt and the Turkish atrocities in response to it and was even more inspired in his rounds of the halls of power in Westminster and Whitehall. He spoke about the Arab revolt and the plight of the Syrians to anyone who would listen. He wrote letters and gave speeches to put more pressure on the War Committee to pay attention to events in the Middle East and see the opportunities it provided. He also reminded them of their responsibility to support the revolt of the Sharif of Mecca, which had been prompted by promises of British support.

On 8 August, in a letter to Maj. Gen. Sir Frederick Maurice,[44] director of military operations at the War Office, Sykes wrote that Syria was critical to the overall situation in the Middle East. He told Maurice of information he received from an informant in Syria that the 50,000 troops stationed there were mostly Arab and unlikely to fight hard on behalf of the Turks. He added the Turks were unable to reinforce the Arabs with Turkish troops, which made it "possible to bring Turkish rule to an end [in Syria] with a very small expenditure of troops."[45]

Still thinking of policy problems in the Middle East, Sykes wrote to Sir Percy Cox in Bushire on 21 August: "We live in considerable confusion here," he wrote to Cox,

> but the war goes well. The situation in London is as follows. The Government suffers from being a coalition and is not given to making decisions nor to adopting very definite lines, nor is it dominated by any one in particular.

As a result, "our affairs [in the Middle East] suffer from a lack of policy, [and] we don't seem to have any definite ideas as to what we are at." Nevertheless, he pessimistically told Cox he doubted there would ever be any reorganisation in future that would make things more efficient and effective.[46] His next words seem almost prophetic. "I believe," he told Cox,

> that if one looks ahead a generation or so that the Arab question will appear to be very important, these people are at present pre-nationalist, but have language, vitality and great capacities, and it rests with us to decide whether as in Persia, India and Egypt, intellectual development is to go against us or with us, if we just drift I think the Sinn Fein element in the Arab mind will go wrong just as it has elsewhere, tho' in Belgium and Serbia we are fighting for Sinn Feinism, and in Hijaz we are using it."[47]

It is not known whether Cox replied or what he thought of Sykes's letter, but it shows Sykes's restless state of mind in the late summer of 1916.

A few days later, on 30 August, Sykes compiled a six-page report entitled "Summary of the Arab Situation" as background for the Imperial War Committee on the Arab revolt. In the first few pages he gave a detailed overview of the region in the Arab Middle East covered by the Sykes–Picot agreement, plus the Arabian Peninsula (with reference to a map he enclosed). He described the indigenous peoples, the areas in which they lived, their lives, whether they were nomads, city-dwellers, Sunni or Shi'a, and made some general comments on their beliefs and attitudes. The remainder of his report focused on the political situation of the area, in which he described the Arab movement as "arising from the progressive educated Arabs of Syria, working on the courageous and disorderly elements of Arabia." He characterised this simply and melodramatically as "The Syrians think, write, and talk, and the Arabians act."[48]

Sykes described the abortive attempts at rebellion in Syria and their brutal suppression by Jamal Pasha. He listed all the rulers in the Arabian Peninsula, noting who was pro-Turk, pro-British or pro-Entente, or neutral, and their relationships with each other. In analysing the "Military Situation bearing on the political situation in Arabia and the Arabic-speaking provinces," he raised the point that "the key lies in the Turkish control of the Area Adana – Aleppo – and the railways to Nisibin [Nusaybin, on the Turkish-Syrian border], Beersheba [Palestine] and Medina." The reason for this was that

> along these lines pass supplies for the Mesopotamian army – for the attacks on the Russian left – for the attacks on Egypt – for Medina – so long as the Turks hold this junction with Constantinople they can hold the Sherif at bay, and dominate the sedentary population, while at the same time operating against Mesopotamia, Persia, the Caucasus, and Egypt.[49]

Given the existing situation, Sykes noted the best Britain could do was "(1) to help the Sherif hold Mecca (2) Help the Arabs ... to harass the Damascus-Medina railway (3) Immobilize a certain number of troops in Palestine by a *POTENTIAL* [Sykes's emphasis] offensive from Egypt and co'operate with the Russians by threatening Baghdad while the Russians threaten Mosul." Should this approach succeed in wearing down the Turks and the Arabs "feel confident enough to rise, we might see a general rising and collapse of Turkish rule" in the area.[50]

He made it clear that the success of the revolt depended on "military success on the part of the Entente forces to be of any great value." Once again, Sykes noted that the existing situation was compromised by the total lack of cooperation "between the Indian political officers and the Egyptian." He emphasised that the "Indian Government does not understand the Arabs and its officers are obliged to act in accordance with its prejudices, although themselves ideal men to foster the Arab movement and take advantage of it if they had orders." As a result, not only are the Arabs "divided among themselves but so are the British."[51]

He further characterised the state of affairs between the Indian government and Cairo as follows:

> Egypt has a definite idea, viz: to make all possible use of the Arab movement, India has no definite idea, dislikes all schemes emanating from Egypt, and contents itself with turning down or watering down any proposals which come from Cairo.

Because of this, Sykes added, "the Arab movement receives the greatest impetus on the political side in that region where our military effort is the most limited and the most obviously defensive." He closed by saying:

I have held and hold most strongly that until the political control is under one hand and follows one policy that the military operations can never reap full benefit for the favourable political situation which [it] obtains in the Arabic speaking countries.[52]

A month later, on 30 September, Sykes was still complaining about problems with the Indian government in Mesopotamia. In a letter to Sir John Hewett, former Lieutenant Governor of the United Provinces in India,[53] he vented his frustration by asking:

Can't they for heaven's sakes take the trouble to read their papers from England and study the Berlin–Baghdad route, if they have taught Indian Moslems to worship Turks in the past let them unteach this nonsense now, the Turks are now led by atheists and anarchists and are employed by the German General Staff, India is the ally of Russia, and the Sharif is the ally of England.

He added:

[A]s for Moslem susceptibilities instead of questioning the action of the Sherif why don't they question the action of the Young Turks who set up German court martial to settle religious disputes, hang every Holy man in Damascus they could lay their hands on, massacred half the Armenian race, and are trying to work Jehad with Oppenheim and Wassmuss!

There is no record of Hewett's response. He was well aware of Oppenheim and German agent Wilhelm Wassmus, who operated in the Persian Gulf area and Persia stirring up trouble among the Persian tribes by claiming to be Muslim and preaching jihad. Wassmuss caused problems there for the British, in particular Sir Percy Cox, British resident in the area, who put a price on his head, captured him, only to have him escape.[54]

A few days later Sykes wrote an extraordinary memorandum to the War Committee on 4 October giving a vastly exaggerated assessment of "the military-political assets of the Allies in the rebellious elements which at present exist in Syria." Based on the premise that Sharif Husayn could hold his own, he listed probable "existing military political assets" among the tribes in Syria at a total of 94,000 – less 15,000 who might desert – should there be an Allied landing in the region." However, he added this was all based on the assumption that

> the Allied force employed is sufficient to engage the main striking force of the Ottoman Army, and further that it will not rely for protection upon the Arabs, whose rising is only contingent on the capacity of the Allied offensive to contain the bulk of the Ottoman Army.

With all this in mind, Sykes suggested that "the best date for the crisis of the regular offensive would be between November 15th and January 15th."[55]

He then gave a detailed list of the all the necessary "arms, munitions, money and supplies [that] should be collected immediately." He followed this with suggestions for coordinated Allied attacks along the Eastern Mediterranean coast, from Palestine to northern Lebanon. He then listed landing sites and strategic military actions to be taken and objectives to be achieved, ending with the routing of the Turks and the occupation of Syria by Allied forces. Confident of his plan, Sykes ended simply by saying that, given all the variables he mentioned, "the above scheme is feasible." Not only that, but

> it will practically break up the Turkish military and civil hold over the country, and contribute material assistance to an Allied offensive. It is submitted that though such guerrilla action can only confirm success and add to its value, that it is well worth the expense.[56]

As the document has no addressee, given its date and the subject under discussion it can be assumed to be a memorandum providing information to the War Committee of conditions in the area, to help their debate at the time whether or not to send British troops to aid the sharif and the Arab revolt. It was followed a few days later, on 12 October, by a similar memorandum, also unaddressed but clearly by both its content and context meant as further information for the War Committee and their discussions. The introductory paragraph made both its provenance and purpose quite clear:

> The following memorandum was prepared under the supervision of the India Office by request of the Prime Minister. Its purpose is to give if possible an inner view of the intellectual and political forces which are predominant in the Ottoman Empire.

More than that, it was a warning about

a traditional friendship or sympathy for the Turkish Empire in the minds of many Englishmen in the past, and this sympathy still survives and tends to foster many misconceptions in regard to the existing state of affairs in the Ottoman Empire.[57]

He took this opportunity to explain the differences between the present-day Ottoman Empire under the CUP with that of the past, about which pro-Ottoman elements at home and in India might not have been aware. After the revolution in 1908, in which the Young Turks came to power through the Army, they extended their power through a political arm, the CUP. In 1909 they successfully crushed the counterrevolution to return Abdul Hamid II to power as Ottoman Sultan. While the supremacy of Constantinople over the Empire remained the same, those in power had changed. Formerly all power was held by the Sultan, supported by "the palace bureaucracy, the inner ring of high functionaries, and the Moslem clergy." Since the revolution

the Sultan has become a puppet, the palace [officials] ... scattered, the older ministers have retired or died off, the Moslem clergy have been completely tamed by terrorism, executions, and the appointment of religious chiefs such as the present Shaykh-al-Islam ... had no connection with religion whatever previous to their assuming office.

Sykes added:

The administration of the Wakfs [religious endowments][58] is now designed with the object of making the religious bodies political slaves of the Government. As a consequence the CUP though small and containing many non-Turkish elements is absolutely supreme and wherever the Turkish flag flies is unassailable.[59]

If the foregoing was not enough to convince its readers, Sykes added that under the Young Turks and CUP

wholesale executions, assassination, delation [accusations by informers], exile, and confiscation have crushed any possibility of a revolution of a real kind, thought it must never be forgotten that the CUP is always ready to engineer a revolution against itself for the purpose of reappearing in another role, or if ridding itself of some cumbersome appendage.[60]

Three days later, along with a covering letter Sykes sent a copy of this memorandum to Lord Hardinge, the former Indian viceroy who was now permanent under secretary at the Foreign Office. After referring to the enclosed papers, he told Hardinge, "It would appear that one way and another we are drifting into touch with very undesirable elements in Turkey." He went on to explain that, "In Paris the financial Turco-phils [sic] are always trying to get into correspondence with the Ottoman state." Then, suggesting the Young Turks would reinvent themselves to obtain any advantage Sykes added:

> [I]n Constantinople there is every evidence that the committee [CUP] is premeditating a dive, viz: an apparent split between Enver and Tala'at and a coalition of Tala'at with the Liberals and old Turks. The objects of such a manœuvre are the splitting of the Entente between England and Russia, and re-attaching the Arabs to the Ottoman Empire.

If such a scheme should succeed, he maintained "there would be great efforts made in Paris and elsewhere to welcome it, with the natural result of complicating the Constantinople situation." In light of this, Sykes suggested to Hardinge, "I therefore submit that we should avoid all contact with the Turkish parties, and stick entirely to the Arabs. Any movement which is led by Arabs is bound to be free of the Committee, any movement led by Turks is sure to get under Committee influence sooner or later."[61]

There is no record of Hardinge's response to Sykes's letter and accompanying memorandum sent in an attempt to curb any contacts with the Turks, personal, political, religious or otherwise. This would be a theme repeated in his reports and memoranda on the war in the Middle East and the Indian government at this time.

Not long after he sent this letter, there was a curious incident involving Sykes and Lord Hardinge. In an entry in his diary for 6 November, Sir Maurice Hankey recorded:

> Difficulty about Sir Mark Sykes who has perpetuated a serious indiscretion in his weekly "Arab Report," and aroused intense indignation in the Foreign Office, so that Lord Grey insists a withdrawal of the Report and an apology. Saw Sykes who was very penitent & promised to make amends.

Without describing what the difficulty had been, Hankey noted in his diary the following day: "Squared Lord Grey about the Sykes difficulty."[62]

In the Hardinge Papers at the University of Cambridge, the following corresponding entries concerning Sykes's "serious indiscretion" can also be found:

7 Nov 1916 – Dear Lord Hardinge

I saw Colonel Hankey last night and he showed me the Secretary of State's minute regarding the first item of the first section of Arab Report 16.

I desire at once in the interests of Departmental Harmony to express my regret that the paragraph was put in the report, which I should have seen could only give rise to friction.

I do not wish to offer any excuse for putting in the paragraph but only to apologise for having done so. I thought that I should give a statement of the events, but I perceive that it was unnecessary and for the reason inexcusable.

My only interest is that the Arab business should be run as well as possible, and I blame myself for doing anything which tends to thwart that object.

If you will allow me to express my regrets personally I shall be very glad to do so.

Yours sincerely,
Mark Sykes

Hardinge to Lord Grey re above (n.d., c. 7 Nov 1916)

Lord Grey

Sir M. Sykes brought me this morning. He expressed penitence & I accepted his apology promising to pass it on to you.

H

Sir Edward Grey wrote the following on Hardinge's note (n.d.):

This may close the incident. The objection to the statement was that it was incomplete and therefore misleading. He did not give the reason why. Lord Hardinge declined to act; it did not state [illegible] was laid before me the next evening; nor did it give the reasons why I [illegible] Lord Hardinge's actions & altered the telegram.

G[63]

There is nothing to be found concerning this incident in the Sykes Papers, the Hardinge Papers or Hankey Papers, nor are there any copies of the offending *Arab Report No.16.* So one can only imagine what might have caused such an upset.

Momentarily chastened, but otherwise undeterred, Sykes continued to chase Turcophiles and do all he could to bring the Indian government in line with Cairo from his desk at Whitehall and his home at 9 Buckingham Gate, across from Buckingham Palace. It appears he was making some headway with the India Office over the future role of the Government of India in Mesopotamia. In a series of papers entitled "The Future Administration of Mesopotamia (SECRET)," with entries from the Viceroy, Lord Chelmsford[64], Sir Thomas W. Holderness[65], the permanent under secretary of state for India (on behalf of the secretary of state), and Sir Arthur Hirtzel, secretary of the Political Department at the India Office, over the period from 18 October 1916 to 3 January 1917, the subject was discussed at length. The main discussion centred on Sykes's repeated contention that the Indian government could not rule Mesopotamia for a variety of reasons, specifically geography, race, language and culture, all of which were broadly discussed.[66]

From this, it seems that Sykes's various memoranda, reports, letters and evidence given to various individuals and committees over time were definitely having their effect at the India Office, as both Hirtzel and Holderness agreed with his position, and contradicted the Indian viceroy. As subsequent events would show, after the war the Indian government would be relieved of its responsibility for Mesopotamia. It would be renamed Iraq after the Arabic name *al Iraq* used locally since the sixth century and made a separate entity under the British Mandate in 1920, with its own government and administration under Sir Percy Cox as its first high commissioner,[67] which Sykes had recommended in his earlier report.

Meanwhile, continued frustration with the Liberal government's policies both at home and in the war had become widespread since Sykes and Amery's speeches in February in the Commons criticising the government. This was especially so in the Cabinet where in the last months of 1916 a revolt was brewing. Over the previous year-and-a-half, after numerous crises and a failing war effort, political loyalties and favours were forgotten and it was time for a change. The change came on 7 December 1916, when Lloyd George deposed his mentor Asquith.

That night, Lloyd George summoned Hankey to the War Office and told him he was now the prime minister. The two men "had a long talk about the personnel of the new Government, the procedure of the new War Committee and the future of the war". Hankey noted in his diary that Lloyd George "and Bonar Law, who was there part of the time, consulted me on many points."[68] As a result of their discussions, in addition to giving advice on a new War Cabinet Lloyd George asked Hankey to establish a War Cabinet Secretariat "to ensure the secretarial services of the War Cabinet itself" and so much more.[69]

What followed was to greatly affect the war effort, by making the workings of the War Cabinet more efficient. It also finally gave Sir Mark Sykes − who had been working in the background all this time − an official position, as one of two political secretaries in the new War Cabinet Secretariat. Leopold Amery was the other. Lord Milner, one of the mainstays of the new Lloyd George coalition had recommended Amery.[70] However, it was Hankey, full of praise for his friend, who recommended Sykes to Lloyd George:

> It is true that you know him mainly as an expert on Arab affairs, but he is by no means a one-sided man, has a considerable knowledge of industrial questions and an almost unique position in the Irish question as practically a conservative Home-Ruler. He also has a most extraordinary knowledge of foreign policy, and has views very similar to yours in regard to Turkey. He has a breadth of vision and a knowledge that may be invaluable in fixing up the terms of peace, which is a task that sooner or later is bound to fall to your lot.[71]

As political secretaries the two men were advisors to the War Cabinet and thus free from parliamentary questioning. It also brought Sykes into the inner circle of what Hankey referred to as the *"Supreme Command,"* once again giving him the authority and clout he had lost with the death of Lord Kitchener, but even more so at the very heart of the government.

At approximately the same time that Sir Maurice Hankey was speaking to Lloyd George about plans for his new government, unaware that a new coalition had taken office, Sykes wrote from home to a friend on the evening of 7 December 1916:

> [P]essimism has been increasing amongst Members of Parliament and civil servants, about the possibility of winning the war. These rumours must be counteracted before they do serious harm.[72]

Sooner than he realised, Sykes would find himself playing a key role in the government's efforts to win the war in the Middle East. Less than a week after his letter he was appointed assistant secretary to the War Cabinet for Political Affairs under his friend Hankey, who continued as cabinet secretary in addition to his new role as War Cabinet secretary. In his new official capacity, Sykes would continue working with Hankey, along with Leopold Amery, the other Assistant political secretary to the War Cabinet. As Amery noted later in his autobiography, the two men "were to be at the disposal of its members and at the same time free, as a kind of informal 'brains trust,' to submit our ideas on all subjects for our chiefs." They also were "to produce a weekly summary of the world situation. For the rest," Amery noted, "I was to make myself useful, as occasion arose, or as I might myself suggest." The same unique situation would also apply to Sykes.[73]

CHAPTER 7

THE ZIONISTS AND A JEWISH HOMELAND

One of our greatest finds was Sir Mark Sykes, Chief Secretary of the War Cabinet, a very colorful and even romantic figure ... He was not very consistent or logical in his thinking, but he was generous and warmhearted. He had conceived the idea of the liberation of the Jews, the Arabs and the Armenians, whom he looked upon as the three downtrodden races par excellence.

Chaim Weizmann[1]

Prior to his departure for St Petersburg in March 1916, in preparation for meeting the Russians Sykes reviewed the draft of the Sykes–Picot memorandum again, particularly the Brown area of Palestine. This time he did it with a new approach in mind – "the accommodation of Jewish opinion." Following Capt. Hall's suggestion in the Nicolson Committee meeting on 21 January that Jewish interest in Palestine should be taken into consideration, Sykes decided to look into the matter. He wanted to get a better understanding of Jewish opinion and its importance, and how it might affect a postwar settlement in the region. So he contacted Home Secretary Herbert Samuel.[2] A prominent Jew, Samuel had submitted two memoranda to the cabinet early in the war, proposing a British protectorate over Palestine after the war that would allow increased Jewish settlement.

In his memorandum "The Future of Palestine," submitted on 25 January 1915, Samuel made a passionate appeal on behalf of world Jewry:

The course of events opens a prospect of a change, at the end of the war, in the status of Palestine. Already there is a stirring among the twelve million Jews scattered throughout the countries of the world. A feeling is spreading with great rapidity that now, at last, some advance may be made, in some way, towards the fulfilment of the hope and desire, held with unshakeable tenacity for eighteen hundred years, for the restoration of the Jews to the land to which they are attached by ties almost as ancient as history itself.[3]

Samuel acknowledged "the time is not ripe for the establishment there of an independent, autonomous Jewish State," and that any attempt to put 90–100,000 Jews among a population of 400–500,000 Muslims would not succeed. Thus, "the dream of a Jewish State, prosperous, progressive, and the home of a brilliant civilization, might vanish in a series of squalid conflicts with the Arab population." So a solution whereby the country was annexed to the British Empire would provide a "solution of the problem of Palestine which would be . . . most welcome to the leaders and supporters of the Zionist movement throughout the world."[4]

He listed many arguments in favour of Britain pursuing such a policy, including an appeal to its greatness, as well as practical benefits: fulfilling its "historic part of civilizer of the backward countries;" adding "lustre even to the British Crown;" avoiding future disputes with Germany, by taking over Palestine instead of German colonies in postwar compensation; providing a "strategic frontier for Egypt"; and winning "for England the lasting gratitude of the Jews throughout the world, whose goodwill, in time to come, may not be without its value." Samuel followed this by listing the alternatives facing Britain if they did not annex Palestine. These included the possible "annexation by France," "internationalisation," "annexation to Egypt" or "to leave the country to Turkey," none of which, he argued, would be to Britain's benefit and some of which might even be to its detriment.[5]

Samuel admitted that "the gradual growth of a considerable Jewish community, under British suzerainty, in Palestine will not solve the Jewish question in Europe . . . [but] some relief would be given to the pressure in Russia and elsewhere." But what it would accomplish was more important:

Let a Jewish centre be established in Palestine; let it achieve, as I believe it would achieve, a spiritual and intellectual greatness; and insensibly, but inevitably, the character of the individual Jew, wherever he might be, would be ennobled. The sordid associations which have attached to the Jewish name would be sloughed off, and the value of the Jews as an element in the civilisation of the European peoples would be ennobled.[6]

However, Samuel's appeal fell on deaf ears. Asquith's response on reading it a few days later showed indifference if not outright disdain. "It almost reads like a new edition of 'Tancred' brought up to date," the entry reads in Asquith's diary on 29 January 1915:

I confess I am not attracted by this proposed addition to our responsibilities, but it is a curious illustration of Dizzy's [Disraeli's] favourite maxim that "race is everything" to find this almost lyrical outburst proceeding from the well-ordered and methodical brain of H.S.[7]

Nothing was done and the matter was dropped.

A year later, after reading a copy of the memorandum given to him by Samuel prior to leaving for Petrograd, Sykes wrote the home secretary on 26 February, "I read the memorandum and have committed it to memory and destroyed it as no print or other papers can pass through the R[ussian] frontier except in the F.O. bag." Unlike Asquith, Sykes grasped the essence of the memorandum, adding in his letter to Samuel, "I imagine that the principal of the ideal of an existing centre of nationality rather than boundaries or extent of territory. The moment I return I will let you know how things stand in [Petrograd]." Concerned with potential difficulties that might arise working with France as partners in a condominium, should Palestine be internationalised and knowing that a British protectorate would anger the French, Sykes suggested Belgium as a possible Entente trustee and administrator. This would keep out the French. It was apparently at this point that he began thinking the Zionists might possibly be of assistance in the problem of Palestine.[8] While he was in Petrograd thinking of Zionism as a possible solution to the impasse with Picot over Palestine, Sykes learned from British Ambassador Sir George Buchanan about the British Jewish Conjoint Foreign Committee's secretary and spokesman Lucien Wolf's formula for Palestine, received at the Foreign Office on 3 March. What he learned was to inspire Sykes further.

On 15 December 1915, Wolf had approached Lord Robert Cecil in a letter with "'suggestions' for pro-Allied propaganda in neutral countries," which he assured Cecil would win over American Jewry to the Allied cause. This could be done if the Entente were to declare its "policy on Palestine." A policy that would "guarantee equal rights for Palestinian Jews with the rest of the population, reasonable facilities for immigration and colonization, a liberal scheme of self-government, a Jewish university, and recognition of the Hebrew language" would, he assured Cecil, "'sweep up the whole of American Jewry into enthusiastic allegiance to the Allied cause.'"[9] While interest in Wolf's earlier proposition had been mixed at the Foreign Office, his subsequent 3 March formula was to attract much more interest. In it he proposed "that Britain could use Zionism to press forward any claim to Palestine in the face of French counter-designs." Anticipating French resistance to any scheme involving Palestine, prior to submitting his 3 March proposal to the British

Foreign Office Wolf had obtained "official endorsements from the [French] *Alliance Israélite*, the *St. Petersburg Committee* and the *Italian Committee*," the major organisations representing Jews in the Allied countries."[10]

Coincidently and unrelated to Wolf's proposal, a month earlier a similar proposal was made on 23 February to Sir Henry McMahon in Cairo, which he forwarded to the Foreign Office. In a letter to McMahon, Edgar Suarès,[11] a prominent Jewish leader in Cairo

> had argued that "with the stroke of the pen, almost, England could assure to herself the active support of the Jews all over the neutral world if only the Jews knew that British policy accorded with their aspirations in Palestine."[12]

Before the end of the Sykes–Picot negotiations and the arrival of Suarès's letter, Lord Grey and the Foreign Office had shown little interest in Zionism. However, upon receipt of Wolf's 3 March formula following Suarès's letter, Foreign Office senior official Hugh O'Beirne[13] minuted on 28 February "that offering the Jews 'an arrangement completely satisfactory to Jewish aspirations in regard to Palestine' would have 'tremendous political consequences.'" In another minute he noted "that 'the Palestine scheme has in it the most far reaching political possibilities.'" Picking up on this, Lord Crewe,[14] who was substituting for Grey while the latter convalesced from an illness, noted:

> We ought to pursue the subject since the advantage of securing Jewish goodwill in the Levant and in America can hardly be overestimated, both at present and at the conclusion of the war. We ought to help Russia realize this.[15]

Now convinced of the importance in seeking worldwide Jewish support over Palestine, the Foreign Office decided Wolf's proposal was inadequate to the task and needed to be improved "to make it more substantial, 'far more attractive to the majority of the Jews.'"[16] Delegated to do this, O'Beirne broadened the scope of the proposal "to hold out the prospect of eventual Jewish statehood once the Jewish colonists had grown large enough in number to 'cope with the Arab population.'" Following up on this, Lord Crewe's letters of 8 and 11 March to Lord Bertie in Paris and Sir George Buchanan in St Petersburg, included the following statement:

> We consider, however, that the scheme might be made far more attractive to the majority of Jews if it held out to them the prospect that

when in course of time the Jewish colonists in Palestine grow strong
enough to cope with the Arab population they may be allowed to take
the management of the internal affairs of Palestine (with the exception
of Jerusalem and the holy places) into their own hands.[17]

After a discussion with Russian Foreign Minister Sergei Sazonov, Buchanan
wired the Foreign Office on 14 March to say that Russia raised "no objection
to the scheme in principle, but sees great difficulties in the way of its
execution." He added that while the "Russian Government would welcome
migration of Jews to Palestine, he doubts whether any considerable number of
them would care to settle there. He will send me [an] answer after a thorough
examination of the question." The following day Buchanan confirmed that the
Russian government agreed to the proposal, with the stipulation that
"the Holy Places were definitely excluded from the scheme and placed under
international control." Thus, with Russian approval, Britain and France were
free to allocate the Arab Ottoman lands between them and Petrograd would
"accept any Anglo-French arrangements, provided that Russian desiderata in
Constantinople and the Straits were met." As for British and French interest
in Palestine, "the Russians hoped that all Orthodox establishments would
enjoy religious freedom and their privileges respected. Otherwise, they had
'no objections, in principal, to the admission of Jewish colonists to the
country.'"[18]

Inspired on learning all this, Sykes developed a scheme of his own.
Although he referred to it as Wolf's plan, it was basically his own creation: "a
Jewish chartered company for land purchase plus full colonizing facilities in
an enlarged Palestine, excluding Jerusalem." By using this approach, he
believed it would win over previously antagonistic Jews. It had been said of
Sykes that "He saw 'Jews in everything,' and attributed to them a corporate
will." This gave them power and influence, and to gain their support Sykes
believed that "the Zionists are the key to the situation – the problem is how
are they to be satisfied."[19]

While he was in Petrograd Sykes shared his ideas with Picot "about how
the Allies might gain Zionist support" for the Allied war effort, and wired
London a suggestion "that a chartered Jewish company might satisfy Zionist
aspirations." On receiving his telegram in London, a startled Nicolson,
"worried about Sykes barging in, fired back a curt wire telling him to keep
such thoughts to himself." Surprised at the rebuke, Sykes responded in a long
letter in which he told Nicolson "his talks with Picot had been harmless, since
he had only told him: 'H.M.G. hate Palestine & don't want it – Zionists want
us for obvious reasons,' which Picot already knew." He pointed out that "the
two of them had come up with the chartered company formula in the hope

that it would give the Zionists more confidence in the Allies." Thus, "If they [the Zionists] want us to win they will do their best which means they will (A) Calm their activities in Russia (B) Pessimise in Germany (C) Stimulate in France, England & Italy (D) Enthuse in the U.S.A." To Sykes, Zionist opinion was now "the key" to acceptance of the Allied agreement of 1916. Strongly confident in his belief, he added in his letter to Nicolson:

> With Great Jewry against us there is no possible chance of getting the thing thro' – it means optimism in Berlin – dumps in London, unease in Paris, resistance to the last ditch in C[onstantin]ople – dissention in Cairo – Arabs all squabbling among themselves – as Shakespeare says, "Untune that string and mark what discord follows."

Posing the question, "is a land company enough?" Sykes acknowledged to the under secretary that while it "might seem rather odd & fantastic … when we bump into a thing like Zionism, which is atmospheric, international, cosmopolitan, subconscious, and unwritten – nay often unspoken – it is not possible to work and think on ordinary lines." He "concluded by reassuring Nicolson that the matter would naturally require 'careful handling' in London, which was 'within reach of Paris.'"[20] Thus, Sykes's growing interest in Zionism coincided with that of

> the British government [which] was suggesting to its allies that a statement on the Palestine question in language to which Zionists would respond would be an effective answer to German propaganda among the Jews in the United States and elsewhere.[21]

After his return to London on 10 April, while the Allied negotiations (renamed because of Russian involvement) were being finalised at the Foreign Office, Sykes's active imagination continued to struggle with ideas on how to resolve the Palestinian issue, this time by involving the Arabs. However, his idea of establishing an Arab sultan as titular ruler of Palestine was rejected both by the Home Office and Cairo.[22] So, once again his thoughts turned to the Zionists as a possible solution and he approached Herbert Samuel at the Home Office. As a non-Zionist, Samuel believed Sykes needed to speak to someone with more knowledge on the subject and arranged for him to meet his friend Rabbi Moses Gaster, a Romanian-born Hebrew scholar and chief rabbi or Haham of the Spanish and Portuguese (Sephardic) Congregation of Great Britain.[23] Gaster was pleased to welcome such an eager pupil into his home.

It was Gaster, Sykes recalled later, who "opened my eyes to what Zionism meant."[24] With Gaster "he discussed questions relating to the future of

Palestine, mentioning (but without disclosing the existence of the Sykes–Picot agreement) the possibility of an Anglo-French condominium."[25] Gaster first met Sykes on 2 May and again two days later. On 10 May he met both Sykes and Picot and noted afterwards in his diary:

> I believe I made them [Sykes and Picot] [see] the importance of [a] Jewish Commonwealth in Palestine and the ideal for which the Republic is at war ... Told him [Picot] of reverence for memory of Napoleon among Jews ... Against positive assurances *we* would do our best for creating public opinion favourable to France."[26]

On 7 July Gaster met Picot and noted in his diary about their meeting:

> Fr. Govt. wishes a manifestation by Jews in favour of Entente. Asked also to define our wishes. I summed them up: Lebanon [word illegible], local autonomy, freedom of exercising civil rights, protection of property and recognition of Jewish enclave so that Jews may develop in peace. He fully understood and seemed satisfied.[27]

Gaster wrote to Sykes over the summer, but there are no entries in his diary to indicate there were any further meetings between them until November.[28] On 23 July he sent Sykes a book entitled, *"Zionism and the Jewish Future,"* which he hoped Sykes would look over to get an appreciation of the way "in which some of us look to the future solution to that great problem."[29] There Sykes's curiosity in Zionism remained, put aside by more immediate matters. That would soon change. Early in 1917 Sykes would become directly involved and enthusiastically assist the Zionists and their cause to establish a Jewish homeland in Palestine at home and abroad at the highest levels of Entente politics.[30]

On 18 October 1916, Sykes met Aaron Aaronsohn, a Jewish agronomist, botanist and Zionist activist from Palestine, who revived his interest in Zionism. In 1906 Aaronsohn had "won international acclaim for discovering a weather-resistant primeval wheat." In 1910, with help from prominent America Jews, he had founded the Jewish Agricultural Station in 1910 at Athlit. Over the next few years at Athlit, Aaronsohn "carried out extensive research on dry-farming techniques." By the beginning of the war, however, even though he was a Turkish citizen and was greatly respected for the work he had done helping the Turkish government with locust infestation, it was hard for Aaronsohn to ignore the evictions and confiscations by the Turks "carried out against the Jews and Arab alike, [and] the horror visited upon the Armenians." This only confirmed his worst premonition that "neither the Land of Israel nor the Jewish settlement there had any future under the . . . brutal Ottoman regime. The Jews' best hope . . . was simply to wrest Palestine away for themselves."[31]

Anticipating a British invasion of the area once the war began, Aaronsohn and his companions at the research station decided to approach the British in Egypt and offer to provide them with information on Ottoman troop movements in Palestine. As veteran settlers, known and respected by the Ottoman authorities for their agronomical work, they "were in a unique position to supply this intelligence . . . [and were] generally permitted freedom of movement throughout Palestine in organizing [anti-locust] campaigns." To put their plan into operation they established a spy network called NILI, taken "from the initials of its Hebrew password, 'Nezach Ysrael Lo Y'shaker' – The Eternal One of Israel Will Not Lie" – to gather the necessary information to pass on to the Allies.[32] However, when Aaronsohn was able to get to Egypt he was met with indifference by British authorities who did not take his offer to spy for the British seriously.[33]

On his return to Palestine, a bitter and disillusioned Aaronsohn learned the Turks were planning a second invasion of the Suez Canal. This time he believed he had to go to London and find someone in the British government who would listen to him. In a trip full of mystery and intrigue worthy of any good spy novel, Aaronsohn travelled in secrecy for five months, crossing and recrossing enemy lines, hiding his purpose with sham papers and outrageous if authentic sounding stories as he bluffed his way through belligerent and neutral countries from Palestine to London. There he was interviewed by Maj. Walter H. Gribbon[34] of the Military Intelligence Directorate at the War Office, in charge of Turkish affairs and deputy to the Director of Military Intelligence Lt. Gen. George Macdonogh.[35]

Alerted to Aaronsohn's presence in London and his interview by Gribbon at the War Office, Sykes was interested to learn what he might have to say. So he got permission to sit in on his second interview with Gribbon on 18 October 1916. At the time Sykes was still a self-professed anti-Semite. He "saw the Jew as the embodiment of international capitalism, and he despised what he saw as the rootlessness of the wealthy assimilated Jews whom he encountered in Britain." While he had some knowledge of Zionism from what he had learned from Herbert Samuel and Rabbi Moses Gaster, it was casual and relatively superficial and had not changed his opinion. However, after meeting Aaronsohn Sykes saw things differently. He found himself intrigued by the man and his total dedication to his people and their cause. His interest piqued, Sykes spent some time questioning Aaronsohn about Zionism.[36]

Like Gribbon, Sykes was greatly impressed with Aaronsohn and the story of what he had accomplished in Palestine. It changed his lifelong negative view of Jews and forced Sykes to consider seriously a possible role for Zionism in the Middle East war. Aaronsohn, in terms of personality and attitude, was the complete antithesis of what Sykes had come to believe about Jews. To the astonished Sykes, he "was no money-grubbing capitalist, but a hardy, hard-working son of the soil, a brilliant scientist to boot, who was manifestly abandoning a career and endangering his life for the sake of his people; and he was not alone." The story he told of his Zichron Ya'akov community in Palestine and the work they were doing at Ahlit "presented a Jewish picture completely new to Sykes."[37] A close friendship developed quickly between the two men. They met three times while Aaronsohn was in London, on 27 and 30 October, and on 6 November 1916, the first time at the War Office and afterwards at Sykes's home, where they discussed Zionism and wartime conditions in the Ottoman Empire.[38] When they met again the following year in Cairo, Aaronsohn noted in his diary on 27 April 1917: "We immediately broached intimate subjects. He told me that since he was talking with a Jewish patriot, he would trust me with very secret matters, some of which were not even known to the Foreign Office.'"[39] Speaking about Aaronsohn to another Zionist, Sykes recalled later "how deeply impressed he was by Aaronsohn, who inspired him with the vision of a Jewish Renaissance in Palestine."[40]

After meeting Aaronsohn Sykes was faced with the very real prospect of helping the Jews to achieve their goal of a homeland in Palestine. However, while he was sympathetic to Zionism and its cause, ever foremost in his mind was his own goal: to find a way for the Allies to win the war in the East. Thanks to his friendship with Aaronsohn, the two causes merged and Sykes began to see in Zionism a solution to his main problem. It helped that he subscribed to the widely held belief about Jews that, through their wealth and political connections, international Jewry had great power and influence. Ever the pragmatist, he saw how the Zionist dream of a homeland in Palestine could help Britain in a postwar settlement. A Jewish colony established in Palestine and overseen by the British would serve as a buffer between the French in Syria and the British in Egypt, while also developing the area without any expense to Britain.

Sykes hoped British support of Zionism would also be helpful in the USA, where Zionism was a growing movement, but where there was also some support for Germany among ethnic German Jews and there was a strong anti-Russian sentiment among a large number of American Jews of Russian origin. The historical mistreatment and persecution of Jews in Russia was not easily forgotten. So while American Jews were generally friendly to the Allies, Russian membership in the Entente caused many of them to resist supporting the Entente early in the war.[41] However, Russia was also the birthplace of

Zionism. So Sykes hoped that British support for Zionism would turn anti-Russian sentiment around in the USA and Zionists there would use their political influence on the government to bring it into the war on the side of the Allies.[42]

In Russia there were Jews among the leaders of the reformers as well as among the revolutionaries. Sykes hoped that these Jews – in appreciation of Allied support of the Zionists – would use their influence to support the Allies and keep Russia in the war. Thus, should there be a revolution, they would be encouraged to keep the vast Russian army fighting on the Eastern front and the German troops tied down, as well as prevent Russian from signing a separate peace agreement and withdrawing from the war. Sykes was not alone in his thinking.

At the Foreign Office, "fears were focused on British and American Jewish bankers, whose German connections and political influence were exaggerated, but whose financial support was [viewed as] vital to the war effort." So when revolution broke out in Petrograd on 23 February 1917, followed by Tsar Nicholas II's abdication on 15 March and the formation of a provisional government, Britain feared that Russia would withdraw from the war and the revolution would spread. *The Times* reported on the Bolshevik sympathies of Russia's newly emancipated Jews as follows:

> Political anxiety in Britain began to focus on the "revolutionary Jew," agitating for the overthrow of the monarchy and parliamentary democracy. One possible way to deter Russian Jewry from supporting Bolshevism, in the eyes of the British government, was to pledge British support for Zionism.[43]

The move to support Zionism began several months earlier in December 1916 with the change in government, when Lloyd George became prime minister and Arthur Balfour foreign secretary. There were two reasons for this. The first was it was never very clear how binding the Sykes–Picot agreement and the Husayn–McMahon understanding "might be at a peace settlement nor even where the exact frontier between French Syria and Northern Palestine might run." The other was that both Lloyd George and Arthur Balfour knew Zionist leader Chaim Weizmann, the acknowledged spokesman for Zionism in Britain. As a result, his influence on the pro-Zionism of both politicians and the newly inspired Assistant Secretary to the War Cabinet Sir Mark Sykes was to reap major dividends for the Zionists.[44]

Balfour had known Weizmann for longer than Lloyd George. The two men first met in 1906, when Balfour was running for Parliament in north Manchester and Weizmann was a reader in biochemistry at the University of

Manchester. A Russian by birth and a naturalised British citizen, in 1915 Weizmann developed a process for a speedier and more economical production of acetone essential in the manufacture of the important explosive cordite needed for ammunition. His scientific discovery coincided with the Shell crisis of May 1915, when it was discovered that British shell supplies had dwindled to such a point there was uncertainty whether Britain could continue in the war. This resulted in a major political crisis in which a new coalition government was formed under Asquith on 25 May, and a new ministry of munitions established with Lloyd George as the minister. In June Weizmann's discovery brought him to the attention of Lloyd George and three months later Weizmann was appointed chemical adviser on acetone supplies to the ministry. On 31 September 1916 he was employed by the Admiralty.[45]

In addition to his scientific work, Weizmann had long been active in Zionist affairs. After many years of working in Zionist circles in England he took up Theodor Herzl's mantle as leader of the Zionist movement ten years after the latter's death in 1904 had left the movement divided and in need of a strong central leadership. Despite many attempts by others to bring the dissident factions together, this was eventually accomplished under Weizmann in England, not in continental Europe, where the movement began. Moreover, at the beginning of the war the World Zionist Organization moved its headquarters from Berlin to neutral Copenhagen, where it could function without pressure and it effectively transferred its allegiance to Britain.[46] By August 1914 Weizmann was one of two vice-presidents in the English Zionist Federation and a member of the World Zionist Organization's Greater Actions Committee and in February 1917 he became president of the English Zionist Federation.

The idea of establishing a modern Jewish state on the site of ancient Israel in modern Palestine has generally been credited to Theodor Herzl, but for millennia at the Passover Seder Jews routinely prayed "Next year in Jerusalem!" and dared to hope that one day the Jews of the Diaspora would return to the Land of Israel. While many fantasised about it and after pogroms in Russia and Central Europe some did immigrate to Palestine,[47] no one had articulated a scheme to make it happen quite like that of Herzl. Born in Budapest on 2 May 1860, Wolf Theodor (Benyamin Zev) Herzl was given a secular education, received a doctorate in law and was admitted to the Vienna bar in 1884. He soon showed more interest in literary pursuits than the law and began writing and publishing plays and essays. By 1887 Herzl was writing more or less regularly for Berlin and Viennese newspapers. In October 1891 the *Neue Freie Presse* appointed him its correspondent in Paris where, among other things, the young journalist covered the Dreyfus trial in 1894–1895. During this time, the virulent French anti-Semitism exhibited during the trial profoundly shocked him. As a result of both this and the long history of anti-Semitism in Europe, Herzl concluded that Jews would never have peace in the world until they had their own homeland. After considering a variety of ways to resolve the so-called Jewish question, he articulated his idea of founding a Jewish state in *Der Judenstaat*, or *The Jewish State: An Attempt at a Modern Solution for the Jewish Question* (its full English title), which was published in Vienna on 14 February 1896.[48]

Far from being a best-seller, *The Jewish State* was generally read only by Jews. Initially, it was not taken seriously in many quarters, especially in the Jewish press. However, it prompted a variety of reactions ranging from enthusiasm to outright hostility and antagonism towards the ideas contained in it.[49] Some Jews in Europe viewed its ideas with fear and apprehension, thinking that espousing such opinions would only cause more problems. Some saw it as a challenge to the assimilation they still hoped to achieve in a Europe, just when some places in Europe opportunities were beginning to open up to Jews as never before. Still others read it with an excitement mingled with hope. The following year Herzl's little book and its ideas led to the first Zionist Congress at Basel in 1897 and the founding of political Zionism with the World Zionist Organization.

A year before Herzl's death in 1904, in April 1903 British Colonial Secretary Joseph Chamberlain proposed to Herzl that Uganda was a possible alternative to Palestine. In what came to be known as the Uganda project, Herzl supported it as a temporary place of refuge for Russian Jews facing persecution and pogroms at the time and he brought it before the World Sixth Zionist Congress in Basel in 1903, where it was approved. However, Russian Zionists opposed it and stormed out of the Congress, which resulted in a split in the movement over the issue. Meanwhile, a group of British colonists already living in Uganda strongly opposed the idea, so the colonial secretary was forced to withdraw the offer. By then, the embattled Herzl "was almost relieved that now the issue had resolved itself, and with it the threat of a conceivably permanent disruption of the Zionist movement." However, the strain of constant work and this latest turmoil was to take its toll on Herzl, who died the following year.[50]

If the controversy over the Uganda project did nothing else, it committed the Zionists firmly to a Jewish homeland in Palestine. But for the first time it also brought Britain into the picture to help the Zionists establish that homeland, which was hinted at in Herzl's original plan. In *The Jewish State*, Herzl described his plan and its organisation:

The plan will be carried out by two agencies: The Society of Jews and the Jewish Company. The Society of Jews will do the preparatory work in the domains of science and politics, which the Jewish Company will afterwards apply practically. The Jewish Company will ... organize commerce and trade in the new country."[51]

In order to do all this, he continued:

The Society of Jews will put itself under the protectorate of the European Powers, if they prove friendly to the plan: We would offer the present possessors of the land enormous advantages, assume part of the public debt, build new roads for traffic, which our presence would render necessary, and do many other things. The creation of our State would be beneficial to adjacent countries, because the cultivation of a strip of land increases the value of its surrounding districts in innumerable ways.[52]

As for the Jewish Company: "[It] will be founded as a joint stock company subject to English jurisdiction, framed according to English laws, and under the protection of England. Its principal center will be in London."[53]

These arrangements would be central to the Zionist's approach to Britain during World War I. In fact, Sykes and Picot were aware of this part of the Zionist plan and had already discussed the matter of the establishment of a Jewish company in Palestine in their conversation in St Petersburg in March.[54]

At the end of December 1916 Sykes attended an Anglo-French conference in London to discuss the offensive by the Egyptian Expeditionary Force into Palestine. Uneasy as to what would come of the military operation, French fears were lessened with the addition of both British and French political officers – Sykes and Picot – to the general staff in the field. Sykes wanted to incorporate the Zionists in his Anglo-French-Arab scheme but the Foreign Office was in no mood to reopen negotiations over Palestine.[55] It was believed such a move might renew French interest in Palestine and allay their suspicion that Britain might only be pretending to be defenders of Zionism as a pretext to annex Palestine for themselves. So, despite the reduction in tensions over the matter, the idea was put aside for the moment.

After his attachment to the secretariat of the Committee of Imperial Defense, Sykes became responsible for providing the Committee with periodical notes known as the Arabian Reports on the situation in the Middle East. He also acted as liaison officer between the various government departments concerned with Middle Eastern affairs, and by late 1916 was also entrusted, in addition to his other duties, with the study of the Zionist question.[56] With the change of government in December 1916 Sykes continued these duties as assistant secretary to the War Cabinet in the Lloyd George coalition government and was also assigned to the Foreign Office in February 1917.[57]

In his dual role as an assistant secretary to the War Cabinet in Lloyd George's coalition government, as well as a member of Parliament, Sykes was now in a special position in relation to both the War Cabinet and Parliament. Along with fellow MP and assistant secretary to the War Cabinet Leopold Amery, Sykes held the status of parliamentary under secretary and was under no obligation to answer for his activities to Parliament.[58] As a so-called brains trust for the War Cabinet, the two men were relatively free to pursue whatever the War Cabinet might ask of them. Or they could also pursue any matter that might interest them other than the political reports assigned to them by Hankey, such as Sykes's Arabian Report. This freedom of action led Sykes to think more about the possibility that Zionism could help with the war effort.

Having been recently sensitised to the Zionist point of view through his meetings with Aaronsohn, Sykes could see the benefit of having "world Jewry behind the Allies," but did not want Zionism to jeopardise the Sykes–Picot agreement and the "promotion of Anglo-French co-operation with the Arabs throughout the Middle East."[59] Two developments in late January 1917 persuaded him to pursue his own personal diplomacy, which led to his first meeting with Dr Chaim Weizmann. These were a proposal for a Jewish Legion to be formed to accompany the British forces into Palestine and a Zionist journal's call for a Jewish Palestine under the British Crown.[60]

Sykes now found himself in a quandary, because other than Samuel and Rabbi Gaster he did not know any leading British Zionists. This was due to Gaster, who according to Weizmann writing later in his autobiography *Trial and Error*, "had a tendency to keep his 'finds' to himself, and play a lone hand." As an example, he wrote that Gaster "did not tell me until after I had met Sir Herbert Samuel that the latter, though not a member of the Zionist Organization, had long been interested in the idea of a Jewish State in Palestine!" Weizmann would eventually meet Herbert in November 1914, without Gaster's assistance.[61] Perhaps Sykes sensed this in Gaster, which was why, instead of asking Gaster, he asked his long-time friend and neighbour James Arootun Malcolm, a well-connected Armenian businessman, to find out for him "who the leading Zionists in London were." With the help of L.J. Greenberg, editor of the *Jewish Chronicle*, Malcolm was able to arrange for Sykes to meet Dr Chaim Weizmann, who was not known to either of them, at Sykes's home at 9 Buckingham Gate on 28 January.[62] At the meeting

Sykes expressed his concern over how raising a Jewish Legion to accompany British forces into Palestine might endanger the lives of Jews already in Palestine and also jeopardise Allied cooperation with the Zionists.[63]

Encouraged by his first meeting, two days later in another meeting with Weizmann Sykes asked him to arrange a meeting between himself and other leading members of the Zionist movement. He also asked Weizmann to prepare a summary of Zionist aims. Just prior to this meeting on 7 February, Sykes received a summarised list of Zionist aims from Weizmann:

> Palestine to be recognized as the Jewish National Home, with liberty of immigration to Jews of all countries, who are to enjoy full national, political, and civil rights; a charter to be granted to a Jewish Company; local government to be accorded to the Jewish populace; and the Hebrew language to be officially recognized.[64]

After his second meeting with Sykes in January, in which he expressed his desire to meet British Zionist leaders as a group, Weizmann sought to address the concerns Sykes had raised. On 3 February 1917, he wrote to Israel Sieff, one of the co-founders of the British Palestine Committee, whose weekly journal *Palestine* had in its first issue on 26 January 1917 greatly disturbed Sykes.[65] Its first banner, repeated in subsequent issues, read: "The British Palestine Committee seeks to reset the ancient glories of the Jewish nation in the freedom of a new British dominion in Palestine." Weizmann wrote Seiff:

> What we desire is to work out the best method for obtaining our aim and you must on careful consideration concede that a fight with the F.O. [Foreign Office] is the last thing we desire ... The matter is entirely in Sir M[ark]'s hands and only when he is ready, it will come before the F.O. All our attention must be therefore concentrated on an endeavour to persuade Sir M. that a joint Protectorate is harmful. I think that Sir M. knows that as well as we do, but he has to move very carefully indeed and it would be *very harmful* at the present stage to start publicly a controversy, which would only make things more difficult ... Above all, I would do nothing *now* until we have seen the results of the conference.[66]

After some resistance, Sieff finally bowed to pressure from Weizmann and eventually conceded in a letter dated 20 February that, "for the sake of 'discipline and unity ... *Palestine* this week will contain a Jewish article which will meet the wishes of Sir M.'"[67]

The meeting arranged for Sykes to meet Britain's Zionist leaders as a group and was held at Rabbi Gaster's home on 7 February. Those in attendance, besides Gaster, Sykes and Weizmann, included Lord Walter Rothschild, Herbert Samuel, James de Rothschild, Nahum Sokolow, Joseph Cowan, Herbert Bentwich and Harry Sacher.[68] After introductions, Sykes explained that he was there in a private capacity without instructions from the Foreign Office or the War Cabinet and, for that reason their discussions must remain secret. He then explained his opposition to the Jewish Legion and why he wanted the Zionist propaganda in the journal *Palestine* to stop. For their part, the Zionists told Sykes they were opposed to the Anglo-French condominium idea and wanted Palestine to be ruled by Britain. Sykes responded that "rejection of the condominium approach brought them up against a problem for which he had no sure solution: France ... refused to recognize that concessions to Zionism might help win the war."[69]

Samuel downplayed any possible French objections, suggesting that arrangements could be made at the peace conference to compensate them for their "pretensions to Palestine."[70] Nevertheless, Sykes strongly recommended that one of the group be appointed to contact the French representative, François Georges-Picot, at the French Embassy in London. After some discussion it was decided that, because of his position as executive secretary of the World Zionist Organization's Executive Board and his fluent French, Nahum Sokolow was the best person to represent the Zionists and Sykes would make the introductions.[71]

Continuing the discussion, Sykes told the group that he sympathised with Zionist aims, particularly a Jewish chartered company, and thought that Britain might guarantee such an undertaking. Eager to ingratiate himself with the Zionists, Sykes exhibited a condescending and patronising attitude towards the Arabs that all too soon would come to haunt the British. He assured the Zionists that "provided the Holy Places were guaranteed, he could foresee no Arab opposition to Zionist colonization in Palestine as long as existing inhabitants were not disturbed."[72] It would do no harm, he added, for the Zionists to strengthen their own case by publicly supporting that of the Arabs elsewhere.[73] For his part, Sykes declared his willingness to promote Zionist aims in the War Cabinet. Before he did so, however, he "suggested they get in touch with Zionists elsewhere, offering to make the War Office telegraph facilities available to them so they could communicate secretly with leading Zionists in Paris, Petrograd, Rome and Washington, D.C."[74]

The day after his 7 February meeting at Dr Gaster's, Sykes introduced Nahum Sokolow to Picot at his London home. He had urged the Zionists to take Picot into their confidence and secure his support of their goals in Palestine and in the hope that this would gain them a hearing with the French government.[75] His domestic French commercial and political interests, as well as his personal experience in the Middle East, proved a potent mixture that made Picot very influential within the Quai d'Orsay and the French government. The Zionists agreed that Sokolow was best suited for the task of meeting such a man as Picot. In addition to being secretary general of the World Zionist Congress, the supreme body of the World Zionist Organization and his fluency in French, it was believed the fact that Sokolow was Russian would enable him to speak for a large portion of the Jewish people. Further, as a layman, he could also "emphasize the fact that the aspirations for the Jewish masses were national as well as religious."[76]

Initially, however, the talks did not go well. Picot was firm in his belief that if the Jews were to given Palestine, the protectorate should go to France. The wisdom of choosing Sokolow to negotiate soon proved itself. He took the opportunity to expound the Zionist cause, explaining that for Russian and Central European Jews there was no alternative to their problems than settlement in Palestine. Avoiding the issue of which country was to receive the protectorate, Sokolow's patient and detailed explanation of the Zionist programme eventually made its impression on Picot. At the conclusion of the meeting, even though he was still unwilling to concede French interests in the area, Picot assured Sokolow he was in no way antagonistic to Zionism and that he was sympathetic to the cause he represented.[77]

Conversations between the two men continued without Sykes the next day at the French Legation in London. At this stage Sokolow and Picot discussed what could be done to make Zionist aims and purposes better known in France. Like Sykes, Picot felt that overt propaganda would arouse opposition. He agreed to explain about Zionism in person to the French government. Picot did have concerns about the reaction of French Jews, but Sokolow assured him once a practical programme was in place any Jewish opposition would cease.[78]

The next day, on 10 February, Sokolow, Weizmann and Sykes met at Sykes's home to discuss the results of Sokolow's meetings with Picot. Pleased with Sokolow's report but urging caution, Sykes suggested that Weizmann and Sokolow did "not emphasize British sovereignty, because it would arouse premature French objections." It was also agreed that support of the Allies was needed and they decided that Weizmann would work with the British statesmen and Sokolow with the French and Italians.[79]

At the meeting Sykes asked Sokolow for information on "existing Jewish ownership of land" and plans for "future settlement of the Jews in Palestine." Two weeks later, after some research, Sokolow personally hand-delivered a three-page note to Sykes. Written in his miniscule hand, Sokolow divided the details between a section titled, "About the territories of Palestine" and one on the "Jewish ownership of land in Palestine."[80] Sykes's subsequent analysis of the material clearly showed the Zionists' ultimate goal was to colonise all of

Palestine, with the exception of the holy places.[81] Knowing this would upset the French, he wrote to Picot at the end of February suggesting that Palestine should become an American protectorate. "Since the USA was still officially neutral, such an offer ... 'would give strong impetus to the Entente cause,'" as well as reduce Anglo-French friction over the area.[82] He also stressed to Picot that he had not mentioned the Anglo-French plans set out in the Sykes–Picot agreement to the Zionists, because they might jeopardise it. They would learn about it much later from someone other than him.[83] He told Picot this was done, because "if the great force of Judaism" felt its aspirations were not met, then he saw "little or no prospect for our future hopes.'" He went on to suggest to Picot that worldwide Zionist resources "would help with the development of the Middle East and, provided guarantees were made to the 'native population' of Palestine, there should be no opposition to Zionist colonization." Picot's response was non-committal.[84]

Meanwhile, events in Russia following the February Revolution continued to capture the newspaper headlines in Britain and Sykes's attention. On 8 March strikes and riots followed by a general mutiny of soldiers in Petrograd began the long-anticipated revolution in Russia. On 16 March a provisional government under Alexander Kerensky was formed after the abdication of Czar Nicholas II the day before. For Sykes the revolution meant that the Zionist supporters of Kerensky's government, who would not fight for the Czar, would now stay in the war for the sake of their homeland in Palestine. That would happen only if the Entente Powers, including Russia, supported a Jewish homeland there.[85] Continued Russian involvement in the war was essential and critical to the Allies' war effort, both in Europe and the Middle East, and he believed Zionist support was the key to Russia's continued commitment. After all, Weizmann had assured the British that "government-endorsed 'Zionism' would deter Russian and East End Jewry from supporting the revolution."[86]

Weizmann also saw this as the time to change the Zionists' approach and tactics. Reacting to Sykes's suggestion that he accompany him to Egypt to be ready to do propaganda work and diplomacy in Palestine once the British occupied it, Weizmann wrote to his friend C.P. Scott, the editor and owner of the *Manchester Guardian*, an early supporter of Zionism on 20 March:

I feel ... that our negotiations must be placed very soon on a more definite practical basis and the plan that I have outlined in my short letter to you before, namely, that I should accompany Sir Mark to the East, enter there into negotiations with the leading Arabs from Palestine and see what can be done almost immediately in the way of acquisition of land in the Palestinian territory already occupied by the

British. It is of the utmost importance that we should *be there* as soon as possible, that the Palestinian people and Jews at large should realize that we mean business and mean to carry it out at once.[87]

However, the trip never materialised for Weizmann, who had too much to do in England. No doubt this had much to do with his recently gained unimpeded access to Lloyd George. The prime minister had told him at a dinner party on 13 March, when asked for an appointment for a further talk, "You must take me by storm. Just come, and if … I am engaged, do not be put off but insist on seeing me."[88]

With this and the help of others, the Zionists' cause continued to gain momentum and support in high places on the continent and on both sides of the Atlantic. Meanwhile, in March Sykes's attention again turned eastward, as he was assigned to the newly formed Mesopotamian Administration Committee as its secretary. With this key position, a new series of challenges and more eastern travel now faced the peripatetic Yorkshire peer. Throughout the rest of the year, whether at home or abroad he remained in constant contact with the Zionists, particularly Nahum Sokolow and made himself available to the Zionists whenever he was needed.

In late March Sykes received a handwritten letter from Capt. William Ormsby-Gore.[89] A fellow Conservative MP and personal friend of Sykes's, Ormsby-Gore had been assigned to the Arab Bureau, where he was responsible for gathering material on economic and agricultural information on Palestine and Zionism. Due to his new areas of responsibility in Cairo he had recently gained an ally when he met Aaron Aaronsohn on the latter's return to Egypt from London in January 1917. This would prove to be an auspicious meeting for Ormsby-Gore. A recent convert to Judaism, he soon joined Sykes as an earnest supporter of a Jewish homeland in Palestine.[90]

In a letter to Sykes on 26 March, Ormsby-Gore told him about the strong anti-French feeling in Palestine and Syria. "The point I want to urge," he wrote,

is that Picot will find *no* friends whatever, except a few clerics, south of the [Orontes] river [in northern Lebanon] and that French influence south of [Beirut] & the Lebanon is nil while Jews (re Dreyfus), Druses (re Maronite War) & Moslems (since France is the protagonist of aggressive Christianity in Syria) are really solidly anti-French now — this feeling has been steadily increasing since the war.[91]

Changing the subject, Ormsby-Gore mentioned Amery had told him "there is a good chance of getting a Jewish Battalion or two if the W.O. will move."

Apparently, he was unaware Sykes was responsible for the delay in the formation of a Jewish Legion and not the War Office. Whether it was his recent conversion to Judaism or the growing relationship with Aaronsohn that influenced his feelings in this regard, he told Sykes:

> [I]t will be a good thing & you would get really useful additions out of the Jewish refugees in Egypt – quite a solid increase to [General Sir Arthur] Murray's force. They would have tremendous esprit de corps in advancing into Palestine & I believe the political effect would be valuable in America and elsewhere.[92]

Shortly afterwards, Ormsby-Gore returned to London and Hankey had him assigned to the secretariat to assist Sykes. With his assignment to the Mesopotamia Administration Committee and the growing activity in the Middle East, Sykes was seriously in need of help. Besides, he was leaving for Cairo with Picot to join General Murray as chief political officer with the Egyptian Expeditionary Force in their invasion of Palestine.[93]

Before leaving for Europe Sykes learned that Lucien Wolf was continuing his anti-Zionist campaign. Joseph Greenburg, editor of *The Jewish Chronicle* and ardent Zionist, wrote to Sykes on 21 March to advise him that an article Wolf had written against Zionism, entitled, "The Jewish National Movement," was to be published in the April edition of the *Edinburgh Review*. Greenburg enclosed a pre-publication copy of the ten-page article he had obtained through the Press Association with his letter.[94] It was clear the Zionists wanted to keep him abreast of anything related to the Zionist movement, in particular, the anti-Zionist activities of the charismatic and formidable Wolf, who had his own contacts in the government. By now, Sykes had become the Zionists' major asset and key advocate in the British government and among its allies.

Sykes and Picot had arranged a visit to Paris so Sokolow could meet Picot and the key members of the French government.[95] Sykes arrived in Paris a few days after Sokolow, on his way to Cairo. Impatient and anxious to make himself available in case his assistance was needed, according to Sokolow, "in spite of his complete confidence in us ... [Sykes] could not refrain from remaining near me, always ready with advice and help."[96] However, by the time Sykes had arrived in Paris Sokolow had already persuaded the French government officials that the Jews in Palestine were to be considered a nationality. Apparently, he had also convinced them that his views represented world Jewish opinion. On 6 April, when the United States declared war on Germany, the French government seemed eager to have American-Jewish opinion on their side.[97]

At Picot's urging Sykes went to see the new French Prime Minister, Alexandre-Félix-Joseph Ribot[98] the next day, on 7 April. He had wanted to make it clear that France would need British assistance in establishing friendly relations with the Arabs in the sphere of influence designated by the Sykes–Picot agreement. In return for this support, he hoped to gain French support of British aims in the area. While the question of Palestine did not come up, Sykes saw the fact that Ribot did not mention it as a positive sign that France would not oppose British sovereignty over the area.[99] The following day he wrote to Balfour of the importance of Zionism and the support of the Zionists "as a lever against the French." He stressed that for the moment

> it would be dangerous to moot the idea of a British Palestine, but if the French agree to recognize Jewish Nationalism and all that it carried with it as a Palestinian political factor, I think that it will prove a step in the right direction, and will tend to pave the way to Great Britain being appointed patron of Palestine ... by the whole of the Entente Powers.[100]

In his 7 February meeting with the Zionist leaders Sykes had told them the French were likely to be the most difficult of the Entente Powers to convince over Zionist aspirations in Palestine. Thus, it would be in their best interests to "find a way to get the French to back down on their claims [to Palestine]," hence, his introduction of Sokolow to Picot and key figures in the French government. It was Sykes's idea that "the Zionists should spearhead the demand that Palestine come under British control in exchange for a commitment that the Zionists would be accorded a privileged position to pursue their goals there under British sponsorship."[101] While he did not specifically mention it to the Zionists, he was clear in his letter to Balfour: use the Zionists to ensure a British Palestine.[102]

On 9 April Sokolow was invited to the Ministry of Foreign Affairs where he outlined the principles of the Zionist programme. He spoke to a group that included Picot, Paul Cambon, the French ambassador to London, his brother Jules Cambon, secretary general of the French Foreign Ministry, and Ribot's chef de cabinet.[103] Once again, Sokolow carefully avoided the issue of sovereignty over Palestine and focused the discussion instead on Zionist aspirations there. Jules Cambon expressed the French government's concern over Jewish influence in two areas: in Russia, where it was hoped that Jewish influence could be brought to bear against the pacifists, and in Italy, where Jewish influence might work toward the "consolidation of the Entente." Sokolow assured Cambon of Jewish support in these areas and, in turn, was assured in turn that the French accepted the principle of Jewish nationality in

Palestine.[104] He was given further assurances that the French government regarded the programme very favourably and was authorised to inform the Zionist organisations in Russia and the United States by telegraph about the results of their talks.[105] It was suggested that Sokolow might contact the Italian government and seek their support as well. In a letter to the Foreign Office Sykes reported that the French felt that by doing this Sokolow might do "some useful work ... towards [the] consolidation of the Entente." Subsequent events were to show that the French wished to ascertain the attitude of the Italian government towards Zionism, using Sokolow to inquire about it before giving their official support.[106]

On 10 April Sykes left for Rome to lay the groundwork for Sokolow. Before leaving Paris for Rome, he wrote to Sir Ronald Graham at the Foreign Office on 9 April, asking him to have the Italian embassy "assist Sokolow in Italy."[107] Contacting the British envoy to the Vatican, Count J. de Salis,[108] Sykes made an appointment to see Monsignor Eugenio Pacelli,[109] assistant under secretary for Foreign Affairs for the Holy See.[110] The next day Sykes and de Salis met Pacelli. Sykes explained that his visit was unofficial and that in seeing the monsignor he was acting in a private capacity. He then raised the shared official concerns of Britain and the Vatican over "the immense difficulties [involving] the question of Jerusalem, the Arab Nationalist movement, the Moslem [sic] Holy Places, Zionism, and the conflicting interests of the Latin and Greek (Churches), besides the aspirations of the various powers."[111]

After discussing these matters in detail with Pacelli, Sykes felt confident the monsignor held a positive attitude towards British control of the holy places and was opposed to that of the French. He then raised the subject of Zionism, explaining its purpose and ideals, and even suggested that the monsignor see Sokolow when the latter came to Rome. While Pacelli's response to Zionism was not enthusiastic, he agreed to see Sokolow. Sykes felt their meeting was important, being aware that the Vatican's primary concern were the holy places in Jerusalem and that they wanted "full assurances that the Zionists had no aspirations in that direction."[112]

Two days later, on 13 April, Sykes had an audience with Pope Benedict XV. While the interview was short and general, lasting only ten to twelve minutes because of the pope's illness, Sykes felt he accomplished a lot. In a letter to Sokolow the next day he wrote about both interviews:

> I visited Monsignor Pacelli and was received in audience by His Holiness. On both occasions I laid considerable stress on the intensity of Zionist feeling and the objects of Zionism. I was careful to impress that the main object of Zionism was to evolve a self-supporting Jewish

community which should raise not only the racial self-respect of the Jewish people, but should also be proof to the non-Jewish peoples of the world of the capacity of the Jews to produce a virtuous and simple agrarian population ... I further pointed out that Zionist aims in no way clashed with Christian desiderata in general and Catholic desiderata in particular in regard to the Holy Places. I mentioned that you were coming to Rome, and I should strongly advise you to visit Monsignor Pacelli, and if you see fit have an audience with His Holiness. Count de Salis, the British representative at the Vatican, can arrange this if you will kindly show him this letter.[113]

Sykes left this letter with the British embassy in Rome and wired Weizmann to tell Sokolow it was waiting for him there.[114] After another meeting with Pacelli, this time with Picot and English Vatican Archivist Francis Aiden Cardinal Gasquet, Sykes and Picot continued on together to Egypt to take up their positions as political officers with the British expeditionary forces.[115]

Before Sokolow left for Rome he spent a month in Paris meeting government officials and becoming acquainted with the leading members of the Jewish community. Unlike Paris, where the French government had been relatively receptive to his message and the Jewish community had lowered its resistance to Zionism,[116] Rome presented Sokolow with a very different situation. There the Jewish community was small and not very influential.[117] Anglo-Italian relations had been strained since 1916 when Italy had been left out of the Sykes–Picot agreement[118] and this was made worse by recent Italian claims to large amounts of Ottoman territory. Even though amendments to Sykes–Picot were allowed by Britain to include the Italians on 21 April in the Saint Jean de Maurienne agreement,[119] the Italian claims to other Ottoman territory were still a bone of contention between the two countries. The British government had questioned these claims, in light of Italy's lack of participation in the Middle East campaign, in a formal written communication to the Italian government on 29 April.[120] Also, the Vatican was wary of the true aims of Zionism. From the outset, Sokolow's task was not an easy one. The Vatican's opposition alone could be a formidable barrier to the Italian government's support.

Here, Sykes had once again paved the way. Arriving in Rome on 23 April, Sokolow was surprised to learn that he was to contact the Vatican as well as the Italian government. To assist him, the presence of the absent Sykes was everywhere, as Sokolow relates:

I put up at the hotel; Sykes had ordered rooms for me. I went to the British embassy; letters and instructions were waiting for me there.

I went to the Italian Government Offices; Sykes had been there too; then to the Vatican, where Sykes had again prepared my way. It seemed to me as if his presence was wherever I went, but all the time he was far away in Arabia, whence I received telegraphic messages.[121]

Sokolow sought appointments with both governments and with the help of Count de Salis he was able to get his first appointments with the Vatican. On 29 April Sokolow had a "somewhat strenuous interview" with Monsignor Pacelli, "who dwelt upon the question of the Holy Places and insisted that they would have to be clearly defined." Two days later, his meeting with Cardinal Gasparri,[122] the papal secretary of state, went more smoothly. Although Gasparri suggested a rather broad definition of the holy places "that included Nazareth and Tiberias, as well as Jerusalem, Bethlehem, and Jericho, but he went on to give a hearty endorsement of Zionist aspirations." Sokolow came away from the meeting with the feeling that Gasparri also preferred British to French sovereignty in Palestine.[123]

Sokolow's meeting with Pope Benedict XV on 6 May was extremely positive. An ardent liberal who had been deeply troubled by the persecution of Jews in Eastern Europe, the pope made his sympathy towards Zionism clear from the beginning of the conversation.[124] With such a supportive listener, it was easy for Sokolow to discuss the aims of Zionism in broad and specific terms. When he asked the pope for the moral support of the Holy See for Zionist aspirations, the pope responded, "Yes, yes, I trust we shall be good neighbors," repeating this several times.[125]

Afterwards, Sokolow wired Weizmann enthusiastically:

Pope declared Jewish efforts of establishing national home in Palestine met sympathetically. He sees no obstacle whatever ... concerning ... Holy Places which he trusts will be properly safeguarded by special arrangement ... His declaration culminated in saying repeatedly "we shall be good neighbors." He spoke almost sympathetically of Great Britain's intentions. The whole impression of honoring me with a long audience and tenor of conversation reveal most favorable attitude.[126]

Flushed with his success at the Vatican, Sokolow once again approached the Italian government seeking an appointment to discuss Zionist aims in Palestine and to gain the government's support. After some difficulty and with the considerable assistance of Angelo Sereni, president of the Italian-Jewish community, Sokolow and Sereni were able to get an appointment with Prime Minister Paolo Boselli, on 12 May. Boselli told them that Italy could not take the initiative in such a matter, but it would offer its support should the Allies make a move in favour of Zionism. "He reiterated the sympathy of the Italian government for Zionism ... and, as a further mark of goodwill, sent out instructions to Marquis Imperiali, the Italian ambassador in London, to give Sokolow a hearing."[127]

Sokolow had kept in contact with the French government during his stay in Rome, informing them of his activities. Apparently satisfied with the results, Sokolow was summoned back to Paris in late May. On 25 May, he met the Prime Minister Ribot for the first time and later that day again met Jules Cambon. In both conversations, firm assurances of French support for Zionist aspirations were given, again prompting Sokolow to press for a statement in writing. It finally came:

Paris, 4 June 1917

Sir,

You were good enough to present the project to which you are directing your efforts, which has for its object the development of Jewish colonization in Palestine. You consider that, circumstances permitting, and the independence of the Holy Places being safeguarded on the other hand, it would be a deed of justice and of reparation to assist, by the protection of the Allied Powers, in the renaissance of the Jewish nationality in that Land from which the people of *Israel* were exiled so many centuries ago.

The French Government ... can feel sympathy for your cause the triumph of which is bound up with that of the Allies.

I am happy to give you herewith such assurance.

Please accept, Sir, the assurance of my most distinguished consideration.

(*Signed*) Jules Cambon[128]

With this success, a crucial part of Sykes's plan for a postwar settlement with the French had fallen into place. It meant that, with the help of the Zionists, he was one step closer to ensuring Palestine would one day either be British, or a British protectorate.

CHAPTER 8

MESOPOTAMIA, ARABIA AND KING HUSAYN

These amateur diplomatists are to my mind most dangerous people and Mark Sykes in particular owing to his lack of ballast.

Lord Hardinge[1]

Over the next year-and-a-half, like a veritable jack-in-the-box the seemingly indefatigable Sykes, by now the government's go-to man in the Middle East and its Mr Fix-it, would seem to pop up virtually anywhere and everywhere. Besides his activities earlier in the year on behalf of the Zionists in London, Paris, Rome, and his trip to Petrograd on behalf of the Entente, April to June 1917 found Sykes once more in the Middle East. While most of his time was spent in Cairo, he took trips to the Red Sea ports of Yenbu, Al Wajh, Jeddah and the Indian Ocean port of Aden, and became directly involved in another area of responsibility: the Arab revolt.

While Sykes was talking to the Zionists in February 1917, Lt. Gen. Sir Stanley Maude raised the siege of Kut and retook the city on 24 February. Once he had secured his hold on the city, Maude marched the Mesopotamian Expeditionary Force north and on 11 March took Baghdad.[2] Thus ended the process begun in December 1914, when Indian troops under Maj. Gen. Sir Charles Townsend were sent to secure British interests in Mesopotamia, but instead were besieged and taken prisoner at Kut al-Amara by the Turks. However, taking the city posed problems on its status and how it was to be administered. Under the Sykes–Picot agreement, Britain would "exercise a *predominant* influence" over Baghdad, while in Basra to the south, in accordance with the Husayn–McMahon correspondence, it would "exercise a *permanent* influence."[3] To resolve potential problems the War Cabinet established the Mesopotamian Administration Committee with Lord Curzon as chairman and Sir Mark Sykes in an unofficial capacity acting as secretary.[4]

In addition to Curzon and Sykes, key officials from the India Office, War Office, Foreign Office and War Cabinet regularly attended the Committee's meetings, as did independent interested parties, like Sir Henry McMahon, former high commissioner in Cairo, recently back in London after being replaced by Sir Reginald Wingate.[5]

Of immediate concern for the Committee was the need to produce an official statement for Lt. Gen. Maude to proclaim to the inhabitants of Baghdad British intentions with its occupation of the city. This task was first assigned to Sir Percy Cox, but because of dissatisfaction with the wording of the statement he prepared it was given to Sykes to improve upon. What he came up with was pure Sykes. Lt. Arnold Wilson, Cox's assistant at the time and later his deputy, was unimpressed. He described Sykes's statement as "ebullient orientalism" written "by a romantically minded traveller," whose "historical references are a travesty of the facts." The proclamation, "pulsated with romantic nationalistic rhetoric," in which "Sykes laid out for the Iraqis a sweeping panorama of their history from the Mongol conquests, since when 'your palaces have fallen into ruins, your gardens have sunken into desolation, and your forefathers and yourselves have groaned in bondage.'"[6]

The proclamation promised "nothing, but state[d] that the British people and Nations in alliance with them desire and hope 'that the Arab race may rise once more to greatness and renown amongst the peoples of the Earth and that it shall bind itself to this end in unity and concord.'" It went further by suggesting that the people of Mesopotamia should

> through your Nobles and Elders and Representatives . . . participate in the management of your civil affairs . . . so that you may unite with your kinsmen in the North, East, South and West in realizing the aspirations of your race.

In the House of Lords, Lord Cromer commented wryly on the proclamation, saying, "it was not necessary for his Majesty's Government to emulate the Hebrew Prophets, but they would have been well advised, before issuing a manifesto, to enlist the help of Muslims in touch with Islamic tradition."[7]

Despite the restrictions put on him, Sykes's suggestion that "the Allies viewed with benevolence the idea of a united or federated Arabia" was at odds with the terms of the Sykes–Picot agreement. It went on to propose a

> Federation in which the Wahhabis of Najd, the "Lords of Koweit and Asir," the Sunni Arabs of Syria and the Shi'ahs of Iraq, not to mention

the usual minorities, would be by some means united to realize their presumed aspirations to govern each other."[8]

Such a scheme was totally unrealistic and unworkable for anyone who knew the history of the region, its rivalries and religious hostilities. The problems inherent in such an idea were apparently lost on the Committee members, who also found other faults to criticise.

Sir Arthur Hirtzel, secretary of the Political Department of the India Office objected to only one line, which was: "It is the desire of the British Government that the Arabs of Irak and Baghdad shall in future be a free people, enjoying their own wealth and substance under their own institutions and laws."[9] The Secretary of State for India Austen Chamberlain also objected to this. "Such a promise, Chamberlain argued, could 'lead to charges of breach of faith in the future' if 'circumstances' made it impossible for the British to give such 'complete freedom,'" that is, "should the Arab State prove to be a failure." Therefore, the sentence was replaced with: "It is the desire of the British Government that the Arabs of Irak and Baghdad shall in future be free from oppression and enjoy their wealth and substance under institutions and laws congenial to them."[10]

Distressed over what he viewed as the India Office "upsetting his Anglo-Arab scheme," Sykes wrote a long memorandum attacking the India Office. "He pointed out that Chamberlain's revision would provoke 'discontent' not only among Syrian intellectuals, but among Egyptian nationalists." Furthermore, he stated that

if the British followed a "White Man's Business" policy in Baghdad, then they would miss an opportunity to "force the hands of the French" to a pro-Arab position in Syria. Wording as loose as Chamberlain's would later give the French freedom to do as they wished in Syria if the British had done so in Mesopotamia.[11]

Nevertheless, the proclamation with this revised wording was sent to Lt. Gen. Maude in Baghdad, who objected to it for other reasons. He claimed that "the revised proclamation did not 'touch on subjects [with] which [the] feelings of communities in Baghdad and Irak are immediately concerned.'" In the end, after the War Cabinet reviewed both Sykes's original and the revision by Hirtzel and Chamberlain the proclamation was turned over to Curzon, Milner, Hardinge and Chamberlain as a committee to decide on its final form. In it, the controversial passage was made even more vague and ambiguous:

It is the hope of the British Government that the aspirations of your philosophers and writers shall be realised, and that once again the people of Baghdad shall flourish, enjoying their wealth and substance under institutions which are in consonance with their sacred laws and their racial ideals.[12]

Sykes was outmatched and gave in for the moment. Elsewhere, he found himself equally unsuccessful in getting the War Council to support his agenda, as the Mesopotamian Administration Committee, guided by its chairman Lord Curzon thwarted his efforts to have the control of Mesopotamia transferred to the Foreign Office from the Indian government.[13] Nevertheless, in the end there was nothing in Lt. Gen. Maude's proclamation that would challenge the general belief that some time in the future, "Great Britain would declare Mesopotamia, like Egypt and Cyprus, to be a British protectorate."[14]

As its name suggests, the Committee was charged with recommending the makeup of the government of Mesopotamia – its administration, personnel, laws and court system – anticipating that it would come under British control after the war. Sykes had much to say on the matter. In his travels in the area for Lord Kitchener in 1915 he had "witnessed first hand the inefficiencies of Government of India control in Mesopotamia." As a result, he was vocal about the inadvisability of the Government of India continuing to control the region after the war in his reports and in a memorandum to the War Cabinet after its meeting on 6 July 1916, at which he gave evidence. Sykes viewed it as a grievous mistake and one that would result in serious problems in the region if the Indian government were allowed to replace the current Turkish and Arab personnel, administration, laws and courts, with Indian personnel.[15]

Eventually the matter was referred to a subcommittee of the Mesopotamian Administration Committee set up for this very purpose. Its members included Sir Thomas W. Holderness, permanent under secretary of state for India, Sir Arthur Hirtzel, secretary of the political department of the India Office, Sir Ronald Graham, assistant under secretary of state at the Foreign Office, George H. Clerk, former first secretary and consul of the British Embassy in Constantinople, Sir Henry McMahon and Sir Mark Sykes. After some deliberation, the subcommittee recommended "that Arabia, and with it control of Haudramaut, South Arabia and the Arabian littoral, should go to the Foreign Office" just as Sykes had suggested. It also recommended that

the proposed new civil service ... might amalgamate with that of Soudan and, later, possibly with that of Egypt. The creation of such a service stretching from Mesopotamia to the Soudan reflected views shared by Hirtzel and Sykes that [it would be] ... a new dependency ... which will [be] a unilingual and unicultural area, from Soudan to the Turco-Persian frontier.

As a result of their deliberations, at the end of March 1917 the Government of India was informed that its "employment of Indians and non-Arabic or Persian Asiatics in the Baghdad Vilayat was 'to be strictly discountenanced' and in the Basra Vilayat, where Indianisation had already occurred, discouraged."[16]

After much discussion and debate over the next year the Committee eventually also decided largely in favour of Sykes's position over the general administration of the region. Baghdad and Basra would no longer to be under the Government of India but under the Foreign Office and a new civil service was to be organised to administer the area. Also, Kuwait was to be made part of the new regional administration. However, not all Sykes's recommendations were followed as the "Government of India was to retain control of Southern Persia, Muscat and the southern and western sides of the Persian Gulf and there was to be no alteration of the 'Cairo sphere of influence.'"[17]

Sykes may have felt his Mesopotamian proposals had been rejected, because the final decisions made by the War Committee were not exactly what he had recommended. However, once again he had played a key role in bringing about fundamental changes to Britain's Middle East organisations and operations. Moreover, he was instrumental in helping resolve the existing problems caused by multiple conflicting centres of authority and lines of communication, which was his intention, just as the previous year his proposal led to the creation of the Arab Bureau in Cairo.

On 3 April Sykes prepared to leave London for Paris, leaving behind its politics, committees and subcommittees, on the first leg of his trip with Nahum Sokolow to introduce him and Zionism to Britain's allies in Paris and Rome. In Paris he would also meet up with Picot for the second leg of his trip to Cairo, where together they would join Lt. Gen. Sir Archibald Murray and the Egyptian Expeditionary Force in its invasion of Palestine. Before leaving, Sykes was summoned to 10 Downing Street to discuss his mission. Although the Foreign Office was well aware of Sykes's plans for his trip, concern had been expressed in certain quarters of the War Cabinet about what he might say when meeting foreign dignitaries or any possible commitments he might make beyond the scope of his instructions.[18]

Present at the meeting at 10 Downing Street, besides the prime minister, were Lord Curzon, who had voiced those concerns, and Sir Maurice Hankey as cabinet secretary, to take notes of the meeting. Asked to describe his plans, Sykes began by saying that his "Political Mission arose out of [the] decision reached at the Anglo-French Conference held at 10 Downing Street on December 28, 1916" and cited its conclusion:

> The British Government agreed that when the British forces now operating in the Sinai Peninsula enter[ed] Palestine a French Moslem detachment should be associated in the operations, and a French Political Officer should be attached to the British Commander-in-Chief. The British Government undertook to warn the French Government as to when [that] ... was likely to occur.[19]

The prime minister then read Sykes his instructions from the Foreign Office, which essentially spelled out his and Picot's "Status and Functions" within the Egyptian Expeditionary Force in Palestine, but little else. After this, Lloyd George asked him "to explain what action he proposed to take?" Known as someone who did not feel limited by his orders, which was the very reason why he had been called in to discuss his mission, Sykes told the prime minister and Lord Curzon his plans. He

> hoped to open up relations with the various tribes in that region, and, if possible, to raise an Arab rebellion further north in the region of Jebel Druse with a view to attacks on the Turkish lines of communication, particularly against the railway between Aleppo and Damascus.

He would then establish himself with the Army and return to Cairo, "and report fully to the War Cabinet."[20]

Anticipating that Sykes might take the opportunity to suggest some unauthorised interpretations of his own in the messages he was to convey,[21] "the prime minister and Lord Curzon both laid great stress on the importance of not committing the British Government to any agreement with the tribes which would be prejudicial to British interests." They were also particularly concerned with the attachment of Picot as French commissioner to the Egyptian Expeditionary Force as it entered Palestine, impressing "on Sykes the difficulty of our relations with the French in this region and the importance of not prejudicing the Zionist movement and the possibility of its development under British auspices." As evidence of this, they noted that

> the attachment of a French Commissioner and of two French Battalions to General Sir A. Murray's force was a clear indication that the French wished to have a considerable voice in the disposition of the conquered territories and the recent negotiations of Colonel Bremond with the King of the Hedjaz showed how pertinacious they were in these matters.[22]

These comments referred to the way that French concern over British control of the revolt and suspicions of their designs in Palestine and Syria had increased with the outbreak of the Arab revolt in the summer of 1916. Matters came to a head when a French proposal to send a token military force to the region "was politely but firmly rebuffed by the British." So the French Defense Ministry decided to send a small military mission to the Hijaz ostensibly to show French support for the revolt. Officially designated as a military mission to Egypt and headed by Colonel Édouard Brémond, an Arabic-speaking French officer and administrator with many years' experience in Algeria and Morocco, it left Marseilles in August 1916. Once in Egypt, in September Brémond announced his intention of accompanying Muslim pilgrims from Morocco on the hajj to the Hijaz. On his arrival in Jeddah he rented a building and "announced the arrival of the French military mission to the Hijaz" and immediately cabled the French Foreign Ministry, "urging the establishment of permanent diplomatic mission to Hussein's adminis-tration." Brémond's request was approved and with this *fait accompli* "the French military and diplomatic presence in Arabia was technically equal to that of the British."[23]

What the British did not know at the time "was that Brémond had a secret agenda; it had been suggested to him in Paris that he should impede the revolt, not assist it."[24] Not long after he arrived, Brémond gave a hint of this over dinner at the French Mission in Jeddah after apparently too much to drink. He surprised his British guests by telling them

> that it was in neither of their interests for the uprising to succeed because "the partisans for a great Arab kingdom seek afterwards to act in Syria, and in Iraq, from where we − French and English − must then expel them."

Over the next year he made every effort to ingratiate himself with Sharif Husayn to the advantage of France. Through various plots and intrigue, Brémond did all he could to limit the Arab revolt to the Hijaz and stop it from moving north into Palestine and Syria.[25]

After difficulties with the French in the region were discussed, the prime minister raised the subject of a Jewish legion, which Sykes had previously taken some pains to resist. Apparently, unknown to Sykes, Lloyd George had breakfasted that morning with Chaim Weizmann and his close friend C.P. Scott, the pro-Zionist editor of the *Manchester Guardian*, who had raised the matter with him. He told Sykes that "he liked [Vladimir Ze'ev] Jabotinsky's idea of a Jewish Legion," noting that "the Jews might be able to render us more assistance than the Arabs." While Sykes agreed with the prime minister "that Palestinian Jews could be helpful," he added, using the argument he had previously used with Weizmann and Jabotinsky: "'it was important not to stir up any movement in the rear of the Turkish lines which might lead to Turkish massacre of the [Palestinian] Jews.'"[26]

The prime minister then made his final point, re-emphasising to Sykes that he "ought not to enter into any political pledges to the Arabs, and particularly none in regard to Palestine." Sykes replied "that he could hardly negotiate at all with the tribes outside the British and French areas without referring to the possible creation of an Arab State." Following the prime minister, Curzon added in his final remarks "that while he criticised the arrangement with the French, he did not quarrel with the arrangement or the establishment of an Arab State. Moreover, he pointed out that this was actually included in the signed agreement from which we could not depart."[27]

Suspicious of French intentions in the Middle East, Curzon in particular saw in this and in their other activities in the region as clear evidence of French intrigue. Before Sykes left the meeting, Curzon "berated ... [him] for the provisions of the Anglo-French agreement – not because of promises made to the Arabs, but because of the cession of important territories to France." It was Curzon's belief, as he told Sykes that, "the French had got 'much the best bargain.'"[28]

Apparently unaffected by the meeting and inquisitorial tone of some of the questioning, Sykes remained "convinced ... that Allied relations could be kept on an even keel. Personal diplomacy would ease the way towards the achievement of his Allied Arab-Armenian-Zionist scheme."[29] Furthermore, he was confident that he was the one to do it. A few days later in Paris he wrote a long letter on 8 April to the foreign secretary. In it, he gave Balfour an overview of French interests in Syria and Palestine, the small clerical-concessionaire cabal that had promoted it at the Quai d'Orsay, and the growing importance and influence of Zionism in helping to resolve the matter in Britain's favour, that is, a British Palestine. He emphasised the importance of Sokolow and his mission to the French, as he "speaks on behalf of U.S.A., Russia, and United Kingdom Zionists." Moreover:

> [I]f the French agree to recognise Jewish Nationalism and all that carries with it as a Palestinian political factor, I think that it will prove a step in the right direction, and will tend to pave the way to Great Britain being the appointed Patron of Palestine ... by the whole of the Entent [sic] Powers."[30]

He went on to describe his interview with French Premier Alexandre Ribot, "who insisted on talking only about Italian claims in Asia Minor." He tried to allay the premier's concerns by alluding to future negotiations and postwar trade-offs over these territories, saying that was only his opinion. Then Ribot asked about "French prospects with the Arabs" and the joint mission Sykes and Picot were about to undertake. Sykes assured Ribot "that very close Anglo-French co-operation would be necessary both now and after the war," because "the Arabs were by nature given to splitting Allies and would endeavour to play us off against each other." This ended the conversation.[31]

While in Paris, Sykes lunched with Boghos Nubar, the Armenian ambassador to the Courts of Europe and son of former Egyptian prime minister Nubar Pasha. An Ottoman citizen, chairman of the Armenian National Assembly, cofounder with several others of the Armenian national movement and president of the Armenian General Benevolent Union, Boghos Nubar lived in Paris and by virtue of his position was very well connected politically. He confirmed Sykes's worst fears "about the French 'Levantine political group' having great influence at the Quai d'Orsay."[32]

After Sokolow reported to Sykes on his successful meeting with the French Foreign Ministry, Sykes wrote to Balfour again on 9 April to inform him of the successful conclusion of the Sokolow–Cambon interview, saying "the Zionist aspirations are recognised as legitimate by the French." He also mentioned that Sokolow told him that, regardless of the Anglo-French relationship

> the bulk of the Zionists want British Suzerainty only; but naturally the moment is not ripe for such a proposal at present, but provided things go well the situation should be the more favourable to British Suzerainty with a recognised Jewish voice in favour of it.

He ended his letter to Balfour by adding, "I am sure the nearer the French get to the Lebanon and Syria proper the less they will care about Palestine."[33]

In a second letter the same day, Sykes wrote to Sir Arthur Hirtzel at the India Office, asking him to intervene in interdepartmental rivalry between Mesopotamia and Cairo. Sir Ronald Storrs, the oriental secretary in Cairo, had been sent to assist in establishing the British administration of Baghdad, which was seen as interference by Cairo in Indian government business. "Try and make [Sir Percy] Cox a little less terrified of the approach of strangers," Sykes wrote to Hirtzel, "Storrs is going out to help him, it was quite unnecessary for Cox to at once begin to say Storrs is to be under Cox's orders, and to await instructions, &c, &c." Hirtzel must have passed on Sykes's request, for Cox warmly welcomed Storrs when the two finally met on 8 May after Storrs' arrival in Baghdad. The two dined that night with Gertrude Bell, formerly of the Arab Bureau in Cairo and now Cox's oriental secretary.

In his *Memoirs*, Storrs would later refer to Cox, as "the outstanding Englishman of the Gulf."[34] Storrs had been sent to Baghdad to help establish the new administration over the region after the conquest of Mesopotamia, and for the months he was there he worked well with Cox, the new high commissioner of Iraq, Bell and Lt. Col. Arnold Wilson, Cox's former Deputy Political Agent, now acting civil commissioner for Mesopotamia.[35] Later, Wilson wrote highly of Storrs in his memoirs of his time in Mesopotamia.[36]

Sykes's third letter that day from his Paris hotel room was to Ronald Graham at the Foreign Office, in which he asked Graham to ask the Italian Embassy to help with Sokolow's visit to Rome. Sykes pointed out that, as Sokolow was Russian, he would need some help because with the revolution in Russia, "his embassy is in a state of some confusion."[37]

Before Sykes and Picot left Paris for Cairo, Sykes told his French colleague that the French were going to have to change their approach towards the postwar Middle East settlement. They would have to come to terms with it being "non-annexationist," in consideration of Italian and Russian claims in the area to ensure the stability of the Entente. It would also involve the British or Americans sponsoring the Zionists and the French the Armenians. He assured Picot that this arrangement would still leave their "Anglo-French Arab scheme otherwise intact." Although Picot made no response, Sykes was surprised at "how this disconcerted Picot, who looked so 'done up.'" Letting the matter rest, he passed this off as Picot's "anxieties over the reaction French imperialists inside and outside the government would have to Sykes's proposals."[38]

A few days later Sykes was in Rome, busying himself with the arrangements for Sokolow's visit. He met Italian government and Vatican officials and was given an audience with the pope, as mentioned previously. After several days Picot met up with Sykes and George Lloyd, who had joined Sykes in Paris and on the train to Rome and was returning to his work at the Arab Bureau in Cairo. The three men journeyed together from Italy to Egypt by ship courtesy of the French Navy.[39]

On his arrival in Cairo Sykes learned the Egyptian Expeditionary Force under Lt. Gen. Murray had failed in its attempt to take Gaza from the Turks. He wrote to his wife that the "atmosphere [in Cairo] was 'very bad' . . . with 'recriminations' rampant as everyone blamed everyone for the failure." From what he heard, he told Edith, word was that "Murray needed to be replaced by a new general 'of character and vitality' under whom, Sykes predicted, the troops 'might yet do great things.'"[40]

With the failure of the Palestinian offensive there was nothing much for Sykes and Picot to do in Cairo other than to report for duty, as their mission was to accompany the Egyptian Expeditionary Force into Palestine. Sykes telegraphed Ronald Graham at the Foreign Office as follows:

[U]ntil a military breakthrough occurred, it would be "necessary to drop all Zionist projects and all schemes involving negotiations with settled rural and urban Arab elements in Syria, whether Christian or Moslem"; otherwise, there was the danger of Turkish reprisal.

He cautioned that, "for the Allies publicly to 'encourage Zionism in London and Paris and foster the Arabs and Syrian hopes in Cairo' would make the British responsible for any "misfortunes to befall these people." He added, "once General Murray broke through, let the Allies profit as liberators of the peoples oppressed by the Turks."[41]

Then, carried away with the moment and in a flight of fancy to which Graham and all who knew him had grown accustomed, Sykes proposed a mass public relations campaign. He suggested that, as a prelude to victory in Palestine and in support of Russia and its continued war effort, meetings of all interested parties be called "in Great Britain, France, Egypt and the USA for Maronites, Syrians, Arabs and Armenians, who could pass resolutions congratulating Russia on her revolution and ask Russia's help in freeing 'races subjected to the Turkish yoke.'"[42] Predictably, nothing came of the suggestion and nothing more was said about it.

Undaunted by the lack of response to his idea in London and having nothing else to do, Sykes asked High Commissioner Sir Reginald Wingate and the Arab Bureau to arrange a series of meetings with Arabs selected for the purpose to discuss "Syrian and Arab desiderata." Three Syrian "delegates" chosen to attend the meetings were introduced to Picot. The Hijazi representative was unable to attend. In the meetings Sykes and Picot reviewed these issues with the delegates. Afterwards, Sykes complained to Graham how difficult it was for him to explain things to the Syrians without being able to use a map; worse yet, not to tell them that an agreement already existed. He was able to report, however, that the Syrians wanted "'an Arab State or Confederation in an area approximating to Areas A and B'" of the Sykes–Picot map. They were "reasonable about Palestine," he added, agreeing that "Palestine presented too many international problems [for an Arab state] to assume responsibility for.'" However, they did insist that should the Jews be "recognised as a millet or 'nation' in Palestine, that 'equal recognition must be accorded to the actual population.'"[43]

Despite Sykes's generally positive report of the meetings, those in attendance must have found the whole situation somewhat artificial and contrived. Picot must have been doubtful about the arrangement and the

Arabs who had been chosen to attend, who were not Syrians of any consequence. The Syrians must have wondered why they were there, and "found Sykes's behaviour extremely strange, as he sailed from one item of the agenda to the next."[44]

Picot may have also noticed what was to be a pattern in British dealings over Palestine; namely, the absence of any Palestinians participating in the discussion of Palestine. Sykes and other British officials would consult on Palestine with Sharif Husayn, his son Faysal and their representatives from the Hijaz, Syrians, Lebanese and even Armenians from Aleppo. However, they would never discuss the matter with any Palestinians about their land, the area under discussion, except to tell them what had been decided for them. Thus, the Palestinians would become the unrepresented people in British policy in the Middle East and this attitude would bequeath a troubled legacy to the region.

Meanwhile in the Hijaz, Sharif Husayn, now the self-proclaimed King of the Arab Lands (an area that included Syria),[45] was becoming alarmed at the reports he was receiving about French designs in Syria. He was worried that Britain and France might be allocating territory without consulting him. He had not yet been informed about the Sykes–Picot agreement. However, his suspicions were aroused by "various oblique and shadowy references" made to an Anglo-French agreement by British officials who were otherwise vague and hazy, if not disingenuous. "Even ... Sir Reginald Wingate would later discount the Sykes–Picot agreement as a dead letter in a cable," when asked directly about it by Husayn. Nevertheless, rumors persisted and grew until the king "became alarmed at published statements which appeared to suggest that the agreement was by no means dead as this threatened him and the Arabs with an imperial *fait accompli*." His alarm increased when he learned that Picot was meeting "in Cairo with groups of Syrian nationalists attempting to elicit their support for an eventual French hegemony over Syria." Husayn knew there was "a strong pro-French lobby among the Syrians and, in particular, among the Maronite Christians" with whom Picot had been working before the war. So, as far as Husayn was concerned, "Picot was just resuming the intrigues he had conducted before the War as a diplomat in Beirut." These had been revealed in documents he left in a safe at the French embassy in Beirut after war was declared. Jamal Pasha had made much of their discovery to show the existence of French anti-Turk activities and their plans in the region.[46]

In retrospect, it has been suggested that the question should not be "was Husain ever 'informed' of Sykes–Picot?'" but "why was he not *involved?*" Like the cavalier treatment accorded the Palestinians, even Britain's key Arab ally was deliberately kept in the dark about their plans, including those directly affecting him. Before things reached a crisis point, Wingate decided it would be a good idea to send Sykes to Jeddah to see Husayn and try to alleviate his fears.[47] So, once again, Sykes would be right in the middle of things, but this time as Britain's Middle East spin doctor.

When Wingate first suggested Sykes's trip to the Hijaz, the Foreign Office expressed reservations about what Sykes might say while he was there. To allay these concerns, Wingate asked for and received Foreign Office approval "for the language which he instructed Sykes to [use] in Jeddah." However, the key concern remained. Would Sykes "speak as clearly and as unambiguously as these instructions required him to do"? This was because, Wingate noted, "he was partial to grandiloquent phrases in addressing oriental personages ... [and] he could not bring himself to treat someone like Husayn seriously enough to warrant speaking to him plainly and precisely." The high commissioner ended his telegram to the Foreign Office by saying "that if Sykes's visit to Jeddah proved successful, a second meeting with Husayn when Picot would be present might be arranged."[48] With his instructions just what to say and a prepared script to follow, Sykes was deemed to be ready for his meeting with King Husayn.

His instructions were simple. When he met the king, Sykes

> was to "reassure" Husayn about French aims in the "interior of Syria": but he was also to tell the king that his rule "cannot be imposed upon peoples who do not desire it"; and, further, he was to "make it clear that in Baghdad and district whilst desirous of promoting Arab culture and prosperity, we will retain the position of military and political predominance which our strategical and commercial interests require."[49]

Sykes also carried with him a friendly, non-committal, greeting from King George V to the King of the Hijaz, as requested by Wingate and telegraphed to him by the Foreign Office before Sykes left Cairo.[50]

The same day that Sykes was given his instructions by Wingate, Sir Gilbert Clayton, Director of Intelligence in Cairo, sent a letter to Lt. Col. C.E. Wilson, British agent in Jeddah, repeating and amplifying Wingate's instructions, since he would be at Sykes's meetings with Husayn. Clayton told Wilson, "Sykes was to 'indicate gently' that special measures had to be taken as regards Baghdad where British military and political preponderance had to be ensured 'at any rate for a considerable time.'" In addition, he was to make clear "what is an undoubted fact, viz: that the Arab movement as represented by himself [Husayn], cuts no ice whatever in Mesopotamia, and that therefore it is quite out of the question to force it upon them."[51] Neither Clayton nor Arab Bureau chief D.G. Hogarth

> thought much of Sykes, despite his exalted position in London as special adviser to the Middle East committee. To them, he was intellectually shallow and hopelessly verbose, pretending to far more knowledge concerning the Middle Eastern affairs than he actually possessed.[52]

However, despite this and because of who he was and what he represented, both agreed that Sykes did have his uses and believed that calming Husayn was one of them. Moreover, it was thought that Wilson's presence at the meetings would ensure that Clayton got a full report of what was said and would thus find out if Sykes went beyond his instructions.

Through arrangements made by Wingate and the Arab Bureau, Sykes left the port of Suez on a naval ship for Jeddah on 30 April.[53] On his journey to the Hijaz, no doubt Sykes reflected on the king's situation when considering how best to approach Husayn in their discussions. As the result of his revolt, the king had been officially removed by the Ottoman sultan from his prestigious post, one of the highest in Islam, and replaced as amir of the Hijaz. Then, after proclaiming himself king of the Arab Lands, a title he assumed the Allies would approve of because of its use in the nebulous promises of the Husayn–McMahon correspondence, he learned to his surprise that instead they would only acknowledge him as king of the Hijaz. So Sykes must have concluded that the insecurity of Husayn's position in the Hijaz, one that was now wholly dependent on the Allies, made him particularly susceptible to all kinds of rumors and innuendo. He heard what he wanted to believe, but fortunately for Sykes, he trusted what he was told by the British. As subsequent events would show, Sykes decided to use this to his advantage.

Midway on his trip to Jeddah, on 2 May Sykes stopped at Wajh, a small coastal fishing village in the Hijaz on the Red Sea some 370 miles north of Jeddah. There he met Amir Faysal, the son of King Husayn and a military leader of the Arab revolt, whose forces had recently captured the village and made it the headquarters of Faysal's Northern Army. In his report to Wingate Sykes gave no details of his discussion with Faysal, other than to say that after "much argument" Faisal accepted the principle upon which an Arab confederation or state was to be based and "seemed satisfied."[54]

Three days later, on 5 May, Sykes met King Husayn in Jeddah. In the three-page report he wrote after the meeting Sykes gave few details of his discussions with the king. Half the report was taken up with the exchange of greetings Sykes passed on to Husayn from King George V and Husayn's response. In the remainder, he described Husayn's concern about being remembered as a Muslim aiding in the overthrow of a Muslim ruler in Syria to replace him with a Christian one. He noted that he had followed his instructions and "explained [to the king] the principle of the agreement as regards an Arab confederation or State." He also impressed upon the king "the importance of Franco-Arab friendship and I at last got him to admit that it was essential to Arab development in Syria, but this after a very lengthy argument." However, he gave no details about what exactly he said to the king. There was also no mention of any maps, boundaries or zones, or of Baghdad. He did, however, get the king to agree to meet Picot to discuss French aspirations in Syria on 19 May.[55]

For a man with a reputation for long and detailed reports on all his activities, the sheer brevity and lack of information in Sykes's report should have raised his superiors' concerns. The following day Sykes sent another letter in which he "claimed to have fully informed Hussein of the British position regarding the future of Baghdad vilayet and the 'position of military and political predominance which our strategical and commercial interests require.'"[56] In the years since, many have questioned his claims and expressed serious doubts that Sykes actually followed his instructions and stressed this latter point with Husayn.

After meeting the king, Sykes had a conversation with Col. Brémond, chief of the controversial French military mission in Jeddah. In Brémond's report of their conversation to his superiors in Paris, he told them Sykes advised him it would be better, when

> discussing Franco-Arab questions with Husayn ... [to] make generous concessions to the king (*il serait bon que le Colonel Brémond aborde ce subject en étant très large au point de vue des concessions*) and not insist too much on precision, so as to allow these ideas to ripen in the head of the Sharif.[57]

This seems to confirm suspicious he did not follow his instructions and shows the disdain Sykes held for Husayn that had concerned Wingate and the Foreign Office when deciding whether to send him to the Hijaz in the first place.

Convinced of the success of his mission to the Sharif and the Arab revolt, a confident Sykes left Jeddah the evening of 5 May and arrived back in Cairo on 9 May, where Picot was waiting for him. Over the next few days in Cairo Sykes had several rounds of meetings, lunches, and dinners with the high commissioner, Clayton, Aaron Aaronsohn, Lt. Gen. Murray and others. On 16 May he left Suez for Jeddah with Picot.[58]

On their way they stopped at Wajh, where Sykes introduced Picot to Faysal. Other than making the introductions, Sykes reported their talks were not productive. Faysal accompanied them to Jeddah, where he joined Sykes and Picot in meeting King Husayn on 19 May, along with Lt. Col. C.E. Wilson, Fu'ad al-Khatib,[59] the king's foreign minister and their interpreters. Sykes reported that the introduction of Picot to the king by Lt. Col. Wilson went well. The king warmly welcomed Picot, who in turn read greetings from the French president to the king, in which he congratulated the Arabs on their movement. However, the discussion eventually became strained when talk turned to the future of the Syrian Littoral (Lebanon). The king repeated what he had told Sykes in their 5 May meeting, that he "could not be a party to ... the handing over of a Moslem population to the direct rule of a non-Moslem state." After further discussion on this and other subjects, talk on the status of French advisors in the area ended the talks. The only agreement reached was to continue the next morning. Sykes reported that Picot was not impressed with the king.[60]

What Sykes did not include in his report of the meeting and the subsequent events that took place afterwards was duly reported to Cairo by others. They gave a slightly different picture from that painted in Sykes's report. The king's foreign minister, Fu'ad al-Khatib, told Lt. Col. Stewart Newcombe,[61] chief of the British military mission to the Hijaz who was not at the meeting, that Picot had proposed to the king "an agreement in Syria with France as you have with Great Britain in Baghdad, which the king refused."[62] Sykes did not mention this in his report. Nor did he mention that that evening he had summoned Fu'ad and asked him "to get the king to agree to Picot's request. Later in the evening [he] sent a message to Fu'ad ... through Wilson, reiterating his request, and repeating his message the following morning." Fu'ad told Newcombe, "it took him three hours to convince the King to accept Sir Mark Sykes's wish."

> He agreed at last because he said that he trusted what the British Commissioner says: He knows that Sir Mark Sykes can fight for the Arabs better than he can himself in political matters and knows that Sir Mark Sykes speaks with the authority of the British Government and will therefore be able to carry out his promises.

Newcombe forwarded this information on to Cairo.[63]

A second meeting was held the following day on 20 May, which was attended by Faysal, Lt. Col. Wilson, Fu'ad al-Khatib and Lt. Col. Newcombe, along with their interpreters. It was a complete turnaround from the day before. Husayn began by having a declaration read in which he expressed

> satisfaction that [the] French Government approved Arab national aspirations ... that as he had confidence in Great Britain he

would be content if the French Government pursued the same policy toward Arab aspirations on Modern Syrian Littoral as British did in Baghdad.[64]

Thus, he agreed to the same situation being established in the Syrian Littoral under the French as it did in Baghdad under the British, where an Arab government would rule with the aid and assistance of Britain while also being part of the Arab Lands and under Husayn's sovereignty.

This was a complete reversal of the king's attitude towards the French proposal the previous day. It appears that Sykes's behind-the-scenes pressure on the king's foreign minister Fu'ad al-Khatib to convince the king to change his position had worked. It also explained the reason for the otherwise inexplicable pro-French declaration with which Husayn began the second meeting. However, the question remains: what did Sykes offer Husayn in return for changing his mind? Although Fu'ad al-Khatib did not tell Lt. Col. Newcombe what this was, it has been suggested that Sykes "must have promised or broadly hinted that at the peace settlement Baghdad would form part not of the red zone (which could be annexed or closely controlled by the British) but of area 'B' (which was to form part of the proposed Arab state)." Nothing short of this could possibly "explain Husayn's sudden readiness to concede Picot's demand which only the previous day he was bent on resisting." Whatever it was, "Sykes's private assurances made Husayn utterly confident in the soundness of his own contention that McMahon [in the Husayn–McMahon correspondence] had promised him Baghdad."[65]

For years afterwards, Husayn was adamant that the discussions on Syria at these meetings were on "whether the Arabs would accept to recognise a French sphere of influence in the coastal regions of northern Syria, that is to say in the Lebanon," referred to throughout the discussions as the Syrian Littoral. Sykes's reports confirm that the discussions on 19–20 May concerning Syria mentioned only the Littoral and made no mention of any discussions on the interior of Syria with either Faysal or Husayn.

Nevertheless, confusion remained. The British agent in Jeddah Lt. Col. C.E. Wilson sent his own report as instructed on Sykes's visit to Brig. Gen. Clayton in Cairo. In commenting on Husayn's pro-French declaration at the meeting, he told Clayton, "it was by no means clear to which territory in Syria it referred." As far as Wilson was concerned, "I am not clear," he wrote, "and probably Picot and the Sherif are not clear, whether Syria i.e. including Damascus etc is meant; or merely the Syrian coast claimed by France; one may have meant Syria the other only the Syrian coast."[66]

So "the troubled Wilson" went to Sykes and asked him directly, "Does the Sharif etc. know what the situation in Baghdad really is?" Sykes's response was, "They have the [Baghdad] proclamation," which, of course, he had a hand in writing for Lt. Gen. Sir Frederick Stanley Maude shortly after the Anglo-Indian forces under Maude had taken Baghdad in March 1917. Momentarily at a loss for words by Sykes's abrupt response, Wilson noted later "the Proclamation said nothing more than asking Arabs to co-operate [with] the Government. Sykes then asked Fu'ad if he had read the Proclamation of Lt. Gen. Maude and Fu'ad said he had and the matter was dropped."[67]

Wilson added that Lt. Col. Newcombe had come to him

> later in the evening (20th) ... and told [him] he had a long talk with Faisal and Fuad and amongst other things told me that it was Sykes who urged Fuad to get Sherif to agree to the [French proposal of Picot] stated at the meeting and that Sykes had told Fuad or Sherif to leave everything to him, this was the first time I heard that Sykes was responsible for the Sherif's action.

In view of this, Wilson felt "very strongly that the Sherif should be told exactly what our interpretation is of our future position in Iraq which [will] ... be much more prominent than that of the French in Syria." If Baghdad

> will almost certainly be practically British ... we have not played a straightforward game with a courteous old man who is as Sykes agrees, one of Great Britain's most sincere admirers, for it means that the Sherif [agreed] verbally to Syria being practically French which I feel sure he never meant to do.[68]

This only served to increase Clayton's concern, because "there was nothing in ... Lt. Gen. Maude's proclamation, to contradict the general belief that at the appropriate moment Great Britain would declare Mesopotamia, like Egypt [on December 18, 1914] ... a British protectorate."[69] Moreover, "consensus among ministers and high officials was that, contrary to what Sykes had advocated, Baghdad should not at that stage be considered as forming part of the prospective Arab state."[70] Therefore, "to equate French influence in Syria with that of the British in Baghdad connoted de facto annexation of Syria by Paris, thereby seriously damaging whatever authority Feisal might later attempt to assert" as King of Syria, and "did nothing but complicate the Arab Bureau's own plans for Damascus and eastern Syria."[71]

If Wilson was confused about the outcome of the meeting, he was not alone. So was the Quai d'Orsay in Paris. Sykes had reported, as mentioned previously, that Husayn "declared he would be satisfied if French policy in the

'Moslem Syrian Littoral' would be the same as that pursued by the British in Baghdad." Apparently picking up on this and further confusing matters, Picot "informed Paris that Husayn would like to see not the 'Moslem Syrian Littoral' but rather 'Moslem Syria' (*la Syrie musulamane*) treated on an equality with Baghdad." The ministry of foreign affairs immediately telegraphed Picot back to confirm that there had not been a mistake in the transmission. Picot replied that there was "no error in transmission." Understandably, "Paris found this news very surprising since they had had no inkling of such a change in British policy."[72]

As early as July 1915, French Premier Aristide Briand had acknowledged in negotiations with the British "the 'importance of the Arab movement' and agreed 'to the towns of Aleppo, Hama, Homs and Damascus being included in the Arab dominions to be administered by the Arabs under French influence.'"[73] This was a major policy change that came from Picot, not from the Foreign Office; hence, it was a complete surprise to the Quai d'Orsay. One can only wonder whether what Picot told his foreign ministry was a deliberated misrepresentation of the facts on his part, whether it came from Sykes, or whether it was just a misunderstanding. Whatever the case, the fallout from these meetings would haunt its participants and their countries for years to come.

Later that summer, in an effort to sway government policy and perhaps cover his actions in Jeddah, Sykes advocated these policies in a memorandum. He suggested "the blue and red zones should be assimilated to areas 'A' and 'B,' retrospectively, and proposed that if the French objected, they should be told that this was the policy the British intended to follow in Mesopotamia."[74] While this was not taken up by the government at the time, it indicates Sykes's opinions, which, perhaps, may have led him to take advantage of his position in his meetings with the king and deliberately mislead Picot, eventually making it *fait accompli* British policy. While this is pure speculation, it does seem plausible, given Sykes's tendency to take independent action based on what he thought best at the time. Nevertheless, after the war France would take over all of Syria based on this misunderstanding.

After the meeting on 20 May Sykes and Picot travelled by ship to Aden before returning to Cairo. While he was there Sykes wrote to Sir Percy Cox in Baghdad on 23 May and shared his thoughts on the Arab revolt, its leader and the agreement with the French with the Gulf resident (whose territory included Aden). While he assured Cox that "the idea of Arab nationalism may be absurd," Sykes told him it would improve Britain's position "if we can say we are helping to develop a race on nationalistic lines under our protection, and if we do not we must hand this race back to an alien oppressor who will crush out the national character."[75]

As for King Husayn, Sykes told Cox:

> I think the main thing is neither to minimize nor magnify his
> importance ... The family has this importance that it has gained
> independence from the Turks, holds a town of first rate importance, and
> is an excellent label and symbol to which the Arab Moslem
> intelligentsia can turn as a mental political rallying point.

He further minimised Husayn's importance to military operations in the
Middle East, saying:

> [T]he King cannot give us much physical help it is true, but where we are
> successful in the Eastern and Central areas, he helps to give moral sanction
> to those who take our side, and later a moral sanction to our control.

It was for these reasons, however, he told Cox:

> [I]t is very important that we should do all we can to promote respect
> for him and[a] good relationship between him and [the] other
> independent chiefs, he should occupy in Arabia proper if possible the
> position of premier Arab and titular leader of the Arab movement.[76]

The not so subtle inference here was for Cox not to promote Ibn Saud, with
whom he was very close, over Husayn as leader of the Arab movement.

As for the Sykes–Picot agreement and the French, Sykes told Cox he
believed that:

> [I]f our aims are realised I think the French will be ready to co-operate
> with us in a common policy towards the Arab speaking people, viz:
> follow in their area a policy similar to our own, as regards language, law,
> education, and administration.

While giving no supporting evidence for this other than wishful thinking, he
added that if this were accomplished then "there is a good prospect of our
benefitting in our areas from the higher standards of life already existing in
the French areas, and the development of a common internal prosperity."[77]

After returning to Cairo and before he left for London, on 5 June Sykes
sent a copy of recommendations to London attached to his report,
"Observations on Arabian Policy as Result of Visit to Red Sea Ports, Jeddah,
Yenbo, Wejh, Kamaran, and Aden."[78] However, they were deemed to be
unrealistic and found little support in Whitehall.

CHAPTER 9

THE ARAB LEGION AND THE FRENCH DIFFICULTY

Sykes was an extremely impetuous man, easily led into enthusiasm, liable to sudden revulsion.

Christopher Sykes[1]

Shortly after his arrival in Egypt, Sykes had a brainstorm. In a letter to Gen. Murray's office on 28 April, he suggested the formation of an Arab Legion made up of captured Arab prisoners of war from the Ottoman army. Not to be confused with the postwar army of the Hashemite Kingdom of Jordan later made famous by "Glubb Pasha," Lt. Gen. Sir John Bagot Glubb,[2] Sykes's Arab Legion was altogether different and a scheme he would fight hard for throughout the summer of 1917.[3] However, it was not his idea. Apparently, he was inspired with the idea by Capt. N.N.E (Norman) Bray, who happened to be in Cairo at the time.

The two first met in London the previous year while Bray was on leave from France. At the time, he was an intelligence officer with the 18th King George's Own Lancers, Indian Army,[4] then serving in France. Previously stationed in Mesopotamia, Bray had also spent time there before as a Directorate of Military Operations[5] officer in the region.[6] While on leave in London from the war in France in June 1916 he read in the papers about the outbreak of the Arab revolt. Believing in its importance and concerned that it would not be appreciated in London, he put aside his vacation plans and went to the Foreign Office to find someone who would listen to what he had to say.

This was an informal time, when intelligence agents like Bray had easy access to policy makers, politicians and bureaucrats.[7] At the Foreign Office he later recalled being "received ... with the greatest kindness" by an official he mysteriously referred to only as "Mr. V – t [Sir Robert Vansittart?[8]]," who, after listening to him, told Bray, "I think, however, that you should in the first instance see Sir Mark Sykes." A call was made and Bray told to go to Sykes's home at 9 Buckingham Gate, where he would be expected. In a meeting that made a lasting impression on Bray, Sykes showed "the greatest enthusiasm for the movement."[9]

Sykes paced back and forth across his study listening intently to what Bray had to say. After he finished, Sykes said, "We must place before the Government every argument we have for giving the Arabs all possible assistance." He was convinced how necessary it was to gain the support of Muslims

> outside Arabia ... in order to counter a vicious propaganda [and] ... send out a mission of Indian officers, carefully selected from the Indian regiments in France, to visit the Hejaz, talk with the Sherif, and learn at first hand his motives for rebelling against his co-religionists.

Once this was done, these men should return to India, where they would tour the country explaining to their fellow Muslims "the true meaning of the revolt." In this way Sykes believed they would "gain for the Arab cause the sympathy and support of our Mohammedan subjects."[10] The excited Sykes saw it as a way to deal with the problems he was concerned about in India and with Indian Muslims.

Sykes hailed a taxi and the two men went straight to the India Office. There they met Sir Arthur Hirtzel, under secretary of state and head of the Political Department, and shared Bray's idea with him. Apparently, Hirtzel was "favourably impressed," as was Sir Edmond Barrow, the military secretary. "We then walked like madmen to the War Office," Bray recalled, "where I was introduced to various officers of that section which dealt with Arabian affairs. Sir Mark rushed me from room to room." When one of the officers wanted to know who Bray was, Sykes said, "Never mind, never mind, he is a pin on my map." Upon reflection, Bray decided he "was more pleased to be a pin on the Arabian map than a bodkin in France. In a few days, the energetic Sir Mark had got all the authorities concerned in the matter to approve the scheme for sending the Indian deputation out to the Hedjaz." Once the Indian officers were chosen for the task from Indian troops in France, Bray gave them instructions on what was expected of them and accompanied them to Jeddah. While waiting in Jeddah for the delegation to return from their meeting with Husayn in Mecca, he was asked by Lt. Col. C.E. Wilson, the British agent in Jeddah, if he would like to "assist him as Intelligence Officer" in the Hijaz. Bray was only too happy to oblige and eventually arrangements were made for him to return to the Hijaz on re-assignment after his mission was completed.[11]

Meanwhile, the fledgling Arab revolt in the Hijaz that was declared in June was already beginning to falter. Despite its recent positive reception at the War Office, the general attitude in Whitehall towards the revolt "ranged from indifference to scepticism to downright hostility." The recent loss of Lord Kitchener at this critical time was devastating. Everything was focused on the war closer to home with little thought given to the war against Turkey in the east. As a result, there was little hope of getting support for "the Sherif's undisciplined and inadequate levies" from London.

In Cairo, with its leader and greatest supporter gone, the Arab Bureau was hard-pressed to provide adequate provisions of food, guns and supplies for the Sharif's forces. By late September few would disagree with the blunt appraisal of Maj. Gen. Sir Arthur Lynden-Bell, Chief of General Staff of the Mediterranean and Egyptian Expeditionary Force, that "that the revolt was 'very much a comic opera performance.'"[12] This judgment was soon to change, as the war had reached a stalemate in France and matters came to a head in the Hijaz in late 1916.

During the summer of 1916 a major confrontation loomed between Turkish forces and those under Faysal at Rabigh, halfway on the road between Mecca and Medina. Rabigh was the only oasis in the area, so the Turks could not slip by on their way from Medina to take Mecca.[13] They had to go through Rabigh. News spread that Turkish troops would soon be on their way at the beginning of the Hajj, bringing with them the Holy Carpet [Kiswah] — the annual gift from the Ottoman Sultan to cover the Holy Kaabah in Mecca — and also Sharif Ali Haidar, newly appointed by the Porte to replace Husayn as Sharif of Mecca because of his revolt against the Ottoman Sultan.[14]

From Jeddah, Lt. Col. Wilson — through Sir Henry McMahon in Cairo — urged London for an immediate decision on reinforcing Rabigh: "Besides having a moral obligation to help the Sherif, [it was a matter of] ... compelling strategic necessity. If Britain did not act, the revolt would probably collapse." Should that happen and Mecca fall into enemy hands, he maintained it would be a serious setback for British prestige in the Muslim world. Not only that, but it would open up the entire Arabian Peninsula to the Turks. "With free access to the eastern shores of the Red Sea there would be trouble in Abyssinia and in Somaliland and the 'whole political and military situation east of Suez would be jeopardized." Wilson believed the situation was serious enough to recommend "the immediate dispatch of two battalions of British troops and a flight of airplanes to block the Turkish advance to Mecca via Rabegh."[15]

However, in Egypt, both Lt. Gen. Murray, commander-in-chief of the Egyptian Expeditionary Force and Maj. Gen. Lynden-Bell disagreed, saying that "Wilson's fears [were] unwarranted." Noting that the Egyptian Sultan Husayn Kamel was against British troops in the Hijaz, Murray pointed out that the War Office "had decided that Britain would concentrate on the Western Front 'every single man we possibly can.'" He added,

> [O]f the 350,000 men he had under his command ... he had 'despatched every single man' he could to France, and left himself 'some mounted troops and four weak and ill-trained Territorial divisions.' He also threw cold water on the request for airplanes.

When the matter was referred to London the War Committee agreed and turned down the request to send a brigade to supplement Faysal's forces against the Turks at Rabigh.[16]

Given this lack of support, Clayton, T.E. Lawrence and those on the spot believed Faysal's tribal forces could hold their own against the Turks if they were given adequate guns and ammunition. This could be done with help from local hill tribes and without the help of British forces. However, as Lawrence noted in his report to Clayton about the situation in Rabigh in November, "when the tribes go over to the Turks, there is no more 'Sherif's movement.'"[17] Meanwhile, Husayn repeatedly told the British "that the Hedjaz was holy territory. 'To send British troops there would have disastrous political effects.'"[18] However, as Lawrence told Clayton, these

> objections to the landing of a British force does not apply to aeroplanes. The Arabs ... are longing for them. They also want instructors in technical matters (artillery, machine guns, bombing ... armoured cars) and are prepared on these grounds to welcome the French contingent and anything we like to send. No European escort is necessary.

He added:

> [T]he policy of not landing a British force at Rabegh should not be made an excuse for doing nothing in the Hijaz. If we spare ourselves this expense and trouble, it is all the more incumbent on us to stiffen the tribal army on which we are going to rely for the defence of Mecca. This stiffening is (by request) not to consist of personnel, but of materials.[19]

Sykes also shared this opinion, as he had spelled out earlier in an appendix to the notes of a meeting in September.[20]

Meanwhile, Wingate disregarded everyone's opinions, ignored the Egyptian Sultan and Sharif and pressed London for British troops to be sent to Rabegh. He agreed with Wilson that the Turks must be stopped at Rabegh and the Arabs would be unable to do it. So the matter came up repeatedly before the War Committee in the last four months of 1916, causing the Chief of Imperial General Staff Sir William Robertson to become totally "exasperated by the repeated appearances of the Rabegh question before the War Committee." In early November, the situation changed:

> Robertson was directed by the War Committee to attend a Foreign Office meeting where Curzon, Chamberlain and Grey agreed that it was "of the highest importance to deny Rabegh to the enemy" and asked Robertson to again report to the Committee on the feasibility of a landing."[21]

The reason for this change in attitude was the result of the arrival of Capt. N.N.E. Bray on 8 November. Sent by his superiors in Rabegh, where he had been with Faysal's Northern Army, Bray came to London to seek Sykes's help for the beleaguered sharifian forces there.[22] As he wrote later, Sykes was

> called "The Mad Mullah" in the War Office, for he would burst into a room, give his views on various weighty matters, demand certain action and rush out of the room in the same tempestuous manner in which he entered it. But he was regarded as a visionary.[23]

He was also on record as supporting the reinforcement of the sharifian forces at Rabegh. So it was agreed that Sykes was the man to see in London for help.

On hearing Bray's story of the situation in Rabegh, Sykes asked Bray whether he would repeat his story to the Cabinet. When Bray said he would, Sykes arranged for the two to meet later that morning at 10 Downing Street. After sitting outside the cabinet room for some time waiting to be summoned, an impatient Sykes got up "opened the door and walked boldly in," leaving a surprised Bray waiting outside. "After an interval, Sir Mark burst out again and I could see he was greatly disturbed," Bray recalled. Sykes told him, "They have put the matter off for this afternoon at three o'clock, when a select committee will discuss the situation." With this, he

> stumped up and down the room in his agitation, impatiently kicked chairs out of his way and then, banging down his hand on the pile of red dispatch boxes [sitting on a table in the waiting room], so that they slithered over the table with a clatter, at which the Prime Minister's messenger's eyes bulged with shocked surprise.

Finally, Sykes said: "'There is not an hour to be lost, not a *moment* − I can't stand this!' and he again burst into the Cabinet room," again leaving an astonished Bray in the waiting room. As Bray recalled, "he soon came out again, this time beaming with delight. We have got our way, every assistance is to be given, and all the men and ships we can spare are to be sent to Rabegh." Then the two men "discussed the terms of the instruction which were to be issued, but just then the Cabinet meeting ended and the ministers trooped out." As they came by, Sykes introduced Bray "to Mr (now Sir Austen) Chamberlain, then Secretary of State for India, who was kind enough to invite me to dinner that evening, so I could inform him more intimately as to what was happening." He also introduced Bray to Sir William Robertson, chief of the general staff, who made a curt comment and "strode off."[24]

Sykes then "plucked me by the sleeve. 'Come on,' he whispered, 'we must get that telegram off,' and then he almost ran to the Houses of Parliament." After they entered the building, Sykes

> hastily drafted a telegram in pencil on a half sheet of paper. This telegram was to be sent on the authority of the Cabinet, to Egypt, instructing the authorities there to hold in readiness a Brigade of British troops, to have ships available, and to render all possible assistance, both political and military.

Sykes then instructed Bray to take it "to the Foreign Office and give the draft personally to Lord Hardinge (then Permanent Under-Secretary of State for Foreign Affairs) and to no one else." Unable to see Hardinge, Bray gave the telegram to his private secretary and reported back to Sykes. That night Bray dined with Austen Chamberlain, "who showed the greatest personal interest in, and sympathy for, the Arab movement, about which he questioned me at length." With his newfound celebrity, thanks to Sykes, before he left London for the Hijaz Bray was summoned to Buckingham Palace to meet the king and to the War Office to discuss assisting the Arabs with Lt. Gen. Sir George MacDonough, director of military intelligence.[25]

The report Bray wrote on the night of 8 November, after meeting the Cabinet and Chamberlain, received a positive hearing and helped change attitudes about support for the Arab revolt in political and military circles in London. In it the government was asked "to support the Arabs as a guerilla army, with all the necessary arms and materiel, but no conventional reinforcements."[26] This was not the first time such a request had been made. Clayton and McMahon had repeatedly argued these very points, as had Lawrence.[27] But now it was receiving a hearing at the highest levels of government.

Bray also "argued for the regional staff's request to locate an intelligence organization at Rabegh itself, headed by 'a capable [British] officer ... who can give us reliable and useful information on improved systems of communication.'" Even Lt. Gen. William Robertson welcomed this proposal. Immediately after receiving Bray's report the following day, in addition to agreeing to provide support for the sharifian forces, the War Committee also approved a British military mission in the Hijaz "to counter the machinations of the French Military Mission under Brémond at Jiddah." Lt. Col. Stewart Newcombe, then with military intelligence in Egypt, was appointed to the post and set out for the Hijaz four weeks later.[28] By January 1917, however, the threat to Rabigh was gone, as the Turks had decided not to leave the security of Medina and in the end there was no landing of troops. However, the importance of the Arab revolt and a willingness to support it had caught on in both Egypt and London and more supplies and materiel were made available.

Six months later Sykes and Bray met again when Sykes went to Egypt with Picot in late April 1917. This time Bray approached him with the idea of lending "his influence to the formation of an Arab Legion, which would give the [Arab] movement a truly national character, which till then it had totally lacked." The idea was originally Clayton's and although it was much discussed between June and October 1916 it failed to materialise because of the refusal of Arab prisoners of war at the time to fight against the Turks.[29] It was also not the first time Sykes heard of it. He had exhibited some interest in Clayton's idea during the summer of 1916, but had not followed through with it and left it for Clayton to pursue. At this time Bray had also supported the idea and had written to his superior in India, Sir Percy Lake, the director of Military Intelligence, India, in October 1916 about it.[30] At the time, nothing came of it.

Taking up the idea again with additional embellishments of his own, Bray approached Sykes on his arrival in Cairo in April 1917 with the idea of an Arab Legion. It was to be made up of Arab former officers in the Ottoman Army taken prisoner and in prisoner-of-war camps in Egypt, as well as recruits from the Arabian Peninsula. It was to be "staffed by British and French officers and to be organized in two full divisions." Bray and Sykes discussed the plan at length, with Bray telling Sykes how disorganised the Arab revolt was. What they needed, Bray believed, was

> a pivot, round which it would revolve in its full strength, and expressed the hope that a striking force, possessing the necessary enthusiasm, would thus be created which might play a serious role on the right flank of the British Army.

Bray recalled that "Sir Mark became enthusiastic over its formation."[31]

In a letter to the chief of the general staff office in Cairo, Sykes immediately passed on Bray's suggestion (without mentioning Bray, or Clayton's similar idea the previous year). After surveying "the Arab situation in Egypt," he wrote, "I have come to the conclusion that it would be of considerable military-political value and of actual military value if steps were taken to raise an Arab legion from prisoners of war and Syrians in Egypt." After briefly outlining the benefits of the proposal in a single-page letter, he asked that the idea be put before Lt. Gen. Murray "for his favourable consideration."[32]

Initially, the idea found interest in some quarters, as the Arab revolt again seemed to have stalled, but it was less welcome in others. This was primarily because Sykes had also enlisted Picot in the scheme, "who, on behalf of his government, had offered to supply half of the necessary capital." At the Foreign Office, Sykes's friend Lord Robert Cecil, normally supportive of his proposals, "considered it 'a very hazardous proposal ... [and] a direct invitation to the French to take further part in the Arabian operations after they have put in a claim for political partnership there.'" While there were those who found some merit in the scheme, Cecil, some at the Arab Bureau, Curzon and Ronald Graham were against involving the French.[33]

Sykes left Cairo on 16 May for his second trip to the Hijaz, this time accompanied by Picot and Lt. Col. Gerard Leachman, a political officer stationed in Mesopotamia who travelled extensively in Arabia.[34] While in the Hijaz, on 18 May Sykes raised the subject of an Arab Legion with Amir Faysal and "ascertained," as he wrote afterwards to Wingate in Cairo, "I. That there is probably useful Arab material for recruitment for Sherif's forces in Prisoners of War camps in India. II. That Hejaz forces would welcome such recruits." After meeting Dr Abdulla, an associate of Amir Faysal, it was Leachman's idea that he and Abdulla "visit the prisoner of war camps in India as he felt convinced that there was good material there if only it could be properly approached." This prompted Sykes to send a telegram to Wingate in Cairo, asking the High Commissioner to forward a message to Foreign Secretary Balfour to "inform Secretary of State for India" of his plans about the Arab Legion. To expedite matters, he had suggested "that Leachman accompanied by two Arab officers proceed to the Indian camps and recruit," noting that Leachman was already going to India in any case. Once the recruits were enrolled, they would be handed over "to a selected British officer for transit to Egypt where men can be trained for service." At this point, Sykes had asked that the commander in chief of India be asked "if he concours with this procedure." Meanwhile, he was taking Leachman and two Arab officers to Aden, "whence they can proceed to India if [the commander in chief, C in C] approves."[35]

By the time he reached Aden on 23 May Sykes had not heard from Wingate about his overtures to the Indian government. So he cabled the secretary of state of India about the Arab Legion. He told the secretary, "C in C Egypt Force [Murray] approved raising of Arab legion in Egypt under Anglo-French auspices. Legion to be raised from Prisoners of War, exiles, and independent tribes, for services in sphere of Arab movement. C in C Egypt Force has taken preliminary steps and has asked London for financial sanction."

Repeating what he had cabled Wingate, Sykes asked the secretary to "please ask if he [the C in C India] concurs in this procedure." He attached an additional page to his cable entitled "Instructions" with "Objects" and "Notes," further outlining the details of his plan.[36]

While he was in Aden, Sykes received several telegrams on the Arab Legion. On 18 May the Arab Bureau in Cairo cabled Wilson in Jeddah, with a copy for Sykes, saying that "General Baghdad [Maude] wires that fifty-one officers and 117 men "will be dispatched shortly to Bombay ... [and proposed] to instruct Baghdad to send them direct to Suez."[37] He also received a telegram from Bray sent on 20 May on behalf of his superiors in India. Bray notified Sykes that the pay schedules for the volunteers was "awaiting sanction of the War Office" and listed the proposed pay for each soldier.[38] Sykes also received a telegram in support of the Arab Legion from Lt. Gen. Murray in Cairo, sent to him in Aden on 21 May.[39]

He finally received a telegram in response to his own sent on 23 May from the Commander in Chief India, Gen. Sir Charles Monro, giving him the approval he had asked for: "We should be glad if Colonel Leachman and two Arab officers would come to visit Prisoners of War camps to enroll recruits for service under Sherif as suggested in telegram from Secretary of State." With this, Sykes sent Leachman and Dr Abdulla to India with instructions on the recruiting efforts he and Picot had developed to guide them.[40]

With these telegrams, Sykes "assumed that the legion was as good as settled and ... made arrangements to begin recruiting at Aden and [Kamaran] with the political officer, Aden." However, despite all this activity and the positive response to the idea in the field, on his return to Cairo a frustrated Sykes learned that London had not yet authorised the Legion. Although "everything was ready to begin work no moves could be made."[41]

Sykes had arrived in Cairo on a special train from Suez at 2 in the morning on 11 June 1917[42] and the next day sent a six-page telegram to the War Office outlining plans for the Arab Legion and giving its details. He described what had been done in raising volunteers for the Arab Legion in prisoner-of-war camps in Egypt, Kamaran Island, Aden, among Arab residents in Egypt and in the prisoner-of-war camps in India and Mesopotamia. From these, he said, there are "good hopes of raising two efficient battalions with adequate drafts at depot by mid-September." He added that "M. Picot has concurred in all steps taken up to present, and is providing advisory officers in proportion to ours." Sykes stressed that

> members of the Legion are engaged to fight for Arab cause on Arab soil where they will serve it best. They will be under the orders of the high command in the sphere selected for them to operate in. Sherif and Faisal have approved of Legion.

He cautioned that while

> half the actual cost is being defrayed by the French on M. Picot's authority. M. Picot has done this on his own responsibility. I beg, therefore that matter be not discussed with French as it is in his hands and he understands his own people.

He implored London not to delay further their decision on the Legion, "otherwise intrigues and trouble will mar inception of scheme. High Commissioner approves from a political point of view and C-in-C from point of view of organisation and administration."[43]

His next few days were spent in a flurry of activity catching up on all his correspondence and a succession of breakfast, luncheon and dinner meetings with officials and friends. Among those whom he met were High Commissioner Sir Reginald Wingate, acting director of the Arab Bureau Commander D.G. Hogarth, director of military intelligence Sir Gilbert Clayton, Capt. N.N.E. Bray, Aaron Aaronsohn and his two close friends Aubrey Herbert, MP and George Lloyd, MP, both currently assigned to the Arab Bureau.[44]

On 6 June Sykes left Egypt on his return to Britain by ship from Port Said. After stops in Rome and Paris he arrived in London on 14 June.[45] Over the next two weeks in London he received telegrams from Leachman in India, Brig. Gen. Clayton (23 June) in Cairo, and Lt. Col. Jacob (27 June) in Aden, all informing him of their successes in recruiting for the Legion. However, approval for the scheme remained stalled in Whitehall.

Meanwhile in Cairo, after Sykes had left for London Wingate was having second thoughts about his recommendations and the Arab Legion. He was particularly concerned with the "dual control" of the Legion with France, because Sykes told him that "the Legion would be used primarily in the Hejaz," an area in which the British did not want the French.[46] So Wingate wrote to the War Cabinet outlining his objections. In London, after reading Wingate's telegrams to the War Cabinet, Sykes responded to his criticisms of his recommendations in a paper to the Mesopotamian

Administration Committee.[47] This shows how utterly convinced Sykes was, no matter how distasteful some aspects of French rule in its colonies were, that maintaining Britain's close relationship with France was as necessary to win the war in the deserts of the east as it was on the battlefields of France.

Wingate emphasised his support of the Sykes–Picot agreement in its "desire to achieve a general understanding and agreement between the two Powers as to [a] broad line of policy each will follow in their respective spheres." As for Sykes's recommendations, he believed they were fundamentally flawed, being based on an unrealistic premise; namely, that Britain and France would "pursue a permanent, identical and co-operative policy in their respective spheres of control and interest." He maintained this could not be done, due to the difference in the inhabitants and conditions in the two areas and, more importantly, because of "the different method of Government by [the] French and ourselves, and [the] present illusory character of an Arabian Federal system."[48]

Not one to be bothered by such inconvenient details, Sykes ignored Wingate's argument about the differences between the areas' inhabitants and local conditions and focused instead on Britain and France as imperial powers. "If I thought that post-war England and France would in any way resemble pre-war England and France I should agree with the High Commissioner." For Sykes, this kind of thinking failed to take into account "the fundamental change which has come over the democracies of the world to imagine that either country can ever return to pre-war Imperial methods, no matter how liberal and enlightened they were." What exactly this change was he did not say, but if it did not occur and Britain and France were "to work toward annexation then I am certain that our plans will sink in chaos and failure." Ever the idealist, Sykes added:

> [I]f on the other hand the two Governments resolve to work hand in hand to revive and re-establish a great people and assist in the development of a new civilization ... then I feel we have before us the prospect of great success, and an opportunity of obtaining for the democracies of England and France the full economic reward which they require.[49]

While agreeing with Sykes on other issues, Wingate further disagreed that "any useful or real similarity of legal or education systems can be attained owing to contrast of French and British methods. I therefore regard a discussion on these points as academic and premature." Sykes disagreed because he was "certain that the French were prepared to revise their previous methods." Perhaps Picot had assured him of this, but he offered no evidence in support of his claim. Then, seeking to end the argument, he pointed out that

> the native personnel, both legal and educational, must be drawn from the same sources, namely, educated Arabs of both spheres, and as education and the law have been identical hitherto, and as the services well probably remain in the hands of the same lawyers and schoolmasters, as were employed by the Turkish Government, I cannot see what there is inconceivable in both Powers agreeing to follow a similar policy.[50]

While this seemed to be a logical assumption, Sykes continued to ignore the differences in the peoples to be governed in the two areas and the well-known history of French colonial rule, guided as it was by its national *mission civilisatrice*, or civilising mission. Under this approach, which was adopted by France in the late nineteenth century, French language, law, education and culture were considered superior and were routinely imposed on the indigenous populations wherever they ruled. Perhaps his blindness to this was due to the fact that it was similar to what British colonial rulers had also done. Nevertheless, Sykes was determined to believe what he felt he must in order to support his proposal, no matter how outlandish it may have seemed to others and, as in this case, to ignore inconvenient facts in order to ensure that everything worked out as he felt it should. An idealist and romantic in a very real world, Sykes faced a stubborn reality and was obliged to succumb to Wingate's logic. Despite his best efforts, Sykes was unable to convince anyone of the feasibility of his recommendations and, although they were officially shelved, they were not forgotten by Sykes.

Years later T.E. Lawrence would characterise Sykes as "the imaginative advocate of unconvincing world-movements," who would "take an aspect of the truth, detach it from its circumstances, inflate it, twist and model it, until its old likeness and its new unlikeness together drew a laugh."[51] While no laughs greeted this example of Sykes's creativity, his recommendations were in some respects a perfect example of how he could be carried away with an unlikely idea and fits perfectly into Lawrence's depiction of him.

The Arab Legion, however, proved less easy to dismiss and had some merit to it. Like the recommendations, Sykes's idea for the Legion was based on a desire for Anglo-French cooperation, but unlike the former, this scheme was developed with the assistance of Picot, who offered both men and the funds to support it. After widely sounding out the idea for several months among military and political leaders at home and abroad and receiving generally positive responses, Sykes officially proposed the idea of an Arab Legion made up from "prisoners of War and Syrians in Egypt" in a paper to the War Cabinet. It was his hope that the idea would challenge their thinking and, as a result, add trained soldiers to supplement the Allied forces in the Middle East from a pool that was already there but was not being used.[52]

In his report, Sykes outlined the process by which he had developed his scheme with Picot's involvement, as well as the input from Sir Gilbert Clayton, director of intelligence and chief of the Arab Bureau, and referrals to both Maj. Gen. Sir Arthur Lynden-Bell, chief of the General Staff of the Mediterranean and the Egyptian Expeditionary Force and High Commissioner Sir Reginald Wingate for review.[53]

Among the key points in promoting the Legion, Sykes listed to "give work for idle hands to do"; promote "Syrian and Arab unity and stimulate enthusiasm"; provide "a rallying point for deserters of the best type, viz: deserters who come over through conviction and not through cowardice"; and provide "a force capable of being used anywhere in the Hejaz, without in any way breaking with our policy of not landing European troops on sacred soil." While these may sound rather superficial and frivolous – "idle hands," "stimulate enthusiasm," "deserters of the best type" – more pragmatic military and political minds saw otherwise. For them there was the benefit in putting the Arab prisoners to use, rather than just feeding and housing them, as well as having trained Muslim troops under their control in the Hijaz with the sharifian forces. Sykes emphasised that these were disorganised, disorderly and untrained, they "waste ammunition, destroy rifles by neglect and devour money and food we send." Not only that, but the sharifian forces were primarily a loose confederation of desert tribes unused to modern warfare. So he believed this plan offered the added bonus of sending trained soldiers and officers to provide the "trained nucleus" that the sharifian forces lacked and by so doing provide the Arab revolt with a more formidable fighting force against the Turks.[54]

Sykes gave a detailed accounting of his efforts over several months on behalf of the scheme with Picot, King Husayn and Amir Faysal, and the military and political authorities in Egypt, Hijaz, Aden and India. He emphasised that while the Legion was to be a cooperative joint Anglo-French venture, once it was fully organised and its troops trained, control over it would change:

In the field [the Legion would] be independent of joint control, but under the control of whatever high command was in authority in the sphere in which it operated, and that no French or British officer should accompany it in the field, and that the executive Battalion authority should be in the hands of Arab officers.

All this he "laid verbally before the C-of-C [Lt. Gen. Murray] and Sir Reginald Wingate."[55]

Two months later, after much persistency and persuasion, Sykes gained the official support of the Foreign Office.[56] A minute describing its prior objections and subsequent approval was initialed by Arthur Balfour, the secretary of state, Lord Hardinge, permanent under secretary, and Lord Robert Cecil, under secretary of state for Foreign Affairs. It was now passed to the Mesopotamian Administration Committee for review,[57] where it was to languish for a while under the disapproving eye and unbending opposition of its chairman, Lord Curzon.

Three days after his return to London in June 1917 Sykes had written and distributed a four-page report on his trip entitled "The Arab Situation."[58] In it he focused on what he saw as the two most important issues in the war in the Middle East: the Arab situation, from which he took the title of his report, and Anglo-French relations. Of crucial importance was what Sykes termed "the French difficulty," which was the main focus of his paper. He characterised it as being based on "French colonial policy of the old school." He further described it by referring to "Colonel Brémond's methods [in the Hijaz], which have as their object the crushing and breaking up of native organisation[s], and the obtaining for France of a special position of advantage by secret petty negotiations ... this difficulty makes the French unsympathetic to the Arabs and irritating neighbours to the British." Despite this, Sykes believed there was hope, as "this school is dying out ... [and] the policy which inspires it is losing ground in Paris."[59]

It was his belief that "French Colonial functionaries like our own Indian and Egyptian officials, are by natural circumstances out of touch with the feeling of the Entente and Central Powers." However, it was "the favourable attitude of the French Government to the ideas of Monsieur Georges Picot in regard to Arab nationalities" that has made a difference. For these reasons, Sykes maintained

> we must not be unduly influenced by the vexations of old school pre-war Frenchmen. We must never allow ourselves to despair of the future on account of them. Most of all we must avoid falling into a similar attitude of mind and should check any tendency to an anti-French attitude in any of our own people on the spot.[60]

As he saw it, the "Arab difficulties" were due to a number of reasons: the Arabs were "scattered"; the "civilised part of Arabistan is in the Turkish occupation"; and further "divided by climate, mode of life and social conditions," ranging from city to desert, and from sophisticated to tribal patriarchies. However, "the Arabs have a sense of race rather than nationality" and having spoken to Arabs of all classes and religions in Cairo, he was "convinced that an Arab National movement is growing and that now is the moment to gain it as an asset of the Entente." Thus, "the great fundamental of our policy should be unity between ourselves and the French."[61]

Here he noted that "the Arabs trust us to a certain degree, are desirous of European protection and co-operation, and intensely desire autonomy and freedom from Turkish control." No doubt recalling his earlier recommendations, he suggested that if the two countries could agree to adopt identical policies in both areas then things would work out well for all concerned. However, if they did not, he believed "the Arab movement will undoubtedly degenerate into a merely anarchical and hopeless form of unrest." It was to this end that he recommended "an Arab Legion be sanctioned and proceeded with under Anglo-French auspices," outlining the terms and conditions of its cost, training and maintenance.[62]

On 2 July Sykes made an appointment to see Curzon in an attempt to bring him around to his way of thinking on the Arab Legion among other things. Before their meeting in the afternoon, he sent Curzon a handwritten note explaining why he had asked for the meeting, respectfully saying he was doing so "in order to save your valuable time and lay before you certain points which I hope to lay before you for consideration." He told the former viceroy, "there are a series of problems which are very closely intertwined and which hitherto lacked a single controlling hand inspired by a defined objective or ideal." He listed the problems that he wished to discuss with Curzon that afternoon: "1. Persia, and future policy towards. 2. The Arab movement in Asia. 3. Anglo-French relations in regard to the Hejaz. 4. Our future policy in Mesopotamia." Here he noted that

the handling of these problems has hitherto fallen to the India Office, Government of India, Foreign Office, and occasional references to the War Cabinet. However ... the Mesopotamia [sic] Administration Committee has under your chairmanship tended to act as ... a co-ordinating instrument

with "both the F.O. and India Office have referred to you for your decisions on matters, which would not otherwise have been dealt with."[63]

Sykes then flattered the former viceroy, commenting favourably on the work of the Committee under Curzon. "The influence of the com[te] under your guidance has been of immense value to speeding up decisions and clearing up situations," he wrote:

[H]owever amongst the great press of work which surrounds you, it is possible that you may not have noticed that as matters are at present the committee is becoming a permanent factor in the conduct of our middle and near Eastern affairs.

He then made a proposal he believed would appeal to Curzon:

If this is to be continued (and it has been what both Clayton and I have longed and prayed for, namely that you should assume control of the whole of the middle Eastern problems), I think that it will be advisable to give a new name for the com^te, to define its scope and work and fix on a regular ordinary meeting day.[64]

He continued:

There should also be some means devised of dispensing of minor questions in order to avoid delay, as the agenda of the committee is liable to become over charged. If the committee resolved that you as chairman should dispose of these lesser questions while keeping them informed of the action taken one weekly meeting would suffice.

Assuming Curzon would find his proposal acceptable, Sykes went on to suggest an agenda for the Committee's next meeting:

I anticipate the next meeting will be heavy and involve considerable discussion, namely: 1. Sir P. Cox's position in Mesopotamia. 2. The Arab Legion. 3. Our Arabian and Hijaz policy. 4. The Persian situation ... Yours very sincerely Mark Sykes.[65]

Unfortunately for Sykes, Curzon failed to succumb to either his flattery or suggestions, nor did he see things the way Sykes hoped.

Several days after his meeting with Curzon in July 1917, a frustrated Sykes vented his exasperation at the anti-French sentiment expressed in a letter he received on 11 July from D.G. Hogarth, former acting director of the Arab Bureau and on assignment with the Admiralty in Cairo at the time. Attached to Hogarth's letter was a paper, "Note on the Anglo-French Agreement about the Near East,"[66] which was an attack on the agreement. In a handwritten note boldly scrawled over Hogarth's letter, Sykes angrily expressed his disgust with the opposition to the Sykes–Picot agreement and the people and politics associated with it. "I am fast growing weary of the Anglo-French-Russian agreement," he wrote in exasperation.

"Particularly of French and English who have not enough guts to tell the Arabs to go to blazes when they flatter them and have not enough common sense to prevent Anglo-Moslem intrigues in Syria and Franco-Arab intrigues in Hejaz." He added:

> Also I am tired of Englishmen who listen to a ridiculous Marmozet [sic] like the King of the Hejaz. The Arabs are weak and divided but manageable by any one who chooses to manage them, and very easily worked by people who prefer to mismanage them ... However, I hope soon to wash my hands of the whole business."[67]

If the Foreign Office was initially concerned about sending Sykes to the Hijaz because of the possibility of his insincerity in dealing with King Husayn, here was evidence that their concerns were probably justified. This also seems to confirm the accuracy of Colonel Brémond's report of his conversation with Sykes after his 5 May meeting with the king and, perhaps, shows the strain of his incessant meetings, reports, negotiations and travel, as well as the lack of ballast referred to by Lord Hardinge.[68] The pressure on Sykes was beginning to tell, especially that which he put on himself.

Later that evening, still upset with Hogarth's letter, Sykes went to see him, as Hogarth was in London at the time. As Hogarth recalled afterwards in a letter to Clayton on 11 July:

> Mark Sykes is very angry and I had a stormy interview with him this evening. Finally he challenged me to produce an alternative draft, and, in my haste, I said I would. But on reflection, I have gone back on the undertaking ... the Western Front situation demands encouragement of the French at any cost, in which case this Agreement must stand.

He noted that Sykes "admits that Curzon's rejection of 'co-operation' in the Arab Legion, had already queered his pitch. My note came on top of that, and 'Alabaster' [Curzon] will get it all his own way."[69]

Hogarth was in London at the Admiralty to present Cairo's view of the Sykes–Picot agreement and Anglo-French relations in the Middle East, both of which were of serious concern to the Arab Bureau, who felt their position was not receiving the hearing it should. In his letter to Clayton, he added:

> [I]f the situation on the western front was dire enough to demand the continued placating of Paris, so be it. If not, "bear skin agreements"

such as the one he [Sykes] had negotiated should be jettisoned, at least until the dust of the war had settled and available options had crystalized. In practical terms, the accord was dead – except in Sykes's fertile imagination.[70]

In another letter to Clayton a few days later, Hogarth reported that he was pessimistic of "the impact of his note on the diplomatic course of events." He told Clayton to tell Lawrence "to be careful not to justify Mark's idea of him as a Fashoda [anti-French] propagandist. I, too, am credited with a 'fashoda mind.' Only you are uncontaminated in Cairo."[71] This "observation ... [lumped] the director of Military Intelligence, Curzon, Milner, and the Admiralty into this Fashodist cabal, each one prepared to support scuttling Sykes's agreement with the French." More importantly, "events were beginning to move too swiftly in Palestine to permit the accord to linger as the primary basis for a postwar territorial settlement."[72]

Two days later, on 22 July, Clayton wrote to Sykes. Among other things in his letter, he assured Sykes, "You need not be afraid of any Fashodism on my part. The indissoluble Entente is everything, and it is the more important that *we* should show that spirit" but he added:

[H]onestly, I fail to see how the French are ever going to make good their aspirations in Syria (all the indications at present available go to show that they are disliked and distrusted by nearly all sections of the people interested in their proposed sphere). It is essential, therefore, that we should give no loophole for any accusation that we have helped towards a failure on their part.

He further assured Sykes, "In the meantime, I am doing my best to muzzle the innumerable talkers here and to impress upon them that they must trust the Entente, whose principles have been announced to the world, and await development."[73]

Clayton commented on Hogarth's paper, saying he generally agreed with it. He told Sykes that

we cannot be a party to any agreement in which our predominant position in the whole of Arabia is not made quite clear. I do not see why this should upset the general spirit of the original agreement provided the French are prepared to be reasonable.

He expressed concern over how far the French would go in this regard, and was unsure how important it may be to keep them contented, but I do think we should make a grave error and lay ourselves open to innumerable difficulties in the future if we did not lay down our position in Arabia beyond any possibility of doubt.

Repeating his concern there were loopholes in the Sykes–Picot agreement, Clayton added, "If we leave any loopholes for any Powers to interfere there we shall lay up a very great store of trouble for ourselves in the future in a region where we can least afford to have it."[74]

In early July Sykes argued his case for the Arab Legion before "a hostile War Cabinet subcommittee that included Hirtzel, Shuckburgh, Graham, and Army representatives; amazingly, his plan was approved, but not necessarily for use in the Hijaz, and not necessarily with joint Anglo-French responsibility." By the end of the month, however, the persuasive Sykes

had managed to obtain an even more favorable compromise: the force was to operate under the orders of the Commander in Egypt for the defense of the Hijaz (and while there was to be nominally responsible to Husain); if elsewhere under the dual Anglo-French control.[75]

At the end of July, despite his earlier gloomy predictions, at the conference in Paris, "Balfour in Paris reached final agreement with France." So a surprised Sykes enthusiastically cabled the officers of the Legion in Cairo news of its success.[76]

However, despite the approvals of London and Paris, in the end the Arab Legion for which Sykes had fought so hard never became a reality. By the end of September 1917, of the 10,000 troops he hoped would volunteer to serve in the Legion the force amounted to "only 500 at Ismailia, and some of these, reported Clayton, were useless. Morale was low, and the officers seemed more inclined to political intrigue than to military efforts." So in the end "it was sent off to 'Aqaba, recently captured by Lawrence's forces, and absorbed into the regular unit under Faisal."[77]

CHAPTER 10

THE BALFOUR DECLARATION

{Sykes} became an enthusiastic Zionist, and his enthusiasm found an entirely new scope when he became a secretary to {the} War Cabinet ... In his new capacity Sykes practically took charge of all the negotiations which led up to the Balfour Declaration. The Zionist movement owed much, at a critical moment in its history, to his infectious enthusiasm and to his indefatigable energy.

Leopold Amery[1]

On his return to London in June 1917 Sykes was surprised to learn of an American effort to negotiate a separate peace with the Turks. Henry Morgenthau, Sr., the former US ambassador to the Ottoman Porte (1913–1916)[2] and a prominent American Jew, with the support of President Wilson and US Secretary of State William Jennings Bryan, a vocal opponent of the war, had arranged to meet Ottoman representatives. While Morgenthau's official purpose was to help relieve the suffering of Jews in Palestine during the war, Sykes saw this as a potentially serious blow to Zionist plans. He immediately contacted Weizmann on how to handle what he believed was an American peace overture to the Turks. He also alerted the foreign secretary, who was considering sending Sir Louis Mallet, the former British ambassador to the Porte (1913–1914) to meet Morgenthau on his trip east. Sykes persuaded Balfour to send Weizmann instead, as he was convinced that Mallet was a member of the "pro-Turk gang" in the Foreign Office.[3]

Meanwhile, a second approach for a separate peace with Turkey was received by the Foreign Office from Switzerland.[4] With some input from the Zionists, Sykes handled this one himself. At the time, Balfour had also referred this to Sir Louis Mallet, who was now working in the Eastern Department of the Foreign Office. In response, Mallet wrote a paper on a separate peace with Turkey, which naturally came to Sykes's attention. Vehemently opposed to any separate peace with Turkey, Sykes wrote a nine-page letter to Lord Robert Cecil on 29 July critiquing Mallet's paper. In his view, Sykes told Cecil, any approach by the Turks, whether through intermediaries like Marmaduke Pickthall, or even a former US ambassador to the Porte, should be viewed with the utmost suspicion. He opposed their attempts and those of others' to cooperate directly or indirectly through intermediaries with the CUP, the Young Turks' political arm and its leaders, in seeking a separate peace. In typical Sykesian hyperbole he characterised the CUP as "evil, corrupt, and hostile, either to this country or the welfare of mankind, and that not one of them desires our future security." All his arguments boiled down to one thing: if a separate peace were negotiated with Turkey, everything would go back to the way it was and the Zionists would be unable to establish a homeland in Palestine.[5] Thus, all his hard work to bring about the peace settlement he envisioned would have been wasted.

At the end of June, as Britain's designated representative Weizmann met the American mission in Gibraltar headed by Henry Morgenthau and Harvard Law Professor Felix Frankfurter before they met the Ottoman representatives.[6] Pressured by Weizmann and Sykes, Balfour came to suspect that the meeting's real purpose was to sound out some of the former ambassador's Turkish friends about a separate peace between the Ottoman Empire and the Allies.[7] Before leaving, Weizmann was given explicit instructions by Balfour to talk Morgenthau out of his mission.[8] After two days of meetings in Gibraltar between Weizmann, Morgenthau, Frankfurter and French representatives Colonel E. Weyl and Albert Thomas[9] between 4 and 6 July Morgenthau was eventually persuaded to cancel his mission. He eventually returned to the United States after a short stay in Paris under Weizmann's watchful eye.[10]

While Weizmann was occupied with matters in Gibraltar and Paris, back in England work proceeded at a fast pace in drafting a Zionist declaration to submit to the government. Harry Sacher began the process on 22 June. On 5 July Sokolow presented his own draft version to Sacher and a group of colleagues, who soon came to be known as the London Zionist Political Committee, or the Political Committee.[11] From this point on, various drafts and wordings were suggested, discarded and amended by the group. In Weizmann's absence, Sokolow had overall responsibility for drafting the declaration but kept in daily contact with the Foreign Office. There, in Sykes's office, Sokolow and Sykes, assisted by junior diplomat Harold Nicolson,[12] reviewed each new draft.[13] It was Sykes's idea for the Zionists to draft a formula that could be sent to Lord Rothschild, who would recommend it to Balfour, who in turn would respond to Rothschild on behalf of the British government. In the end it was essentially Sykes and Sokolow who drafted the Zionist's formula.[14] It was approved by Lord Rothschild and, after a final review and approval by Sykes,[15] was submitted to Balfour with a cover letter on 18 July:

> Dear Mr. Balfour,
> At last I am able to send you the formula you asked me for. If His Majesty's Government will send me a message on the lines of this formula, if they and you approve of it, I will hand it on to the Zionist Federation and also announce it at a meeting called for that purpose ...
>
> Yours Sincerely,
> (Signed) Rothschild

Listed were two draft declarations:
1. His Majesty's Government accepts the principle that Palestine should be reconstituted as the National Home of the Jewish people.
2. His Majesty's Government will use its best endeavours to secure the achievement of this object and will discuss the necessary methods and means with the Zionist Organization."[16]

The Zionist declaration (as it would be called) submitted by Lord Rothschild was startling in its audacity. It read that the Zionists were asking for all of Palestine to be turned over to them and that the British government would do all they could to ensure this happened.[17]

In retrospect, it is incredible to think that while he was helping to write the Zionist declaration Sykes actually believed that the Palestinians would have accepted this giving away of their land, *and* without being consulted. However, this was never a consideration on his part or on that of the Zionists. Both had their agendas and any real concern about the Palestinians was never an issue. For his part, Sykes did not want to jeopardise what he believed was the best solution for a postwar Middle East through the implementation of the Zionist plan. All his efforts could be undone by mentioning what he knew would be an inconvenient fact: the Palestinians would be against it and would resist it. For this reason, as in the past, he purposely did not consult any Palestinians on the matter. Besides, his immediate goal was for the declaration to generate much-needed Jewish influence and support in the United States and Russia for the Allied war effort.

Sykes believed that after the war a Jewish homeland in Palestine would be strategically important for Britain in the region and lead to the flowering of the undeveloped and backward Middle East. In short, Sykes's approach to Zionism and Palestine – far from being pro-Zionist – was to use Zionism to achieve his own goals. However, his motivations were based on several flawed assumptions. These were "a belief in Jewish unity and power, the conviction that Jews were largely pro-German, and that they also constituted a leading force in pacifist and Russian Revolutionary circles."[18] He was not alone in this; nor were the Zionists above using the beliefs of their would-be benefactors to further their own cause. In fact, they counted on it.

Balfour responded to Lord Rothschild's letter the next day:

My dear Walter,

Many thanks for your letter of July 18th.

I will have the formula which you sent me carefully considered but the matter of course is of the highest importance and I fear it will be necessary to refer it to the Cabinet. I shall not therefore be able to let you have an answer as soon as I should otherwise have wished to do.

A. J. Balfour[19]

There the matter rested for several weeks.

Meanwhile, in Cairo the Palestinian offensive continued to remain stalled despite General Murray's replacement by General Edmund Allenby in June 1917. The problem was that in order to take Palestine and Syria Allenby needed more troops. Since Sykes's Arab Legion had not provided the hoped-for 10,000 troops and no troops were available from Europe, where the war was at a crucial stage, other options had to be considered. In early 1917 Sykes had resisted efforts by Zionists Vladimir Ze'ev Jabotinsky and Joseph Trumpeldor to establish a Jewish Legion to support the invasion of Palestine. At the time he felt a Jewish Legion accompanying British forces into Palestine might bring repercussions from the Turks against Jews in Palestine.[20] However, now that his Arab Legion had failed to materialise, Sykes looked for other alternatives and reconsidered the Jewish Legion.

After Murray failed to take Gaza, the Turks knew that the British would try again to take Palestine. As for Sykes's fear of retribution against the Jews in Palestine because of the formation of a Jewish Legion, Jabotinsky pointed out that Jamal Pasha had already expelled 10,000 Jews from Jaffa in December 1914 for no reason at all. So, with or without the formation of a Jewish Legion, he did not need a reason to take revenge on the Palestinian Jews.[21] This was all Sykes needed to hear to change his mind. He "now supported the recruitment of Russian Jews in London for a 'Jewish Regiment' with its own distinctive badge in the form of King David's shield" to accompany Allenby's Egyptian Expeditionary Force into Palestine.[22]

However, resistance to the plan came from an unexpected quarter. Edwin Montagu, the newly appointed secretary of state for India and the only Jewish cabinet member in the government, was an anti-Zionist. The second son of the first Lord Swaythling[23] and Herbert Samuel's first cousin, Montagu was appointed secretary of state for India in mid-July.[24] On learning about the formation of the Jewish Legion, Montagu protested in a memorandum to the Cabinet that "a Jewish Legion makes the position of all Jews in other regiments more difficult and forces a nationality upon people who have nothing in common."[25]

Montagu was not alone in his opposition to the formation of the Jewish Legion. Another prominent Jew, the Lord Chief Justice Lord Reading, also voiced his opposition. Several other deputations of prominent Anglo-Jewish leaders also made their feelings known, with some objecting more to the name than the Legion itself, while others argued in favour "of a return to Palestine, but refuting the means desired by Jabotinsky." However, on 12 September, 1917 in Army Council Instruction 1415, the War Office authorised the "Formation of Battalions for the Reception of Friendly Alien Jews, confirming their establishment ... [under] the regimental name ... the [38th] Royal Fusiliers."[26] Now the Jewish Legion, long opposed by Sykes, who had relented, was a reality under another name.

By the middle of August the Zionist declaration prepared with Sykes's help and submitted by Lord Rothschild to Balfour was submitted with minor alterations to the War Cabinet. Prospects for its approval seemed good, as most of the seven men in the War Cabinet were favourable to Zionism. These included Lloyd George, Andrew Bonar Law, Lord Milner, Lord Curzon, George Barnes, Sir Edward Carson and General Jan Smuts.[27] While the Prime Minister's pro-Zionist feelings were well known among his colleagues, South Africa's General Smuts, an unknown entity recently appointed to the War Cabinet, showed a similar strongly pro-Zionist stance.[28] Milner, Barnes and Carson were supportive as well. However, Bonar Law and Curzon soon showed mixed feelings and some antipathy towards Jewish aspirations in Palestine. Balfour, as foreign secretary, frequently attended the meetings and, although he was not a member of the War Cabinet and had no vote in the proceedings, he could be counted on to give his utmost support to put the declaration through.[29]

However, strong opposition to the declaration came from outside the War Cabinet from the regular cabinet. This again came from Edwin Montagu, secretary of state for India, who claimed his Judaism "meant little to him as race or religion, [but] meant absolute hostility where Zionism was concerned." Montagu was a member of the Asquith government at the same time as his cousin, Herbert Samuel. At the time he had characterised Samuel's Palestine Memorandum of March 1915 to Asquith as "disastrous policy." In the summer of 1916, when asked by the Foreign Office for his views on Zionism, Montagu responded with "a diatribe against Jewish nationalism, which he called 'horrible and unpatriotic' and characterized as 'pro-German.'"[30]

On 23 August 1917, Montagu presented a lengthy memorandum titled "The Anti-Semitism of the Present Government," in which he warned that in making a pro-Zionist declaration the government would be advocating a policy which is "anti-Semitic in result and will prove a rallying ground for Anti-Semitism in every country in the world." He went on to say, "Zionism has always seemed to me to be a mischievous political creed, untenable by any patriotic citizen of the United Kingdom." He added, "I have always understood that those who indulged in this creed were largely animated by the restrictions upon and refusal of liberty to Jews in Russia."[31]

Then in eerily prophetic statements, Montagu wrote that "a declaration calling for the reconstitution of Palestine as the national home of the Jewish people [would mean] . . . that Mahommedans and Christians are to make way for the Jews, and the Jews should be put in all positions of preference and that Turks and other Mahommedans in Palestine will be regarded as foreigners." And, "perhaps also citizenship must be granted only as a result of a religious test." His final point was "that gentile support for Zionism in England was based on anti-Semitism, on a desire to get rid of Jews and send them to Palestine, which would thereby become the world's Ghetto." Ending his memorandum, Montagu wrote:

> I feel that the Government are asked to be the instrument for carrying out the wishes of a Zionist organisation largely run, as my information goes, at any rate in the past, by men of enemy descent or birth, and by this means have dealt a severe blow to the liberties, position and opportunities of service of their Jewish fellow-countrymen.
>
> I would say to Lord Rothschild that the Government will be prepared to do everything in their power to obtain for Jews in Palestine complete liberty of settlement and life on an equality with the inhabitants of that country who profess other religious beliefs. I would ask that the Government should go no further.[32]

At the 3 September meeting of the War Cabinet when the declaration on Palestine was first considered, both of its major supporters Lloyd George and Balfour were on vacation. So was Sykes.[33] So, when it came time to consider the declaration those members present – Bonar Law, Milner, Carson and Smuts – reviewed three documents: the proposed Balfour letter to Lord Rothschild, the Montagu memorandum, and an alternative draft to Balfour's note. The latter was prepared two weeks earlier by Lord Milner, who felt Balfour's draft declaration was too similar to the Rothschild proposal and, as such, was too strongly Zionist.[34] The key difference between the two was the substitution of "every opportunity should be afforded for the establishment of a home for the Jewish people in Palestine" in place of the Zionist declaration that "Palestine should be reconstituted as the national home of the Jewish people." Montagu was present at the meeting at the cabinet's request, and shared his views with the cabinet members. After some discussion, the Under Secretary of State for Foreign Affairs Lord Robert Cecil, sitting in for Balfour, was instructed to advise the United States government that the British government was being asked to "make a declaration of sympathy with the Zionist movement, and to ascertain their views as to the advisability of such a declaration being made."[35]

To virtually everyone's surprise, especially the Zionists, President Wilson's response to Cecil's solicitation of support was not positive. It had been assumed from discussions with the Americans throughout the year that Wilson would be in favour of a declaration. This included Balfour's visit to the United States in April and May and meetings with both Wilson and Brandeis, and Weizmann's constant communications with Brandeis, Frankfurter and other prominent American Zionists.[36]

While the Foreign Office continued to seek Wilson's approval through official channels, Cecil suggested that Weizmann should contact Justice Brandeis for his personal assistance in obtaining a presidential affirmation.[37] However, Weizmann included the Rothschild (Sokolow/Sykes) Zionist declaration draft along with his message, not the Milner draft favoured by the cabinet at its meeting.[38] Unlike Cecil's official overtures, Weizmann's proved successful. In two separate telegrams, Brandeis wired back on 24 September confirmation of Wilson's support for the more strongly Zionist declaration, but also suggested he contact the French and Italian allies for their opinion. Apparently, the coolness of Wilson and his chief advisor Colonel House two weeks before was due to a concern over British sovereignty in Palestine, which had not been spelled out in the declaration but was implied in it.[39] For them, the Allies' attitude towards this issue had assumed more importance than the declaration itself.[40] So, with the French and Italian approvals already received, this response assured American support for a declaration on Palestine. On 26 September, Brandeis wrote to Weizmann:

> From talks I have had here with the President . . . I can answer you that he is in entire sympathy with declaration quoted in yours of the 19th, as approved by the Foreign Office and Prime Minister. I, of course, heartily agree.[41]

However, Montagu had not given up. On 14 September he wrote a long, impassioned letter on Zionism to Lord Robert Cecil, which he had printed and circulated among Cabinet members. He pointed out that Zionism was foreign in origin and that the majority of Zionist leaders in Britain were of foreign birth, and "at least half of the Jews in this country" were anti-Zionist. In his view, the declaration the War Cabinet had approved and shared with President Wilson

> to help the Allied cause in America . . . implies there is a Jewish people in the political sense and that any Jew who happens now to live in England, France, Italy, or America is an exile in belonging to the English, French, Italian and American people among whom he dwells at present . . . Such a declaration would be felt as a cruel blow by the many English Jews who love England, the birthplace of themselves and their ancestors for many generations, who wish to spend their lives in working for her, and whose highest aspiration is to continue to serve here.[42]

By this time, Sykes had returned to London in mid-September, determined to revive the stalled momentum in the War Cabinet on the declaration and meet the criticisms levelled at it, particularly Montagu's. He quickly wrote a memorandum, "Note on Zionism," in which he

> [c]haracterized Jewish opposition to Zionism as including both the "assimilated Jewish case" – Jews like Montagu, who were totally Westernized – and the "international and cosmopolitan Jewish case" – Jewish socialists and financiers who cared more for Karl Marx or capitalism than for a homeland in Palestine. Zionists, according to Sykes, only wanted recognition and continuance of colonization in Palestine with privileges equal to "the various religious and racial nationalities in the country," viz. the "Latins, Orthodox, and Moslems."[43]

However, other matters proved more instrumental than Sykes's memo in getting the government to declare its support for Zionism. There was growing unease in the Foreign Office over British failure to commit to the Zionist programme. It was believed that continued inaction would not only leave the government open to charges of failing the Zionists, but "would upset not only leading Zionists in London but the Jews in Russia and America. Furthermore, the Foreign Office received a report from Weizmann that Germany was on the verge of making a similar declaration, which "would steal Allied thunder." Later evidence would show, however, that this was something suggested in the press rather than being given serious consideration by the German government.[44]

Montagu's was not the only voice at this time raising concern about Britain's support for a Jewish homeland in Palestine. In August Sykes had received a letter from Antoine F. Albina, a Christian Arab. In 1903 Sykes had met his father, Joseph Albina, who was a prosperous Syrian contractor in Jerusalem. The younger Albina had served as Sykes's translator on his trips in the Middle East.[45] He periodically wrote to Sykes with observations on Middle East matters that he felt were important and did occasional translation work for him.

In a three-paged typed letter sent from Cairo dated 10 August 1917, Albina had much to say on Zionist aspirations in Palestine:

> The rumour spreading about and the propaganda undertaken in the Press that the Jews will be given possession of Palestine, or at least will be granted extensive privileges there, are causing a feeling of fear and great anxiety amongst the Christian and Moslem Arabs.

He went on, "Jerusalem, as I have pointed out in former reports, is the most sacred city in the world to Christians and the third holy place to Moslems. The Jews have nothing there but memories of the past." Albina continued:

> How can the Allies [reconcile their efforts] of freeing small nationalities, by imposing upon the Palestinian Arabs, who are the original settlers of the country, the rule of a foreign and hated race, a motley crowd of Poles, Russians, Roumanians, Spaniards, Yemenites, etc, who claim absolutely no right over the country, except that of sentiment and the fact that, their forefathers inhabited it over two thousand years ago?

He added, "the introduction into Palestine of Jewish rule, or even Jewish predominance, will mean the spoliation of the Arab inhabitants of their hereditary rights and the upsetting of the principles of nationalities." Albina warned Sykes, "Politically, a Jewish State in Palestine, will mean a permanent danger to a lasting peace in the Near East. Besides brewing discontent [it will create] a spirit of rebellion amongst the populations."[46]

There is no record to show that Sykes replied to Albina's letter. Given its content and being from Albina, a Middle East source he had frequently relied on in the past, it would not be something he would normally ignore. At the end of this document in the Sykes Papers, there is a note in small, neat handwriting: "The interesting thing here is that Albina put the anti-Z pt of view strongly. It was known to MS [Sykes], & therefore to AJB [Balfour]." This was probably written by Sykes's son, Christopher Hugh Sykes (1907–1986), while researching for his book *Crossroads to Israel: Palestine to Bevin.*

The author of the note adds that he had met Albina: "A very dark man he seemed to me. I suppose he was Syrian. He was a pious man and used to serve mass at Sledmere." Both are similar to the handwriting found on another document in the Sykes Papers with Christopher Hugh Sykes's initials after it.[47]

On 21 September, Sykes sent Balfour a seven-and-a-half-page memorandum, "Note on Palestine and Zionism," apparently as an aid to guide the foreign secretary and the War Cabinet in their decision on the Zionist proposal. Originally a six-page document simply entitled "Note on Palestine," Sykes had given it to Weizmann to read, according to a handwritten note on an early copy of the document found in the Sykes Papers (again in the handwriting of Sykes's son, Christopher Simon Sykes). The Zionist leader's handwritten comments can clearly be seen in the margins of the document, identifiable from other handwritten letters in the Sykes Papers. Subsequently, the name of the document was changed to "Note on Palestine and Zionism" and Sykes added an additional page-and-a-half of material to the final document, which also included Weizmann's suggestions, before giving the final version to the foreign secretary.[48]

In this document Sykes set out the basis for the government's policy on Palestine. He pointed out that it was purposely set apart from the areas included in the Sykes–Picot agreement. "Although Palestine is chiefly inhabited by Syrian Arabs the questions of Zionism, the Christian and Moslem holy places, oblige us to consider it as a category apart."[49] The reference here to the inhabitants of Palestine as "Syrian Arabs" instead of "Palestinian Arabs" was blatantly disingenuous and part of his continuing campaign to categorise them as non-indigenous peoples; hence, foreigners that had no special rights in the area. By thus disenfranchising the native population, Sykes was able to make the so-called Palestinian problem resolvable on a strictly religious basis and promote the Zionist agenda. This was consistent with his earlier meeting in Cairo in May 1917, where he spoke to Syrian Arabs and not Palestinian Arabs about Palestine.

He went on in "Note on Palestine and Zionism" to make religious identification of paramount importance as far as Palestine was concerned, expanding on the interests of the different Christian churches, pilgrimages and shrines in Palestine in about a page. He suggested that because of these, there existed "a general feeling that Palestine should be treated as a region consecrated to religious memories, and as little as possible involved in the hurly-burly of politics."[50] In fact, however, it could not have been more political.

Muslims were given a single paragraph in the second page of the document, in which Sykes mistakenly noted, "the mosque of Omar [in

Jerusalem] is regarded as next to Mecca itself the most sacred site in the world."[51] However, the Mosque of Omar is not among the three most important mosques in Islam, which are, in order of precedence, the Great Mosque in Mecca, the Prophet's Mosque in Medina and the Al-Aqsa Mosque in Jerusalem. The latter is on Temple Mount along with the most prominent feature of the Jerusalem skyline, the Dome of the Rock, from where Muslims believe Muhammad made his midnight journey to Heaven in the *mi'raj*. Neither of these was mentioned, which seems to suggest that what Sykes knew of Jerusalem, despite his many visits there and his much-vaunted expertise in Islam, was exaggerated. Otherwise, why would he make such a glaring error about something learned by virtually every tourist in a guided tour of the city?

In the remaining six pages of the document Sykes focused almost entirely on the Jews and Zionists, emphasising that "in regard to Palestine and its holy places [the Jews] require careful consideration." He began by explaining the religious and historical importance of the return to Jerusalem by the Jews of the Diaspora. He then divided worldwide Jewry into two categories: the assimilated Western Jews and non-assimilated Eastern European Jews. The former lived in large numbers in England, France, Italy and the USA, and, although they were assimilated, they remained religiously apart from the communities in which they lived. These contrasted with the "immense masses of Jews in East Europe & England & America of East European origin, who do not tend to assimilate into the race among which they reside." It was among the latter, who have "a special sense of Jewish nationality which in some cases survives strict adherence to the Jewish faith." It is among these Jews, he maintained, that

> there is an instinct to revive the Jewish nation once more in Palestine, and there exists a Zionist organization whose members are of opinion that it is only by fostering this idea that the masses can be redeemed from the squalor and degradation which centuries of oppression, racial isolation, and enforced abstention from manual labour has plunged them.

Here he added, "it is further believed by leading Zionists that this national inspiration can only be drawn from the soil of Palestine."[52]

Sykes went on to describe two types of Jews who oppose Zionism: the assimilated Jews and the cosmopolitan Jews. The former, he explained, "feel they have attained nationality," while the latter "do not desire nationality, but feel that rather they would desire to see the whole world as cosmopolitan as they are themselves." The latter he referred to as "an influence or mental tendency, it has its foundations in ...

A. The Extreme Socialist Jew of the underworld who regard Karl Marx as the only prophet of Israel and who work toward the destruction of the present Nationalistic basis of the world and the setting up of a world state.

and

B. the fact that — There are a few magnates of Jewish birth whose connexion with international finance has been so engrossing as to identify them mentally rather with finance as a cosmopolitan affair than with any nationalist cause, be it that of the Nation in which they are assimilated or Zionism itself."[53]

"These two influences are readily understandable," Sykes explained:

[S]ince they arise from the conditions of life under which Jews live at either end of the social scale, viz: those who are engaged in world finance which is cosmopolitan and those whom persecution and isolation have driven down to the depths which in misery squalor and poverty is also cosmopolitan.

It was the "Cosmopolitan Jewish case," he maintained, "which consciously or sub-consciously impels a few Jews into the Anti-Zionist Camp." As for the assimilated Jew, he continued, their lack of interest in "the scheme for Palestine" can be attributed to the attitude that "the country is already populated and will not hold the Jews whom it is proposed to send there," or that "the experiment of Zionism will strengthen anti-Semitism by distinguishing Jews politically and racially from the rest of the community, and will place already assimilated Jews in a dangerous and false position and retard the assimilation of other Jews."[54]

Sykes then used an approach that he believed would appeal to Balfour and the War Cabinet and remove any reservations that might remain on the Zionist proposal. In doing this, he also countered the anti-Zionists' arguments. The Zionists would not "flood Palestine with Jews," he said, but instead "stimulate Jewish agrarian colonization in Palestine, and thus to build up gradually the Jewish community, Jewish language, local Government, and culture." Moreover,

[F]ar from strengthening anti-Semitism [this] will combat it by (a) showing the world that urban Jews are capable of being transmuted into an industrious peasantry (b) by stimulating in the Jews who at present have no nationality with a sense of national consciousness and so improving their moral[e] which has been impaired by ages of wandering and aloofness.[55]

If all this were not convincing enough, Sykes then spelled out exactly what the Zionists wanted and what they did not want, some of which appears to be deliberately misleading. If he had read Herzl's *The Jewish State*, upon which political Zionism was founded, he would have known this, and there is no reason to doubt that he had. At the very least, it appears that Sykes was willing to say or do whatever it took to ensure his vision of a postwar Palestine in which a thriving Zionist community would not only act as a buffer between British-controlled Egypt and the French in Syria, but also bring peace, prosperity and stability to the region. He listed the following:

I. What the Zionists do not want is: – To have any special political hold on the city of Jerusalem, itself or any control over the Christian or Moslem Holy places.

II. To set up a Jewish Republic or other form of state in Palestine or any part of Palestine.

III. To enjoy any special rights not enjoyed by other inhabitants of Palestine.

I. On the other hand the Zionists do want: – Recognition of the Jewish inhabitants of Palestine as a national unit, federated with national units in Palestine.

II. The recognition of [the] right of bona-fide Jewish settlers to be included in the Jewish national unit in Palestine.[56]

Sykes continued by giving every assurance "that Zionism is a permanent and positive force in world Jewry." He admitted that the number of Jews in England, Italy, the USA and France "were a small and comparatively prosperous body [that] has assimilated itself gradually into the local nationality." However, such was not the case in "Russia, Poland, Roumania and East Europe generally. The Jews of those countries are in such compact masses, are so numerous and so distinct in social habits, occupation, appearance and language, from the nations among whom they reside," that the possibility of their assimilation in the near future is unlikely; adding, "there exists a mutual hostility based on long traditions between the Jewish mass and local nationality." (Jews and non-Jews alike had long referred to this situation in Russia and Eastern Europe as the Jewish problem. While this term was not used by Sykes, the inference was clear and the term was in common use at the time.[57])

This brought him to Palestine. Without actually saying it would be a place for these "masses of un-assimilated Jews," instead Sykes let the reader make that connection. He helped this along by adding, "it may therefore be assumed safely, that so long as there are large un-assimilated masses of Jews in East Europe Zionism will be a positive force with supporters in all Jewish communities throughout the world."[58] In other words, it would resolve the so-called Jewish problem by establishing a Jewish homeland in Palestine and housing Jews there. In doing this, he assured the ministers that "the Zionists are not working for the establishment of Jewish supremacy in Jerusalem but for the building up of a regenerated Jewish race in Palestine."[59]

In his closing remarks, Sykes said that "the Palestinian problem [was] . . . extremely complicated," because "we must consider the questions of the Holy Places, the local population, and Zionist aspirations." He then spelled out what he believed the British government must do:

(i)   The Government would have to perform the following Offices: – Guarantee order in the Christian Holy Places.
(ii) Hand over the Mosque of Omar to a Moslem body commanding [the] respect of the Moslem world as a whole, and guarantee the Mosque of Omar from violation.
(iii) Guarantee respect of the recognised Jewish Holy places.
(iv) Provide an arrangement for safeguarding the remaining holy places shared by various bodies.
(v)   Devise a system of equitable land purchase, setting a mediator between a willing buyer and a willing seller.
(vi) Devise a means of constitutional Government recognising the various religious and racial nationalities in the country viz: the Latins, Orthodox, Jews and Moslems, and according equal privileges to all such nationalities.[60]

Once again Sykes mentioned the Mosque of Omar and left out the most sacred Muslim places in Jerusalem – Al Aqsa Mosque and the Dome of the Rock – about which he would certainly know from his many trips to Jerusalem. This was doubtlessly done in deference to the Zionists, because of their being situated on Temple Mount the site of the Jewish Temple destroyed by the Romans in 70 CE and the most holy site for Jews. To his credit, however, the military authorities and later the Mandatory government in Palestine would follow some of Sykes's suggestions. In fact, some remain in force to the present day. Unfortunately, much more needed to be done in order to ensure the peaceful integration of the various religions and communities. As subsequent events would soon show, this could not be done by marginalising the native population in favour of others from outside the area.

In early October, Lord Milner, dissatisfied with the wording of the Zionist declaration, asked Leopold Amery to devise another formula for consideration at the 4 October War Cabinet meeting that combined the previous drafts.[61] In it were added two provisions: that "nothing shall be done which may prejudice the civil and religious rights of the existing non-Jewish communities in Palestine," or "the rights and political status enjoyed by Jews in any country who are contented with their nationality."[62] This did nothing to appease Montagu. The next day, after it was decided that the declaration was to be on the agenda for the War Cabinet's meeting of 31 October, Montagu wrote to Lloyd George. Speaking for himself and the majority of British Jews, who viewed themselves as British and having no other nationality, he told the prime minister:

> You are being misled by a foreigner, a dreamer and idealist ... who sweeps aside all practical difficulties ... If you make a statement about Palestine as the National Home for the Jews, every anti-Semitic organization and newspaper will ask what right a Jewish Englishman, with the status at best of a naturalized foreigner, has to take foremost part in the Government of the British Empire ... The country for which I have worked ever since I left the University – England, the country for which my family have fought, tells me that my national home, if I desire to go there, therefore my natural home is Palestine.[63]

On 9 October, Montagu circulated a second memorandum simply titled "Zionism," and in its four pages he argued passionately against the passage of the Zionist declaration. He included a list of prominent British Jews who were opposed to Zionism, together with the text of anti-Zionist statements made by well-known Jews in France, Italy and the USA.[64]

By this time, the War Cabinet, in addition to having decided to vote on the proposed declaration at its 31 October meeting, again decided to consult President Wilson as well as to seek the views of both Zionist and British anti-Zionists. The Zionists included Chaim Weizmann, Lord Walter Rothschild and Nahum Sokolow and the anti-Zionists interviewed (all on Montagu's list) were Claude Montefiore, Sir Leonard Cohen, Stuart Samuel and Sir Philip Magnus, MP.[65]

Noticeably absent from the latter group, however, was the Zionist movement's most outspoken opponent, Lucien Wolf of the Conjoint Foreign Committee. While there is no documentation to prove whether this omission was deliberate or not, it probably was, as Sykes and Weizmann had suggested whom to invite. Nevertheless, without its leading spokesman, the esteemed and highly regarded leaders of the Jewish community chosen to give evidence before the War Cabinet were assured a relatively benign contribution from British anti-Zionists. The War Cabinet also consulted Herbert Samuel, who, although he was not a Zionist, favoured a declaration as soon as a military victory was achieved. On 16 October, President Wilson sent a favourable reply and the stage was set for the declaration to be considered. All Montagu's attempts to convince the War Cabinet that it was a serious mistake to support the declaration had failed. Discouraged and bitter, he left shortly afterwards on a previously scheduled state trip to India.[66]

At the end of October, taking up Montagu's lonely place in opposition to the declaration, Lord Curzon openly declared his opposition to Zionism, which he called "a dream." Sykes answered Curzon's criticism with an anonymous note attached to a Foreign Office memorandum reviewing what Zionist colonisation had done in the past and predicting greater achievements in the future. His note concluded, "If the Zionists do not go there ... some one will, nature abhors a vacuum."[67] The latter remark was absurd. The region was far from empty, as he knew well from his travels in the area and writing about its diverse populations. (A study of Ottoman records shows that in 1914 Palestine had a resident population of 736,000.[68]) This seems to show just how far Sykes was willing to go in pursuing his agenda by ignoring the existing population, the unrepresented people, and thereby deliberately misrepresenting the true situation in the halls of power to achieve his goals.

At the 31 October War Cabinet meeting Balfour reviewed all the arguments against making a declaration to the Zionists. Particular focus was on just what the Zionists meant by a national home. Sykes and Balfour assured the ministers it meant "not the early establishment of an independent Jewish State, but rather Allied guarantees that the Jews might work out their own salvation" and "build up, by means of education, agriculture, and industry, a real centre of national culture and focus of national life."[69] After some more discussion, the War Cabinet agreed "to authorize the Secretary of State for Foreign Affairs to take a suitable opportunity of making a declaration of sympathy with Zionist aspirations in the terms of a slightly amended version of the draft submitted by (Lord) Milner on 4 October:"[70]

His Majesty's Government accepts the principle that every opportunity should be afforded for the establishment of a home for the Jewish people in Palestine, and will use its best endeavours to facilitate the achievement of this object, and will be ready to consider any suggestions on the subject which the Zionist organisations may desire to lay before them.[71]

While the Cabinet was in session approving the final text, Weizmann waited anxiously outside. He recalled the scene many years later, when Sykes brought the document out to him from the Cabinet room and exclaimed: "Dr. Weizmann, it's a boy!" According to Weizmann, "I did not like the boy at first. He was not the one I had expected. But I knew that this was a great departure."[72] Two days later, on 2 November, Balfour officially presented Lord Rothschild with a letter signed by him expressing Great Britain's support for the Zionist goals in Palestine.

To all outward appearances, the meeting on 2 November 1917 at the Foreign Office was little more than a social call by a distinguished peer of the realm on Britain's foreign secretary. Walter Rothschild, the second Lord Rothschild, head of the British branch of the Jewish banking family and a major force in British financial circles, was paying his respects to Arthur Balfour. Such was far from the case, however. After greeting his guest in his office and sharing a few pleasantries, Balfour handed Lord Rothschild a note. A few more words were exchanged between the two men, after which Lord Rothschild left Balfour's office with the note firmly in hand. It read:

Foreign Office
November 2, 1917

Dear Lord Rothschild,
I have much pleasure in conveying to you, on behalf of His Majesty's Government, the following declaration of sympathy with Jewish Zionist aspirations which has been submitted to, and approved by the Cabinet.

"His Majesty's Government view with favour the establishment in Palestine of a national home for the Jewish people, and will use their best endeavours to facilitate the achievement of this object, it being clearly understood that nothing shall be done which may prejudice the civil and religious rights of existing non-Jewish communities in Palestine, or the rights and political status enjoyed by Jews in any other country."

I would be grateful if you would bring this declaration to the knowledge of the Zionist Federation.

Yours,
Arthur James Balfour[73]

At the time, the Lloyd George government believed this was little more than a small, symbolic gesture, albeit an important one, for which little was given but for which much was expected in return. This was not to be. As events turned out, the British would get nothing but grief from both the Arabs and Jews over the next thirty years, as the result of the Balfour Declaration and eventually denounce it in the White Paper of 1939 (Command Paper No. 6019) and debate in the House of Commons on 23 May 1939[74] and give up the Palestinian Mandate under Zionist pressure in 1948. However, for the Zionists in 1917 led by Weizmann, it was the first major step toward bringing the Jews of the Diaspora back home after almost 2,000 years in exile, and establishing a Zionist state. It was also the beginning of unrest and turmoil unseen in the region for more than millennia, leading to several wars and bloodshed that continues to this day.

In Mecca King Husayn was greatly disturbed on hearing the news. He asked the Cairo government "for a definition of the meaning and scope of the Declaration." In response, Commander Hogarth of the Arab Bureau was dispatched to Jeddah at the beginning of January to see the king. Hogarth was apparently able to mollify Husayn, which was important at the time for the Arab revolt. He gave the king "on behalf of the British Government ... an explicit assurance that 'Jewish settlement in Palestine would only be allowed in so far as would be consistent with the *political and economic freedom of the Arab population.'"*[75]

The foregoing was written by Arab historian George Antonius, who in 1931 was given complete access to King Husayn, then in exile in Jordan, and "a 'jealously guarded chest' containing hundreds of documents, including the McMahon–Hussein correspondence of 1915–1916," which would form the basis of his book on the Arab revolt, *The Arab Awakening*.[76] Antonius learned that Hogarth delivered his message orally, "but Hussain took it down, and the quotation I have just given is my own rendering of the note made by him in Arabic at the time." Antonius noted, "the phrase I have italicised [above] represents a fundamental departure from the text of the Balfour Declaration which purports to guarantee only *the civil and religious rights* of the Arab population." He further maintained that,

> [H]ad the Balfour Declaration in fact safeguarded the political and economic freedom of the Arabs, as Hogarth solemnly assured King Hussain it would, there would have been no Arab opposition, but indeed Arab welcome, to a humanitarian and judicious settlement of Jews in Palestine.[77]

Unfortunately, the coming together of East and West – Middle Eastern Arab and European Jew – proved to be an unfortunate clash of civilisations and was not the hoped-for harmonious merging of peoples, religions and cultures promoted by Sykes and the Zionists. Each had their own agendas, in which Arab rights and welfare were not a priority. And, as events would soon reveal, things were not what Husayn was led to believe. After the war it became clear that what Husayn was getting in return for leading the Arab revolt against the Turks was far less than he thought. It was not the Arab State he envisioned but what he already had: the Kingdom of the Hijaz. In fact, what he believed the Husayn–McMahon correspondence gave him, the Baghdad Declaration, the Sykes–Picot agreement and the Balfour Declaration took away. Whether he ever learned that Sir Mark Sykes had played a key role in the last three is not known, particularly the latter, of which it has been said that "but for [Sykes], there would have been no Balfour Declaration. For he was the all-important connecting link between the War Cabinet and the Zionist Organization, the one who had the ear of both and the confidence of both."[78]

The announcement of the Balfour Declaration came as a surprise to many besides Husayn, especially the British officials in Cairo, including Hogarth, who shortly afterwards would have to defend it. While there had been hints about it during the year, even Clayton, chief of British Intelligence in the region, who oversaw the Arab Bureau and the Arab revolt, Wingate's second-in-command and Hogarth's boss, who had worked closely with Sykes, knew nothing about it: "There is no evidence ... [that] Clayton was aware of the efforts of Weizmann and his colleagues in 1916–1917 to secure British sponsorship of a Jewish national home in Palestine." Nor does it appear he had any knowledge of the serious consideration Whitehall was giving to the Zionist programme.[79]

In December 1916, when Aaron Aaronsohn had returned to Cairo from his trip to London, Clayton recalled he was "eager to pursue 'schemes which he had started on [in England] under Sir Mark Sykes's instructions' for sending agents to Russia and appealing to Jewish labour parties in Russia and America." While he was "interested in Aaronsohn's proposals, ... [he] had no idea whether they should be encouraged." As late as July 1917, he wrote to the Cairo residency, "we are in ignorance of the policy which H.M.G. is adopting in regard to the Jewish question and Palestine .... Our complete ignorance ... is handicapping us considerably in dealing with the local Jewish community."[80]

In August Clayton wrote to the Foreign Office when "Cairo censors held up publication of a pamphlet that had appeared in Egypt that contained excerpts from a speech by Weizmann in London on 20 May." In it, the Zionist leader was reported as saying that

"Palestine will be protected by Great Britain" and that, under such protection, "the Jews will be able to develop and create an administrative organization which, while safeguarding the interests of the non-Jewish population, will permit us to realise the aims of Zionism."

Weizmann then said, "I am authorized to declare ... the H.M.G. are ready to support this plan.'" Apparently, this was "Clayton's first intimation that London was about to announce support for the Zionist programme."[81] It also shows Weizmann's confidence that the government would support the Zionist programme, saying publicly in a speech that it was authorised five months before it was actually approved, and thus speaking as if it were already *fait accompli*.

A concerned Clayton also immediately wrote Sykes, telling him that the

lack of any knowledge of the policy, if any, decided upon as regards the Jewish question makes it increasingly difficult to deal with Aaronsohn and other Jews here who are becoming restive and impatient. If no definite line has been settled we can quite well keep them in play – but we ought to know.

He added:

I am not sure that it is not as well to refrain from any definitive pronouncement just at present. It will not help matters if the Arabs – already somewhat distracted between pro-Sherifians and those who fear Meccan domination, as also between pro-French and anti-French – are given yet another bone of contention in the shape of Zionism in Palestine as against the interests of the Moslems resident there.

He warned Sykes that the "more politics can be kept in the background, the more likely are the Arabs to concentrate on the expulsion of the Turks from Syria, which if anything will do more than anything to promote Arab unity and national feeling."[82]

Clayton had every reason to be concerned about being kept out of the loop of any apparent policy change in London. It directly affected him. He and Wingate were "running an Arab revolt predicated on the notion that the Arabs were fighting to liberate Arab territory from the Turks. If Palestine was now to be turned over to the Zionists, he ought to be informed." After all, "having sponsored the Arab movement for more than a year, London ought to consider what effect the Zionist programme would have on that movement."[83]

Others were also concerned with Sykes's plans in support of the Zionists in Palestine and the French in Syria, but none more so than T.E. Lawrence.

In September 1917 he wrote Sykes a five-page handwritten letter from Aqaba, where he was with the Arab Army under Faysal, after having taken the Turkish outpost after two months of heavy fighting. In it, Lawrence demanded to know exactly what Sykes's plans were for "Near Eastern Affairs ... since part of the responsibility of action is inevitably thrown on to me, and unless I know more or less what is wanted there might be trouble." Addressing the issue of the Zionists, he told Sykes there was a difference between the Palestinian Jews and colonialist Jews:

> The former speak Arabic, and the latter German Yiddish. [Faysal] is in touch with the Arab Jews ... and they are ready to work with him, on conditions. They show a strong antipathy to the colonialist Jews, & have even suggested repressive measures against them. [Faysal] has ignored this point hitherto, & will continue to do so. His attempts to get in touch with the colonialist Jews have not been very fortunate. They say they have made their arrangements with the Great Powers, & wish no contact with the Arab Party. They will not help the Turks or the Arabs.[84]

Lawrence continued, saying that Faysal

> wants to know (information had better come to me for him, since I usually like to make up my mind before he does) what is the arrangement standing between the colonialist Jews (called Zionists sometimes) and the Allies ... What have you promised the Zionists, & what is their programme?

He went on to say that he had spoken to Aaronsohn in Cairo, who told him that "the Jews intend to acquire the land-rights of all Palestine from Gaza to Haifa, & have practical autonomy therein," which prompted him to ask Sykes,

> Is this acquisition to be by fair purchase or by forced sale & expropriation? Arabs are usually not employed by Jewish colonies, and are often forbidden to enter a colony or purchase in its shops. Do the Jews propose the complete expulsion of the Arab peasantry, or the[ir] reduction to a day-labourer class?

Under these circumstances, he told Sykes he could "see a situation arising in which the Jewish influence in European finance might not be sufficient to deter the Arab peasants from refusing to quit – or worse!"[85]

As for the French, Lawrence wrote, "You say they [the Arabs] will need French help afterwards in the development of Syria – but do you really imagine anyone in Syria (bar Christians) wants to develop Syria? Why the craze for change?" Moreover:

> If the French put a tangible price on their help, the Sherif might pay for it. He is not going to sell "spheres of influence" etc. for gold or mountain guns. If the French want to annex a province, in return for "X" materials, let them say so. You can't buy gratitude by a secretly-conditional gift.[86]

Pointing out that the Sykes–Picot agreement was outdated, Lawrence said that with Britain's continued help, Husayn would be successful in their independent fight against the Turks and end up taking parts of both British and French areas [under the Sykes–Picot agreement] for his Arab state. To prevent this from happening, Lawrence sarcastically suggested "it seems to me that England & France can either take their areas first, or turn the Arabs out by force, or leave the Arabs there, or leave the Turks there (by ceasing to pay the Sherif's subsidy)."[87]

Lawrence ended his letter by saying:

> You know I'm strongly pro-British, & also pro-Arab. France takes their place with me: but I quite recognise that we may have to sell our small friends to pay for our big friends, or sell our future security in the Near East to pay for our present victory in Flanders.

Finally, he asked Sykes, "If you will tell me once more what we have to give the Jews, and what we have to give the French, I will do everything I can to make it easy for us." He added:

> [T]he future seems to me all over thorns, since military action by the Arabs, independently, was not in our minds when the S.P. was made, and if it's to be a Mede and Persian decree, [i.e., cannot be repealed] we are in rather a hole: please tell me what, in your opinion, are the actual measures by which we will find a way out.[88]

Lawrence sent his letter to Clayton to read and send to Sykes in London, but Clayton did not send it. It remains to this day in the Clayton Papers in the Sudan Archives at the University of Durham with Clayton's handwritten "Not Sent" on it. In a typed letter written a few days later, Clayton told Lawrence he had decided not to send his letter because Hogarth had told him that Sykes had "dropped the Near East just now and the whole question is, for the moment, somewhat derelict. All the better, and I am somewhat

apprehensive lest your letter to Mark may raise him to activity." As for the
Sykes–Picot agreement:

> From all I can hear the S-P agreement is in considerable disfavor in most
> quarters viz: Curzon, Hardinge, D.M.I. [Directory of Military
> Intelligence, General George Macdonogh] ... etc. The change is the
> Russian situation has wounded it severely and the general orientation of
> Allied policy towards "no annexation, no indemnities, etc" militates
> still further against many of its provisions. I am inclined, therefore, to
> think that it is moribund.[89]

He added:

> At the same time we are pledged in honour to France not to give it the
> "coup-de-geace" [sic] and must for the present act loyally up to it, in so
> far as we can. In brief, I think we can at present leave it alone as far as
> possible with a very fair chance of its dying of inanition.

As for his questions to Sykes about H.M.G.'s policy, Clayton wrote:

> I am entirely at one with you and I should like them to go on to him.
> I have endeavoured repeatedly to extract some definite statement of
> H.M.G.'s policy in this question, but without success. Your query may
> elicit some information beyond the meagre assertion that we are
> working with the Zionists which is all that our present information
> amounts to.[90]

Sykes would never learn of Lawrence's letter. One can only imagine what he
would have done if he had. Hogarth was quite wrong about Sykes "dropping
the Near East." He was busier than ever working to ensure the passage of the
Balfour Declaration and the establishment of a Jewish homeland in Palestine.
Also, neither man was aware that Sykes had already addressed some of the
issues Lawrence raised a month earlier in his Memorandum on the Asia-Minor
Agreement and sent to the Foreign Office, where apparently it languished for
lack of interest.[91]

With the Declaration secured, in November a triumphant Sykes turned his
attention to other matters he had also been working on. One of the most
serious matters was that some members of the War Cabinet led by Lord
Milner were once again considering "a separate peace with the Turks ... Sykes
was furious and put his reasons for opposing Milner's view before the War
Cabinet." He assured them "any such Turkish peace initiative was a sham; as

soon as one leader in Constantinople made peace, another would come to power and declare war again." Not only that, but

> any British responsiveness would be interpreted by the Turks as a sign of weakness and would encourage them to stiffen their terms. Now was no time for the British to act; they should wait until General Allenby had accomplished his task.

He then reminded the War Cabinet

> of British moral responsibilities to Zionist immigration, Armenian liberation and Arab independence. "Zionism is the key to the lock. I am sanguine that we can demonstrate to the world that these three elements are prepared to take common action and stand by one another."

He further "argued that when 'the Turks see the Zionists are prepared to back the Entente and the oppressed races,' then they 'will come to us to negotiate.'" The prime minister agreed and no further action was taken on peace negotiations with the Turks.[92]

While Sykes may have had a point about power struggles in the Young Turk leadership and letting Allenby establish the Allied forces in Palestine and Syria, his emphasis on the importance of Zionism to the Turks and the strength of relationships between the Zionists, Armenians and Arabs was pure fabrication. The goal of such talk was merely to dissuade the ministers from considering peace talks with the Turks at this time, and it had the desired effect. There was no evidence to substantiate any of these claims. A month later, perhaps in an act of collective guilt over the Balfour Declaration and 24 November revelation of the secret Sykes–Picot agreement by the Bolsheviks, on 14 December the War Cabinet increased King Husayn's subsidy.[93]

CHAPTER 11

PALESTINE

0, what a tangled web we weave when first we practice to deceive!
Sir Walter Scott[1]

From Suez, across the Sinai and into Gaza, General Allenby's Egyptian Expeditionary Force successfully pushed back the Turks, advancing into Palestine to take Jerusalem and entered the city on 11 December 1917. On hearing the news, "Sykes collected messages from Arabs, Armenians and Zionists, congratulating the British on their victories over the Turks, and distributed them to the War Cabinet." As Allenby approached Jerusalem the War Cabinet designated Curzon to write the proclamation for the general to make to the inhabitants of Jerusalem and consulted Sykes in its preparation.[2] Sykes was also behind Lloyd George's suggestion to Allenby that he should dismount and enter the city on foot through the Jaffa Gate, in contrast to Kaiser Wilhelm II's triumphal entry on horseback through the same gate almost twenty years earlier in 1898.[3]

Although he was not asked to write the proclamation himself, an elated Sykes took the opportunity to let his imagination run wild in describing possible reactions of the populace in his 13 December *"Eastern Report"*. In it, he enthused that "Christians – whether 'Bible-reading,' 'Roman' or 'Greek' – would be deeply moved because Jerusalem was 'so profoundly impressed as a name and an idea on all who have been reared in a Christian environment.'" He continued:

[A]s for the reactions of "Jewry" ... "Wherever there are Jews there are Zionists, in theory at least, and ... no matter what views these may have held about the war up till now, henceforth the goal of their ambitions rests in Entente hands."[4]

As for the Islamic world at large, Sykes believed the capture of Jerusalem would be a great loss of prestige for the Ottomans and their German allies and result in an increase in pro-British esteem. However, he admitted Arab reaction would be less enthusiastic:

> [He] conceded that "Our adoption of Zionism and our capture of Jerusalem will tend to a certain extent to somewhat abate Arab enthusiasms." Yet this lack of Arab support could be countered, Sykes maintained "if we take an early opportunity of showing that we are behind the Arabs, appreciate their assistance and desire their liberation."[5]

What he meant by this or how it would be done, he did not say. However, given his actions and attitude towards the Arabs during the war he probably meant little by it.

The Times and the *Manchester Guardian* carried Reuter's account of Sykes's speeches at Zionist meetings at the London Opera House on 2 December and at the Manchester Hippodrome on 9 December. At Manchester, he told his audience that the Zionist ideal was to ensure a future world where Jews were no longer persecuted and consigned to ghettos, but to give them "a world higher position than ever before." He reminded them, however, that the success of the Zionist plan depended on "a Jewish-Armenian and Arab *Entente.*"[6]

On 12 December the *New York Times* published an article reporting on his Manchester speech under the headline "Sees Great Future for Jew and Arab." Under the headline were several subheadings: "But Brotherhood and Conciliation Must Animate Revived Nations, Col. Sykes says"; "Responsibility on Zion"; and "Jerusalem Must be Vital Heart Healing Europe's Scars and Calling Asia Back to Life."[7]

On 16 December Sykes gave an interview published in the *Observer* on the importance of the capture of Jerusalem. Entitled "Jerusalem and its Future," the reporter wrote that he

> brought away ... from [his] interview with Lieutenant-Colonel Sykes that Jerusalem ... [was] a new Light of the World, shining out on all men and upon all nations, and bidding them, when the war is over and peace once more restored, take up their lives again with hope reawakened and faith restored.

He quoted Sykes as saying that, since Selim the Grim's[8] conquest of the area 400 years ago, "forces ... had used Jerusalem for the purpose of fomenting

discord in Christendom, of holding Jewry at arm's length and promoting war
and ill-will among men." Once the war was over, he said, "and Jerusalem out
of the devastating hands of the Turk," what will the future bring?" Answering
his own question, Sykes told *Observer* readers, "In this future we are
conceiving, the world is at peace, and also, we presume, the Zionist movement
is having an immense influence. That is one thing." He continued:

> Another is after this war . . . we may well see a spiritual revival in Islam,
> of a nature totally different from anything we have ever have seen before.
> The unintellectual Turk will play a lesser role. The intellectual,
> speculative, and spiritual Arab and Indian will play a great role.
> Consequently, we may see a tendency among Moslems to think more of
> the Word and Book than of Dominion and the Sword.[9]

One hundred years later we can see just how wrong Sykes's predictions
were.

He ended his interview by saying there must be a League of Nations to
ensure there would "be no more wars in the future [and] some force
which will control nations." And "the physical force of a League of Nations
must be at the call of a moral force," which Sykes saw as being in
Jerusalem. There, he told *Observer* readers, existed moral forces "stronger
than any man could imagine – the moral force of Calvary and Sacrifice, the
moral force of Zion and eternal hope, the moral force of Islam and
obedience."[10]

While blatantly playing both the religious and romantic cards for public
consumption, Sykes received a large dose of reality about his Middle East
dreams from Cairo. In a letter to Sykes dated 15 December, Clayton bluntly
told Sykes:

> I quite see your arguments regarding an Arab-Jew-Armenian combine
> and the advantage that would secure if it could be brought off. We will
> try it, but it must be done cautiously and, honestly, I see no great chance
> of any real success. It is an attempt to change in a few weeks the
> traditional sentiment of centuries.

He added:

> [T]he Arab cares nothing whatsoever about the Armenian one way or
> the other – as regards the Jew the Bedouin despises him and will never
> do anything else, while the sedentary Arab hates the Jew, and fears his
> superior commercial and economic activity.[11]

While claiming not to know how much influence the Zionists carried in America or Russia, Clayton said that by giving them all they asked for and promoting their plans for Palestine "as hard as we appear to be doing, we are risking the possibility of Arab unity becoming something like an accomplished fact and being arranged against us." Besides,

> whatever protestation Jews like Sokolow and Weizmann may make and whatever Arabs, whom we may put up as delegates, may say, the fact remains that an Arab Jewish entente can only be brought about by very gradual and cautious action.

This was because "the Arab does not believe that the Jew with whom he has to deal will act up to the high flown sentiments which may be expressed at Committee Meetings." Furthermore:

> [I]n practice [the Arab] finds that the Jew with whom he comes in contact is a far better business man than himself and prone to extract his pound of flesh. This is a root fact which no amount of public declarations can get over.[12]

One can imagine that Sykes was not pleased to read this, in light of his pro-Zionist campaign and also because of the high regard and respect he held for Clayton, which made his words hard to ignore. As Lawrence would later write of him:

> [A]ll of us [at the Arab Bureau] rallied round Clayton, the chief of Intelligence, civil and military. Clayton made the perfect leader ... He was calm, detached, clear-sighted, of unconscious courage and assuming responsibility ... He never visibly led; but his ideas were abreast of those who did ... he impressed men by his sobriety, and be a certain quiet and stately moderation of hope.[13]

The wise words in his letter to Sykes showed all this was true.

Clayton had more advice for Sykes: "We have therefore to consider whether the situation demands out and out support of Zionism at the risk of alienating the Arabs at a critical moment." In addition:

> [T]here is also to be considered the mass of sentiment which is bound to be called forth in every Christian country by the fall of Jerusalem into Christian hands, and which might easily be offended by a wholesale pro-Zionist policy.[14]

In contrast to Sykes's positive and upbeat *Observer* interview on the effect of the capture of Jerusalem on the Arabs and Islam, Clayton further warned Sykes there were "indications of considerable revivalist movement on Wahabi [*sic*] lines in Central Arabia, such as has occurred in the past when the prestige of Islam has fallen low."[15] While admitting he was not in a position at the time to assess properly "the strength of this movement, but the defeats which Turkey has suffered, the lack of a temporal head in Islam, and finally the fall of Jerusalem conduce to fostering it." He further advised Sykes that "it may modify the whole situation considerably and give rise to yet another complication." At the end of his typed letter, Clayton added a handwritten PS: "Do not think from this above that we are not trying to act on the lines you suggest, but I wish to point out clearly the dangers & difficulties which exist here."[16]

Clayton lived to see his predictions come true as Ibn Sa'ud's fierce Wahhabi Ikhwan ousted King Husayn from the Hijaz in 1924 once British patronage was withdrawn from both men;[17] the Islamic revival movement it would foster would take another 100 years to materialise fully and further destabilise the Middle East and the Muslim world beyond. Meanwhile, the war and resulting peace had let the genie out of the bottle and would result in far worse complications than either man could have imagined. However, Clayton's warnings and those of other officials in the Middle East who expressed serious concern over the newly revealed secret Sykes–Picot agreement and Balfour Declaration failed to dim Sykes's dreams of a postwar Middle East that corresponded to his vision.

Meanwhile, Picot had returned to Cairo and announced that "he was to be the French representative in a joint Anglo-French provisional administration which was to govern occupied enemy territory in Palestine until the end of the war – when some sort of international arrangement would be made." Clayton told Sykes he knew nothing about this and could not "protest too strongly against any such unworkable and mischievous arrangement. The country is under martial law and under martial law it must remain for a long time to come – probably until the end of the war." He stressed that

> The administration must be military and the Commander-in-Chief supreme until such time as he can inform H.M.G. that he considers the situation such as to allow of a civil government being established. That moment has certainly not come yet, nor do I see it, even in the distance.[18]

While Sykes may have been surprised at Picot's announcement, he was not surprised at his appearance in Cairo, because he had made the arrangements for Picot to travel there from Europe with Oriental Secretary Sir Ronald Storrs who was returning to Egypt after home leave. Apparently, during the trip Storrs and Picot became friends. Shortly after their arrival, Storrs wrote to Sykes:

> When I found the incredible knots into which they appear to have tied themselves here [in Cairo] over Picot I gravely feared a row, and was glad to be able to explain quite fully his side of the question before the matter was discussed. It would indeed be too foolish to complicate relations and problems at this very critical moment.[19]

He attached a copy of a letter he sent the same day to Sir Ronald Graham at the Foreign Office about Picot and hoped that Sykes would approve of his advice to Graham. In it, Storrs had told Graham of his recent trip from Europe with Picot, noting that "Towards the end he spoke with extraordinary sincerity and frankness; and I am convinced that so long as things are fully and firmly explained to him, he will prove not only an ally but a help." He added:

> There appears to be some apprehension as to his exact status and functions but upon fuller explanations and still more the receipt of your very satisfactory telegrams, rather an awkward incident and a certain amount of bad blood was avoided.[20]

As letters to Sykes from Clayton, Wingate and others seem to indicate, for some time this was not to be the case.

It was this and other matters that prompted the Foreign Office to send Sykes in late December visit to Paris. His good relations with the French were seen to make him the ideal candidate to smooth over any difficulties with Britain's closest ally. The hope was that he would be able to resolve any outstanding issues that might weaken the Anglo-French relationship and hinder the Allied war effort. Specifically, he was instructed to improve "the coordination of Allied policy in the Middle East by removing ambiguity in French policy ... [and] as a subsidiary matter ... raise the question of the Arabian Peninsula." With Picot in mind, Sykes was also "to insist that the administration of Palestine must remain in the hands of the [British] military authorities." The Foreign Office maintained the Sykes–Picot agreement "did not provide for any such immediate dual administration" and to insist that France cease attempts to promote its banking operations in the Hijaz.[21]

These talks, conducted by the pro-French Sykes, was motivated by the belief at the Foreign Office that a strong Anglo-French relationship was the linchpin to British Foreign Office policy and the war effort. It was what Sykes worked so hard for in his dealings with the French, and for which he had often been criticised.[22] So hearing contrary views and attitudes expressed by senior figures in Whitehall, in particular Curzon, as well as Wingate and Hogarth in Cairo, despite the existence (or perhaps because) of the Sykes–Picot agreement, did not help matters. These negative sentiments, so exasperating to Sykes and others at the Foreign Office, can be found in Wingate's dispatches to the Foreign Office in early July[23] and are repeated again in his November and December dispatches.[24] They were not what the Foreign Office wanted to hear; hence, they would turn on the messenger – Wingate, in this case, speaking for others – and his message, calling it "Fashodism," that is, giving in to the French. The inference here was that these men and others like them had an imperial mind-set and believed that Anglo-French co-operation in the Middle East was impossible. For them, only unrestricted British supremacy in the region was acceptable.[25] This did not help to promote Anglo-French relations.

On his return from Paris after talks with the French, where he had smoothed over difficulties and given assurances, at the end of 1917 Sykes decided that he had done all he could at the War Cabinet Secretariat. He saw a move to the Foreign Office as one where he "could be in a better position for keeping an eye on British wartime administration in the Middle East, and for laying the ground for Allied post-war settlements." Six months earlier, on his return from Egypt, Lloyd George had offered him a position as assistant secretary at the Foreign Office. At the time Sykes had turned down the prime minister's offer, believing he would have more influence on Middle Eastern foreign policy in the Secretariat.[26] Now that the Balfour Declaration, a central part of the government's Middle East foreign policy, had been approved by the War Cabinet he felt his work at the Secretariat was done. Sykes believed that once he was at the Foreign Office he would be on top of breaking events as they occurred and in a position to give advice and direction as needed to those "officers on the spot" as events happened. He believed that at the Foreign Office "he would be in a better position ... to encourage Arab, Armenian and Zionist delegations who sought clarification of British policy in the Middle East." He also hoped that, like the outcome of his recent trip to Paris, by being at the Foreign Office "he would be able to bring Allied Middle Eastern policies into line with those of Great Britain," which he had helped create.[27] He believed he would be in a better position in the Foreign Office to influence policy and ensure that all parts of his vision for a postwar Middle East would be in place by the end of the war.

When he approached his friend Lord Robert Cecil with the idea of moving to the Foreign Office, the parliamentary under secretary of state for Foreign Affairs "was delighted." The two men had been friends since before the war, when Cecil had been a frequent luncheon guest at Sykes's London home. More recently, they "had been closely associated in making Middle East policy and in backing Zionism." They were also "in general agreement that the Turks must be defeated, that old Allied notions of imperialism must end and that peace in the future could only come about through national independence under the auspices of the League of Nations."[28]

When Sykes approached Cabinet Secretary Sir Maurice Hankey about transferring to the Foreign Office from the Secretariat, to his surprise Hankey promptly approved his request without objection. He asked only that Sykes continue providing the War Cabinet with his weekly Eastern Report. Expecting some resistance to his request and receiving none, Sykes was slightly miffed. Later, on learning this Hankey would apologise. Once at the Foreign Office, however, Sykes found that far from being "an Under-Secretary looking down on the War Cabinet Secretariat ... his official title would be 'Acting Adviser on Arabian and Palestine Affairs,' and his office was located in the Foreign Office basement." Neither of these mattered much to Sykes, as titles meant little to him, and in the basement he would be free of the stifling bureaucracy he disliked so much. Most importantly, "given the loose manner in which Balfour headed the Foreign Office, and the large measure of discretion left to his cousin, Cecil, Sykes expected to run everything his own way."[29] However, the New Year would bring its share of disappointments for Sykes, whose star had been in the ascendant over the past two years and who considered himself irreplaceable to the government in matters dealing with the Middle East.

At the Foreign Office Sykes continued his independent ways with Cecil's support, injecting himself into any and every matter he believed required his particular expertise. After only a few days in his new position in January 1918 Sykes sent "several telegrams to Russia and Persia telling British officials there how to combat Bolshevik and Pan-Turanian propaganda." Lord Curzon came across these telegrams in a routine review of Foreign Office telegrams on behalf of the War Cabinet in his capacity as chairman of the Middle East Committee. Furious at what he read, he wrote immediately to Cecil:

> This morning I read an astonishing telegram to Buchanan ... suggesting that we should stir up the Russian Moslems by laying stress on *female suffrage* (Good God!) and progressive self-government for Oriental peoples! Who in your office is doing all these things I cannot imagine.[30]

Of course, Curzon knew full well it was Sykes, if not a collaborative effort concocted by Sykes and Cecil together. Their close friendship and agreement on foreign matters was well known, and the two had worked closely together before Sykes joined the Foreign Office. Cecil also was the chair of the Foreign Office's Russia Committee, created after the Russian Revolution in December 1917. Besides, Cecil's well-known independent streak and confidence "in his own judgement were such that he took to drafting telegrams without seeking advice and making last-minute changes to dispatches written by more seasoned Foreign Office hands."[31]

Another telegram that upset Curzon was pure Sykes, the self-styled Ottoman expert. It was a "telegram to Persia denouncing the Turanian movement. 'The Turanian movement was not to be attacked,' Curzon reminded Cecil."[32] It was an Ottoman-led political movement to unite all Turkic-speaking peoples of Central Asia, including those outside the Ottoman Empire. In late 1917 the British Department of Information published a report on the movement, in which it warned that it could be "a dangerous instrument in the hands of the Young Turk leaders," if they were able to "create a Turkish-Islamic state there, in alliance with Persia and Afghanistan. India would be directly threatened. 'It would create a vast anti-British hinterland behind the anti-British tribes on the North-Western Frontier'";[33] hence, Curzon's alarm.

The former viceroy then came to the point of his letter to Cecil: "There used to be a Middle East Committee of the Cabinet of which I am or was Chairman and of which Sykes is or was secretary. We used to have frequent meetings and all the earlier Mesopotamian and Hejaz policy was formulated by us." He continued:

> The Foreign Secretary was of course present whenever he desired. Now I observe that no questions are referred to us. We have not been summoned for some 2 months & the Foreign Office policy as regards these countries is formulated and published without any reference to us at all.

He noted that since

> the Middle East Committee was created by the Cabinet to promote coordination between Foreign Office and War Cabinet ... I do not think we ought to be given the complete go by and I am sure that is the very last thing that you yourself would desire.

Here Curzon added a not so subtle cautionary note, no doubt in reference to Sykes: "Both in Palestine and Arabia and Central Asia we ought to be very circumspect and there are considerable advantages in the consultation of men who know."[34]

While this might serve as evidence why Sykes wanted to be out from under Curzon's stifling control, it also showed how far Curzon's oversight and influence reached. Even the Foreign Office was not free of his watchful eye and for this reason Sykes was not alone in his aversion to the former viceroy. Furthermore, there was no love lost between Cecil and Curzon, as "each viewed the other with suspicion, continually suspecting plot and intrigue as they vied for influence within the Foreign Office itself." In a letter to Balfour two days after Curzon's letter, Cecil made his feelings about Curzon clear when told his cousin the foreign secretary "of his desire to 'smother decorously' the work of the Persian Committee and the Middle East Committee." When he was unable to do so, Cecil chaired the meetings himself in Balfour's place "often with only the most reluctant approval of Curzon."[35]

So it must have been both galling and humiliating for such a proud man as Curzon to learn much later that the month before he made his complaint to Cecil in January 1918, the Foreign Office had sent Sykes to Paris on a mission involving the Middle East without his knowledge. At the time Sykes was still secretary of Curzon's Middle East Committee. Not only that, but no one, including Sykes, had told him about it. On 13 March 1918, after learning of Sykes's trip, an irritated Curzon minuted:

> It seems to me a very extraordinary thing that though I happen to be the Chairman of the Middle East Committee I was neither informed of Sir Mark Sykes's visit to Paris, nor was ever shown his report upon it, nor even heard of his speech until it was long over.
>
> This must be an accident: for it is of course impossible for a War Cabinet Committee to do its duty when even its chairman does not get vital papers.[36]

With this and other recent events, the Foreign Office – whether willingly or unwillingly – was clearly succumbing to the realities of war and the politics of a changing world scene in which old style imperialists like Curzon and the imperialism of the past that he represented was being replaced by an internationalist spirit, one of multilateralism and international cooperation. That is not to say that actions on behalf of His Majesty's Government in matters of national self-interest did not occur, but as with the Sykes–Picot Agreement, these were more apt to be done in cooperation with other nations.

No one embodied this desire to change more than Sir Mark Sykes, as Britain's man in the middle, the government's propagandist and fixer in Middle East matters. A wealthy landed aristocrat, member of

Britain's privileged ruling elite, an ardent and outspoken modern imperialist, and Francophile, Sykes was gifted with an easygoing, friendly, persuasive and charismatic personality. He represented in word and deed the new diplomacy of the changing times just as Curzon represented the past. However, as events would soon show, Curzon was not out of things yet.

Several days after his letter to Cecil and five months after the Middle East Committee last met, Curzon reconvened the committee on 12 January. Cecil attended the meeting on behalf of the Foreign Office but "only to suggest that the committee dissolve and its place be taken by a new Middle East Department in the Foreign Office run by Sykes." Needless to say, his proposal was not well received by Curzon and it was not accepted.[37] Cecil would continue in his efforts to wrest control of Egypt and the Middle East from Curzon and the Committee throughout the year. Despite Cecil's best efforts, however, Lord Curzon maintained his iron grip over Middle East policy. As a result, Sykes would never again be in a position to have any major influence over Middle East policy, as he had on the Middle East Committee and at the Secretariat.

Although he had lost the central position he had held in policy making the previous two years, Sykes continued his usual frenetic pace in pursuit of numerous schemes and projects in the Foreign Office. According to Zionist leader Nahum Sokolow, who spent much time there, Sykes's two-room office in the basement of the Foreign Office was a hub of constant activity. It housed Sykes, his official secretary Mr Dunlop, and his personal secretary from Sledmere, Sgt. Wilson. Sokolow noted that Sykes ignored the lift and took the stairs to the main floor of the Foreign Office "about twenty times daily at a lightning speed, which made it impossible for me to keep pace with him in spite of my most strenuous efforts." Reflecting on the many schemes of its occupant, his office saw "a constant coming and going of Foreign Office men, MP's, Armenian politicians, Mohammedan Mullahs, officers, journalists, representatives of Syrian Committees, and deputations from philanthropic societies."[38]

Sokolow also noted that "in the midst of this busy world Zionism maintained its prominent position." As the primary author of the original Zionist proposal that became the Balfour Declaration – aided and abetted by Sykes – and other Zionist activities, Sokolow spent much of his time with Sykes in his basement offices. "Often and for long periods at a time," Sokolow recalled, "my work, indeed, required my attendance there more than at the Zionist offices, and sometimes I had to go there three times a day and to remain there till late at night." This was because

[i]n order to avoid confusion and divergence of effort [over Zionist activities, Sykes] insisted upon what was readily conceded him, namely that he should pass an opinion on every question and every detail, and in this there was no hesitation, no delay.

Sokolow noted several instances where Sykes was able to get things through official channels to help the Zionists. Following one of his requests, after much difficulty Sykes was able to get official approval to authorise the Zionist Organisation "to protect the Jews of Palestine and Syria ... who were technically alien enemies [as Turkish citizens]." This was a major achievement, in Sokolow's view, as it gave "official recognition of the Zionist Organisation as competent authority."[39]

As evidence of Sokolow's claim about his Zionist focus, two weeks after the announcement of the Balfour Declaration Sykes wrote to Hankey that British support for Zionism must be made abundantly clear to the world. "We are pledged to Zionism, Armenian liberation, and Arabian independence," he wrote the Cabinet secretary: "Zionism is the key to the lock. I am sanguine we can demonstrate to the world that these three elements are prepared to take common action, and stand by one another." In Sykes's view there was good reason for this: "If once the Turks see the Zionists are prepared to back the Entente and the two oppressed races, they will come to us to negotiate with the real situation clearly in our minds."[40] Although Hankey supported the Zionist movement and saw it as central to a postwar peace in Middle East, there is no evidence that he agreed with or responded to Sykes's bizarre claim that such an announcement would alarm the Turks enough to panic them into negotiating a peace settlement on Allied terms. Nor was there much support from Hankey or others for the unlikely Zionist–Armenian–Arab coalition Sykes continued to promote. Zionist and Armenian organisations had worked together in England for some time, but neither were close to any Arab groups.

Sykes had persuaded Clayton to attempt to bring about a meeting of minds between Arab leaders and Zionists in Cairo. On 20 December 1917 Clayton cabled Sykes to say he had

interviewed various members of Arab committee in Cairo and also Jewish leader [Joseph "Jack"] Mosseri,[41] urged combination between them on lines already suggested and brought them into touch with each other. Arguments advanced appeared to impress Arabs but they are still nervous and feel that [the] Zionist movement is progressing at a pace which threatens their interests.

Clayton added, "Discussion and intercourse with Jews will doubtless tend to curb their fears, provided latter act up to liberal principles laid down by Jewish leaders in London, but they must be careful not to frighten Arabs by going too fast." At the end of his telegram Clayton added: "Hope to arrange for a Moslem Arab delegate to proceed to London and serve on Sir Mark Sykes's Committee and King Hussein is being approached." At the bottom of the telegraph, which had been forwarded to him in Paris, Sykes wrote "Show W[eizmann] & S[okolow] very important. We should go after I return London with Billy [William Orsmby-Gore]."[42] However, the trip to Cairo did not materialise.

Soon after Sykes had moved to his new position in the Foreign Office in early January, Assistant Under Secretary Sir Ronald Graham sent a warning to Wingate in Cairo in a private letter. Commenting on Sykes's view of the situation in Jerusalem, he noted: "Mark Sykes has been very dissatisfied with the way in which matters in Palestine have been running." He apparently felt so strongly about it, Graham told Wingate, that Sykes "went to the Prime Minister and the War Cabinet about it and has had himself attached as a sort of Head of Department to deal with the subject." As a result, "he will have nothing to do with Egypt or, indeed, anything but the Arabian movement and Palestine."[43]

A couple weeks later in a letter to Wingate, Graham reconfirmed Sykes's new position:

> Sykes is now installed in the Foreign Office and has taken over Palestine and Arabia, so that I am no longer responsible for any telegrams that reach you on the subject. He relieves me of a good deal of work, but [from my personal experience I find that] at the same time he makes a good deal for other sections of the Foreign Office.[44]

Wingate passed this news on to Clayton in a letter on 1 February, commenting: "[G]enerally speaking, I should think that it was a move in the right direction and Mark Sykes's energy and varied knowledge of the countries and people concerned should be very helpful."[45]

Meanwhile, Sykes had received a letter from Clayton in Jerusalem in which he told Sykes he hardly had a moment to write or do much else because of his duties as political officer. "Political and administrative problems arise every instant and it takes time to build up an organization to deal with them." He had no problems with Picot, because of their good personal relationship, and he sympathised with the Frenchman, whom he felt was in a difficult position because his government had sent him "to play a certain part and . . . on arrival he [has found he] cannot play it." However, he told Sykes "that P

[icot] is far from sympathetic towards our present Zionist policy which he thinks will not be favourable to French interests either here or in Syria." He learned from

> several independent sources – Arab (Christian and Moslem) and Jew – that [Picot's] attitude when talking with Arabs on the Zionist question is not calculated to do away with their feelings of uneasiness or to promote that Arab-Jew-Armenian sympathy which I have – with some success – been at pains to promote.

Otherwise, Clayton was hard-pressed with "the restoration of normal conditions, the opening of trade with Egypt, the currency question, and above all the feeding and supply of Jerusalem."[46]

In his new position at the Foreign Office, Sykes's daily routine was spent doing everything he could for Clayton and others in Palestine and elsewhere in the Middle East. His time was spent reviewing and responding to dispatches and letters from Clayton, Wingate, Storrs and others, listening to petitions from various groups, hearing complaints, mediating when he could, offering advice and seeking to resolve crises brought to his attention. He also attended numerous meetings, wrote memoranda, produced his monthly Eastern Report for the War Cabinet and met Zionists, Foreign Office personnel and parliamentary colleagues. Sykes worked tirelessly day and night in his office and at home, as new problems and challenges found their way to his desk following Allenby's successful campaign in Palestine. He was still at the centre of actions in the Middle East, but now it was from behind a desk and without his previous influence. This was a major change from his activities of the previous two-and-a-half years, when he was more often than not a key player at the centre of events as they happened and was even involved in making them happen. Nevertheless, Sykes still believed he could make a difference, although he was far from the action and was no longer a member of either the Secretariat or Middle East Committee.

Through it all, as Sokolow noted, Sykes's primary focus remained on Zionism. In a letter to Wingate at the beginning of March he wrote: "With regard to Zionism, the important point to remember is that through Zionism we have a fundamental world force behind us that has enormous influence now, and will wield a far greater influence at the peace conference." As a further emphasis, he added, "If we are to have a good position in the Middle East after the war, it will be through Zionist influence at the peace congress that we shall get it."[47]

In March 1918 the Middle Eastern Committee was again renamed. This time, as the Eastern Committee, it subsumed under its ever-expanding wing both the former interdepartmental Persian Committee and Cecil's Russian Committee at the Foreign Office. With this change, the War Cabinet was relieved of the responsibility of an increasingly complicated Middle Eastern situation, which now was under a single authority that could provide it with "with some much-needed consistency and regularity." This was what Sykes had recommended to Curzon in July 1917 – that the several committees dealing with the Middle East be combined under Curzon into a single entity for the sake of efficiency and policy coordination.[48] However, when doing so Sykes had expected to be a member of the expanded committee. Nor was the Committee to be under Sykes in the Foreign Office, as Cecil had tried so hard to achieve. Under Curzon, the new Eastern Committee was composed of "the heads and chiefs of staff of the Foreign Office, War Office, India Office and Board of Trade," with General Smuts as the permanent member representing the War Cabinet. Sykes was not even an unofficial member of the committee, as he had been previously. He was invited to its meetings only on matters specifically concerning Palestine, Arabia or French Syrian claims.[49] While this may have been disappointing for Sykes, he was finally out from under Curzon's thumb and able to work freely in the Foreign Office. He still had access to the War Cabinet with his monthly Eastern Report, through which he could attempt to influence Middle East policy and strategy.

After much prompting and prodding from Sykes, early in the year the Eastern Committee finally gave its approval for the Zionist Commission to go to Jerusalem. With Weizmann as its chairman, the Commission consisted of Zionist leaders from all the principal Allied countries except the United States, which had not yet entered the war against Turkey, and Russia, whose representatives were unable to join the Commission in time "for political reasons" before it left for Palestine.[50] As Sokolow noted, it was to "represent the Zionist Organization," and also

> act as an advisory body to the British authorities in Palestine in all matters related to Jews, or which may affect the establishment of a national home for the Jewish people in accordance with the Declaration of His Majesty's Government.[51]

The Commission's departure was set for 8 March 1918, but as Weizmann recalls in his autobiography, *Trial and Error*,

> a few days before that date Sir Mark Sykes, who was responsible for collecting and organizing us, and making our travel arrangements ... suddenly had the idea that it would be useful for the prestige of the

commission if I, as the chairman, were to be received by His Majesty the King before we left.

On the morning of his audience with the king Weizmann, top hat and all, went to the Foreign Office where he was to meet Sykes and go from there to the palace. At the Foreign Office, he was met by

> a very confused and apologetic Sir Mark Sykes, who informed me that he had just received some "very disquieting" telegrams from Cairo, to the effect that the Arabs were beginning to ask uncomfortable questions ... He was inclined to think that it might be better to cancel the audience.

However, after much discussion and referral to the foreign secretary for a decision, Balfour telephoned the palace and made arrangements to reschedule the audience, which was held on the day of their departure.[52]

With Maj. William Ormsby-Gore as their official Foreign Office escort, the Zionist Commission left London as scheduled on 8 March for Paris. From Paris they travelled to Rome and on to Taranto, where they caught a ship for Alexandria and travelled overland to Cairo.[53] In Cairo, before leaving for Jerusalem, Weizmann met High Commissioner Sir Reginald Wingate, who recalled in a letter to Hardinge at the Foreign Office his "long and interesting talk" with Weizmann. "I had to warn them," he wrote, "to go a little slow in this country [i.e., Palestine] as it is, above all things, Moslem and Pan-Islamic and, as such does not view Zionism too favourably." Wingate told Hardinge that he "recommended them to feel their way carefully and do all in their power to show sympathy and good-will to the Arab and Moslem peoples with whom their future must lie." He also "warned them to be very careful in regard to their discussions on acquisition of land, etc., but you will probably hear all this from Sykes with whom I know Ormsby-Gore is in full communication."[54]

After a delay in Cairo, the Commission arrived in Palestine, where they met General Allenby at the general headquarters of the Egyptian Expeditionary Force. Afterwards, Clayton was effusive in his report to Sykes in praise of Dr Weizmann. Long talks were held with the Zionist leader, he wrote, and

> we are all struck with his intelligence and openness and the Commander-in-Chief [Allenby] has evidently formed a high opinion of him. I feel convinced that many of the difficulties which we have encountered owing to the mutual distrust and suspicion between Arabs and Jews will now disappear.[55]

Clayton told Sykes that Weizmann's "conversations in Cairo with Dr. Nimr[56] and others had satisfactory results and I feel sure that the same will be the case when he meets the leading Moslems in Jerusalem." Weizmann had not yet met Picot and Clayton wrote that he

> was not extraordinarily anxious to do so, but he will have to see him within the next day or so. What the result of their conversation will be remains to be seen, but I think Weizmann can be trusted to manage the interview with discretion.[57]

Changing the subject to the Administration of Occupied Enemy Territory, Clayton went over the plans and suggestion he had received from Sykes, adding they

> were very much those we have been working on and I am extremely glad to get as full an explanation from you of the general trend of policy. We have separated as far as possible the three main questions with which we have to deal, e.g. –

> (a) The Holy Places and religious questions.
> (b) Palestine and its local population.
> (c) Jews and the Zionist movement.

Clayton expanded on each of these at length, before turning to the "Anglo-French Agreements," that is, Sykes–Picot. "Your clear statement of the state of affairs in regard to this question has helped me greatly. I felt that it was so, but until now I had not been quite sure that H.M.G. had come to a definite decision in the matter." He told Sykes that he presumed that

> nothing has up to the present been said to the French Government [about any changes] and that for the moment it is intended to let the agreement die gradually until the time comes to administer the "coup de grace" which may not even have to be given at all.[58]

As far as Picot was concerned, Clayton saw "no sign ... he has any idea that such a policy is in contemplation and he still regards the agreements as his bible – at least such is the impression which he conveys to me." In an attempt to find out just how much Picot knew, Clayton proposed to get a reaction out of him by raising the subject, including giving him some of his own opinions that the agreement was out of date and observing his response. It was clear to Clayton that the situation had changed so much that the agreement was out of date and would not "hold a drop of water at any conceivable peace conference." If Picot," Clayton continued,

> could get definite instructions from his Government to act on these lines and renounce, when talking to Arabs, any idea of interference except in popular demand, it would have the best of effects, but I am not sure whether things have progressed far enough to make such a consummation possible.[59]

Thus, Sykes and Clayton set into motion a scheme to revise the Sykes–Picot agreement so that it would relate better to existing conditions and potentially be more acceptable at any peace conference. If successful, the two men who made the original agreement – Sykes and Picot – could then recommend the changes.

Meanwhile, Clayton told Sykes not to concern himself "with the small questions which arise regarding seats of honour at religious ceremonies, candles, guards for Holy Shrines, etc. They are of little importance and we can settle them locally." Knowing that numerous such complaints were finding their way to Sykes's Whitehall office, he suggested that

> when such matters are brought up by foreign Governments they might be told that the actual administration is in the hands of the local military authorities who are doing the best they can to meet the wishes of all concerned in so far as military exigencies permit.[60]

While this was some relief to Sykes, he continued to review these complaints in order to keep his finger on the pulse of the events both large and small that were taking place in Palestine.

A few days later, Ormsby-Gore wrote to Sykes from Tel Aviv (Jaffa) on the progress of the Zionist Commission and his observations of the situations in Cairo and Jerusalem. Overall, he reported, the Commission was doing well. He was concerned, however, about British personnel recruited from the Sudan for duty in Palestine. Their "experience in the Sudan does not make for a ready realization of the very wide questions of world policy which affect Palestine." It was quite noticeable, he wrote, for those recruited from the Sudan and India "to favour quite unconsciously the Moslem against both the Christian and Jew. Still thanks to Clayton for whom my admiration is continuously increased I think things are proceeding on sound lines." As for the Palestinian Arabs, Ormsby-Gore noted that they "are being very tiresome about Hebrew and are trying to get Arabic acknowledged as the only official language of the country ... [specifically] ... about Hebrew signboards over Jewish shops."

Furthermore, "the Arabs in Palestine are ... showing their old tendency to corrupt methods and backsheesh [bribes] and are endeavouring to 'steal a march' on the Jews. The only really decent ones are in Jerusalem." At this point Ormsby-Gore added, "Picot and I had a friendly talk to [sic], but only about generalities. He is not exactly 'loved' here – but Clayton does his utmost to work with him."[61]

Another concern Ormsby-Gore shared with Sykes was the continued connection between Palestine and Egypt, and he pleaded for Sykes's help in resolving what he viewed as a serious problem:

> I hope you will do all you can to make the political cut between Palestine and Egypt a clean cut economically and financially. Egypt certainly has the idea that Palestine is a sort of Egypt irridenta [sic] and Egyptian business & banking interests are anxious to get in.

In his view, "the soonest after the war we can secure a separate Palestinian currency and direct British trade the better." In closing, Ormsby-Gore remarked that "the Zionist Commission is very small" and would soon need additional members. He gave Sykes a name he was given as a possible addition to the group and asked him to inquire about him.[62]

Letters and dispatches like these from Ormsby-Gore, Clayton, Wingate, Storrs (now governor of Jerusalem) and others in Palestine and Cairo continued to flood daily Sykes's desk in his basement office at the Foreign Office. So, too, did occasional reports from the political intelligence officer in Jerusalem reporting in elaborate detail on the activities of the Jews and Muslims in the holy city, although those reports found among the Sykes Papers primarily concerned the Jews.[63]

Sykes reviewed everything sent to him and if advice, assistance or action were needed, he did whatever he could to provide it. During this time, he was also requested to give speeches on the Middle East. Of particular note were two talks he gave during the summer of 1918. On 21 June, he spoke to the Manchester Syrian Association and two weeks later, on 7 July, he spoke to the Zionists in Hull. The two speeches could not have been more different in content or tone.

To the Syrians, his theme was "that everybody who has been oppressed by the Turk must pull together – the one thing necessary before all things. I have done all I can to promote that unity." He then read a telegram from Picot, in which the Frenchman apologised for not being able to be there and to convey to the meeting his "wishes for the liberation and prosperity of Syria under the supervision of Great Britain & France." Then, without commenting on Picot's message or about France and Syria, Sykes posed a rhetorical question about what "the the future "British Government's policy in regard to Syria and the Middle East would be."[64]

Answering his own question, Sykes repeated what has come to be known as the "Declaration to the Seven." The week before, seven Syrians of the newly formed Syrian Unity Party had petitioned the Arab Bureau in Cairo for a clarification of Allied plans for postwar independence. They hoped it would be "complete independence" with a "decentralized Arab government" that would have administrative autonomy within an Arab Kingdom.[65] The government's response given to the Syrians in Cairo on 11 June 1918 – written by Sykes[66] – was deliberately vague and ambiguous, and Sykes shared it with his Hull audience.

He told them the region was to be divided into four broad categories and described the government's policy towards each. Of the first and second regions, "countries like parts of Arabia which were free and independent before the war" and "countries emancipated from Turkish rule by the action of the Arabs themselves ... His Majesty's Government recommends the complete and sovereign independence of the races inhabiting [them] and will support them." As for the third category, those

> countries formerly under the Turkish dominion now occupied by the Allies as the result of operations during the present war ... His Majesty's Government's intentions ... are embodied in the declaration of the general Offices of Command at the capture of Bagdad [sic] and Jerusalem.

As for the fourth and final category "Races still under Turkish control" (which at this time included Syria), he said, "it is the wish and desire of H.M. Government that the peoples of these races should obtain their freedom and independence and to the attainment of this they continue to labour." To this end, he added, "our object is to do the best for these people as well as we can, and help them to help themselves."[67]

Sykes then returned to his theme that the Arabs must put aside their differences and work together, for it was these differences the Turks had used to divide, control and oppress them for 500 years. Referring to the Turks as "those Blackgardly people," he reminded his audience that, "most of your people are hostages in the hands of the enemy, that is a fact." Once the Turk was gone, he told his audience "the result will be like a fairy tale." He described how Jerusalem and Jeddah were now quiet places compared to the "pillages, murder and outrages committed when those places were held by the Turks."[68]

Again he told them, "Now I must ask you to pull together. I do entreat you to do this work for your country." He suggested several ways they could do this, including direct military service, propaganda and sending "out a good Medical unit," and to get in touch with all Syrians and Arabic-speaking people and have them join them in their cause. Sykes left them with the example of the Sharif of Mecca who, in the last two-and-a-half years of fighting, had begun his fight against the Turks with a small number of men and, with the help of the British, "[had] killed, wounded or taken prisoner 40,000 of the enemy and captured about 220 field guns." He ended by saying, "I hope you realise that the object we have to keep in view is to get the Turks out of Syria and something else in their place." Exactly what was to take their place, he did not say.[69]

Sykes's speech to the Zionists two weeks later in Hull on 7 July could not have been more different than his speech to the Syrians. Considerably less vague and far more specific in detail, it was essentially the same material found in his September 1917 memorandum "Note on Palestine and Zionism," which he had prepared for the foreign secretary and War Cabinet to assist them in their deliberations of the Balfour Declaration with Weizmann's help.[70] Unlike his speech to the Syrians, a typed copy of which was sent afterwards to Sykes by the Syrian Association, only the abbreviated notes of his Hull speech can be found in the Sykes Papers.[71]

Sykes was joined on the speakers' platform by Nahum Sokolow, who heard his friend relate the trials and tribulations of the Jews of the Diaspora well known to the audience. While it is impossible to know exactly what he said because there is no record of it, by using his notes it is possible to reconstruct what he may have said. From the notes it appeared to be an inspiring speech full of passion and high-flown rhetoric, meant to appeal to the hearts, minds and souls of his Zionist audience, and it was doubtlessly well received.[72]

It was because of the long history of Jewish suffering and persecution that every year the Jews of the Diaspora traditionally proclaimed at the Passover Seder, "Next year Jerusalem!" Many did so through "[cries] of pain [and] anguish," as they envisioned a dream "so impossible [and] remote." However, it was this dream that led to the "gradual growth [of the] Zionist ideal [of] Palestine [which became] a Jewish national ideal." It was there, where

Jews [could] develop [a] national life [with] Jewish art, Jewish service, Jewish literature, Jewish peasantry, Jewish national existence, a Jewish centre of racial pride, a storehouse of traditional inspiration. [As a] result, every Jew in [the] world [would be] elevated; every Jew in the world more truly [a] citizen.

This would result in the "disappearance [of the] feeling of yearning [and] replaced [by a] feeling of participation in world life [as] never participated [in] before."[73]

Here, his notes appeared to indicate Sykes was to pause and stress dramatically that "unless Jewish aspirations [are] fulfilled [the] war [will] not [have] achieved [its] full purpose, [the] final settlement [of the] Jewish question [the] necessary element [in the] world." This settlement would be "two fold ... equal Jewish rights in all countries & recognition [of] Palestine [as the] centre of Jewish national ideal." There was to be "no question of Jewish dual nationality, [a] Jew [would be] no less British, [or an] American."[74]

Sykes ended his speech describing his hope, dream and vision were that after the war, tyranny and militarism would be banished, there would be a "League [of] free peoples bound by mutual [values?] ... [no more] subjection [of people], [no more] treaties [of] convenience ... & [a] moral centre [of the] world [in] Jerusalem." Here he was to add dramatically that there would be "[The] call [of] Mount Zion. Mount Calvary. Mosque of Omar. Christendom Islam. Jewry. Bound [together through a common] belief [in] the same God. [Each] following [their] interp[retation and] common sense." While there are many other notes, the foregoing gives a gist of what Sykes said, or intended to say.[75]

Two speeches. For the Jews, Sykes offered the fulfilment of their hopes and dreams, while for the Arabs he offered only vague and nebulous promises. In these speeches, we see the attitude that would be played out again on the world stage at the Paris Peace Conference the following year. But for the present, the capture of Jerusalem brought with it new questions and new problems.

CHAPTER 12

FINAL DAYS

He had an extraordinarily difficult and complicated policy to run, namely, to reconcile the Arabs to an arrangement with the French for which he was largely responsible; to reconcile the Arabs and the French to an arrangement with the Jews, for which he was also largely responsible; and to reconcile alternatively English military and political opinion with both.

Everard Feilding[1]

On 2 September 1918, Sykes wrote a three-page letter to the prime minister. In it he expressed his concern over a change that he believed would seriously affect the important work he was doing at the Foreign Office. This was the transfer to France of Maj. Gen. Sir George Macdonogh director of military intelligence at the War Office. With this promotion Macdonogh would be raised to the rank of lieutenant general and made adjutant general to the Forces, the second-highest ranking military member of the Army Board, responsible for developing the Army's personnel policies and supporting the troops. However, Sykes was concerned that "unless some arrangements are made for some person to take over Sir George's Oriental work and that that person has the requisite knowledge and authority ... dire confusion will ensue." He told Lloyd George that he had worked closely with Macdonogh over the previous eight months as political advisor on Eastern Affairs, and as a result of their good working relationship, "the military-political developments have been fairly satisfactory. Our Arab, Syrian, and Palestinian policy has not landed us in any great difficulties and has on the other hand given us a considerable return in prestige, booty and enemy casualties." Not only that, but "we have friendly populations, native allies, and good material and moral assets for a peace conference should one occur at any time."[2]

This had been possible, Sykes told the prime minister, because

Sir George understood the Eastern situation in relation to the war so well that every detail, name, move and event has its proper significance in his mind, and consequently he could take necessary action without more ado when circumstances arose.

He was concerned that Macdonogh's replacement would be not be able to learn enough for some time for Sykes to be able to give him the support he needed. He had

Zionist agents scattered all over the world, Arab chiefs at each others throats, French military and civil officials to contend with, Italian representatives to pacify, Syrian colonies to keep in touch with, affairs and agents to move hither and thither.

With Sir George's help, "up to now a phone message, a word on a bit of paper, a mere initial has been sufficient to do the work."[3] Sykes's pleas were to no avail and Macdonogh assumed his responsibilities in France. General Sir William Thwaites, fresh from the Western Front, where he had been general officer commanding 46th (North Midland) Division in France, took over the position of director of military intelligence at the War Office.

More changes were in store for Sykes. On 7 September he received a letter from Lord Robert Cecil confirming a conversation the two men had the day before about Sykes's reassignment in the new Foreign Office Middle East Department. Sykes was to continue as political advisor on Arabian and Palestinian Affairs, but instead of being under Lord Hardinge his new superior was to be Sir Eyre Crowe, who was to be permanent head of the new department.[4] On hearing the news, an obviously upset Sykes pointed out to Cecil in a letter that this was a demotion for him. Previously, he had

advised Lord Hardinge who passed the stuff on to the Secretary of State. Under the present arrangement I advise Sir Eyre Crowe who advises Lord Hardinge, and when the stuff comes back it will have to go back to Sir Eyre Crowe again. This seems of course double delay both ways.

Nevertheless, a disappointed Sykes admitted he realised he had no choice in the matter. "I don't see any way out of the difficulty, until Sir Eyre Crowe has really mastered the complexities of the situation, which will take at least 3 months."[5]

To add insult to injury, Cecil also told Sykes to postpone his negotiations with Picot for the moment, because he wanted to meet Picot the next time he was in London and get to know him. It was much better, he told Sykes,

> if it can be arranged, that negotiations should take place after, and not before, our troops have entered the French sphere. We must never forget that, internationally, the French are a grasping people, and we shall have a much better chance of getting reasonable terms out of them if they come to us in the first instance to get some thing which they want.[6]

With these changes and no longer a member of the Eastern Committee, it was not lost on Sykes that he was being eased out of his previously held key positions of influence in Middle Eastern affairs.

Sykes made no comment on Cecil's request about Picot, but did tell him that he believed it was premature for a visit to the United States to see President Wilson that Cecil urged him to make. He told Cecil this should be done only when the American government had decided on its postwar policy towards Turkey.[7] Although it had declared war on Germany on 6 April and joined the Allies in the war in Europe, as well as recalled its ambassador to Turkey, the United States had not joined the Allies in declaring war on Turkey. Complicating this was Wilson's insistence that after the war all peace negotiations over Turkey were to be discussed with the twelfth of his Fourteen Points in mind. This stated in part that

> The Turkish portion of the present Ottoman Empire should be assured a secure sovereignty, but the other nationalities which are now under Turkish rule should be assured an undoubted security of life and an absolutely unmolested opportunity of autonomous development.[8]

This could seriously impact upon British and French aspirations in the region, Sykes told Cecil. So the United States must decide on its postwar position on Turkey and to do so it must first reach a clear understanding with France and "the Arabs, Syrians, and Armenians." Once this was accomplished "we could do the Zionist and Syrian work at the same time."[9] Otherwise, any such meeting should be postponed.

There matters rested and Sykes's trip to Washington never materialised. Nevertheless, he was officially one step further away from policy and decision-making than before. His new superior, Sir Eyre Crowe, was a long-serving key senior member of the Foreign Office but he had no experience at all in Eastern matters. From 1895 to 1905 he had served in the African Protectorates Department, after which he had served in the Western Department, where he became under secretary of state and the Foreign Office's specialist on German affairs.[10] However, this did not deter Sykes. He was not out of the picture yet and would not let Crowe have any effect on his activities. He was determined that little would change.

A month later, Sykes wrote directly to Cecil without mentioning or involving Crowe, Hardinge, or the new director of military intelligence. Apparently, after a conversation with Cecil about the current situation in the Middle East talking about some of the ideas and future plans he had for the region, Cecil had asked him to write it all down and send it to him in a letter. Sykes dutifully responded with eleven handwritten pages, which were subsequently typed and sent to Cecil on 12 October 1918. In his letter Sykes gave a brief overview of the war in the Middle East from the beginning to the present, as well as an analysis of the current situation. He focused primarily on Palestine and Syria, which were his areas of responsibility, and he included plans for what he believed should be done there. He also included detailed plans for his return to the Middle East at the head of a mission to ease the transition from war to peacetime and to implement his plans.[11]

Believing that peace, or at least an armistice, was imminent, Sykes felt it was vitally important for Britain to do all it could "to obtain as satisfactory a settlement of the Middle Eastern questions as military and political circumstances will permit." He believed that everything should be done so that when it came time to choose a mandatory power for Palestine Britain would "be the most likely candidate" by general acclamation of its population. He also told Cecil that it was important for Britain to "establish our political influence in the Mosel [sic] area" and do all it could to "be on good post-war terms with the French, so that the Entente may continue as a permanent factor." He added that it was essential that "we should hold the confidence of the Arabic-speaking peoples of Asia."[12]

Sykes included several long lists in his letter itemising what he termed were "general world interests, which as Great Britain is a world partner are British interests [and are] desirable." These included fostering and reviving "Arab civilization and promot[ing] Arab unity, with a view to preparing them for ultimate independence," promoting "the permanent settlement of the Armenian question" and pursuing "a policy in Palestine which will take into equal consideration" several objectives. Among them were: "safeguarding ... the rights of the indigenous population," nurturing "the wise and practical development of the Zionist movement" and "safeguarding ... the various interests in the Holy places," as well as establishing an effective "administrative organisation of Palestine [so] that the post [war] conference administration can take over these various problems in a condition which will make it easy for this policy to be continued." His final recommendation was that the United States become involved in "Middle East problems and . . . [be] associated in their settlement."[13]

Sykes stressed that everything was well in hand in the Middle East and as a result "we have magnificent prestige" there. So everything was set to put these plans into effect. However, to do so successfully everything had to be coordinated, and he suggested that he be "sent out as a special commissioner for Arabian Affairs . . . [as he] could do a good deal to bring about this result." He listed eight general objectives for the special commissioner:

1. To minimise friction & secure practical co-operation between British and French.
2. To assist in composing Arab-French and Arab-Syrian disputes with the object of providing a workable Syrian political unit.
3. To co-ordinate Mesopotamian and East Palestinian policy in consultation with Mesopotamian authorities.
4. To develop Palestinian administration.
5. To consolidate on the line of the tri-part interests political parting [of] N. Mesopotamia in Mosul area.
6. To prepare Arab case for peace conference.
7. In consultation with [listing of military and colonial authorities in Egypt and Mesopotamia], French representative to evolve a co-operative policy in regard to the Arab movement to the raising unity and uplifting the Arabic-speaking peoples.
8. In consultation with above [authorities and French] to do what is possible with regard to Armenian refugees with a view to their repatriation.[14]

Confident his recommendation that he be sent as the special commissioner would be accepted, Sykes then listed just what the "purposes of my mission should be." In two pages he enumerated grandiose and elaborate schemes, in which he gave himself additional duties and authorisations as special commissioner (summarised below).

1. Subject to the authority of General Allenby, to organise the Anglo-French liason [sic] on political and administration affairs in Syria proper and in the area of General Allenby's operations exclusive of Palestine.
2. Subject to the same authority to assist in promoting good relations between Arabs and French.
3. To organise a liason [sic] between the political services in General Allenby and General Marshall's[15] spheres subject to the concurrence of both commanders.

4. In consultation with General Allenby or his delegate to report on the reorganisation of the provisional administration of Palestine with a view
to preparing the main lines of the future government of Palestine.

5. In consultation with Political authorities in Mesopotamia and the High
Commission in Egypt to evolve a scheme or policy for unifying the
Arabic-speaking peoples, turning the Arab racial ideals into sound and
progressive channels; and form a basis of permanent friendship and
alliance between Arabic-speaking people and the British Empire.

6. In consultation with the same authorities to submit a scheme of
organisation for carrying the above policy into effect.

7. In the event of the War Cabinet approving the schemes of policy and
organisation, to set up the proposed organisation in co-operation with
the above authorities.[16]

He followed this by describing the staff he needed to support his mission,
saying that each member was to be assigned a military rank and suggesting
specific individuals with ranks and pay grades assigned them. Among the ten
people he listed for the mission staff were the special commissioner (himself)
with the rank and pay of major general, Foreign Office Second Secretary
Harold Nicolson as counsellor with rank and pay grade of a lieutenant colonel,
his Foreign Office secretary and clerk Mr Dunlop as head clerk with the rank
and pay of captain, Nahum Sokolow as honorary Zionist civilian advisor with
no pay, Lt. Antoine F. Albina as his Arabic interpreter with the rank and pay
of captain and several others. Furthermore, he expected to be provided with
housing and transport along with domestic staff for the mission. As the
special commissioner, Sykes would be given "full liberty of movement subject
to the approval of local military authorities." He would also "correspond
directly with the Eastern Committee through the Foreign Office, but repeat
his telegrams and despatches to Mesopotamia, Egypt Force [Allenby] and Sir
Reginald Wingate."[17]

On 30 October 1918, the armistice ending hostilities between the Allies
and the Ottoman Empire was signed. The ceremony between Ottoman and
British military officials took place on board HMS *Agamemnon* in Mudros
harbour on the Greek island of Lemnos; hence its name, the Armistice
of Mudros.[18] On the same day Sykes and his mission left London for Paris,
Rome and the Middle East. While the Foreign Office and the Eastern
Committee had agreed the mission needed to be sent, it was far less grandiose
than what Sykes had proposed. The small group leaving London that day led
by Sykes numbered five men. Including Sykes, these were Major Ronald
Gladstone, an old Yorkshire friend of Sykes's whom he had asked to come
along to keep a record of the mission, two former Ottoman Arab officers

whom Sykes had taken in hand and had trained in England, and Sgt. Walter Wilson, his confidential clerk and private secretary. Others would join them along the way.[19]

Sykes was not promoted to major general as he recommended, nor were the others as he had recommended. A household entourage was not assigned to the mission, nor was Sykes to act independently but was under the authority of Lord Curzon and the Eastern Committee. Despite any disappointment this may have caused him, at least Sykes was free from the London office routine he found so stifling. Once again, he was in his element. He was to represent the British government in the capitals of its European Allies and in the East, where he would meet with key politicians, senior government officials, military, civilian and local community leaders, dispense good will on behalf of His Majesty's Government, and give advice to one and all.

In Paris, Sykes met the French Foreign Minister Stephen Pichon and discussed the Turkish armistice and the future of Armenia. While at the Ministry, he also saw Jean Gout, chief of the Asiatic Section of the Foreign Ministry and discussed the current Arab situation. Before leaving Paris, Sykes saw Boghos Nubar Pasha and the Armenian delegation, which expressed their concerns about the terms of the armistice with Turkey and its effect on Armenian territorial ambitions.[20]

In Rome he met the British minister to the Vatican Count de Salis, with whom he discussed safeguarding the Holy Places in Palestine, as he did later with Cardinal Pietro Gasparri the Vatican secretary of state. At the British embassy he spoke to the ambassador, Sir Rennell Rodd, and also met Maj. Gen. Sir Charles Townsend. Townsend was on his way back to England after being released from two years of Ottoman captivity. A prisoner of war since his surrender to the Turks at the disastrous Siege of Kut Al Amara on 29 April 1916, Townsend had recently witnessed the Ottoman surrender to Allenby's Egyptian Expeditionary Force and been a witness to the signing of the Mudros Armistice.[21]

In Rome, Sykes met the Italian Armenian Committee and attended a meeting at which he was asked to speak. He gave his speech in French, telling his rapt audience about the future plans for Armenia and Turkey. Sykes acknowledged the harsh conditions the Armenians had lived under Turkish rule and atrocities they had suffered and left them with hope for a better future.

Two more members joined the small mission as it left Rome. These were Lt. Albina of the headquarters staff of the Egyptian Expeditionary Force, who joined the mission as interpreter, and French Capt. Louis Massignon. Having recently recovered from a bout of flu, Massignon was on his way to Port Said, from where he would travel to Beirut to join Picot.[22]

On his arrival in Palestine, Sykes reported to General Sir Edmund Allenby, commander-in-chief of the Egyptian Expeditionary Force, and discussed his mission. From there, he went to Jerusalem where he met his close friend Sir Ronald Storrs, the military governor, who provided accommodation for Sykes and his mission. While in Jerusalem, Sykes met local community leaders and others, including the pro-British Muslim leader and Mufti of Jerusalem Kamil al-Husayni,[23] Rabbi Moshe Segal[24] of the Zionist Commission and Maj. Gen. Arthur Money, the newly appointed Administrator of Occupied Enemy Territory.[25]

Over the next few days, Sykes and his group visited the sights and holy places in Jerusalem and travelled to Al-Salt via Jericho, Ramallah and Jaffa, from where they travelled to Tel Aviv. In Tel Aviv, Sykes met the Zionist Commission and its leader Dr Montague David Eder, a prominent London psychoanalyst[26] who had come with Dr Weizmann and remained in Palestine. After meeting Dr Eder, he returned to General Headquarters and reported on his meeting with the Zionists to General Allenby and Brig. Gen. Gilbert Clayton, the chief political officer. He made another trip to Tel Aviv for a second meeting with the Zionist Commission before leaving for Syria with Picot, who had come from Beirut to meet him and update Sykes on the situation in Syria.[27]

On 19 November, Sykes left Palestine for Damascus. He had learned that Amir Faysal was preparing to leave for Paris and the Peace Conference and wanted to see him before he left for Europe. After meeting Faysal in Damascus, Sykes travelled with Picot to Beirut where he stayed the night with Lt. Gen. Sir Edward Bulfin,[28] an old friend from the Green Howards. While in Beirut, Sykes also met Dr Howard Bliss, president of the Syrian Protestant College (later the American University of Beirut), from whom he learned more about the situation in Syria.[29]

After these meetings Sykes left Picot in Beirut and returned to Damascus, where he met British military officials and reported on the political situation in Damascus and Lebanon. After this the mission was to travel north to Aleppo. Before they left Damascus, as Gladstone noted, "Sir Mark saw innumerable people and gave them advice in regard to the Governing of the City which we left on Nov. 24." As their motorcade approached the city of Hama north of Damascus where the local governor knew Sykes from previous visits, he received a grand welcome. Gladstone recorded that

> a mounted escort of Arabs came out to meet him, and they cantered in front of the cars and conducted the Mission to the local Serai (town hall) where a Guard of Honour of local Arab troops was drawn up at the Present to the strains of an Arab Band, playing "God Save the King."

Sir Mark inspected the guard and expressed his approval at their appearance. From there he was taken to the government office building, where "he was received by the local Civil Governor and notables of the City."

The mission was afterwards quartered in the governor's house as his personal guests. That evening a reception was held with all the notables in attendance in honour of Sykes, who was greeted by each one and welcomed with a little speech, translated by Lt. Albina. At the dinner Sykes thanked them all

for their great welcome [and] addressed them with words of encouragement and gave them advice in regard to their future, pointing out how necessary it was for the whole community of Hama to work very hard in order to show the Allied Powers their ability for municipal self government, a speech that was greeted with great applause.

As they left Hama on 26 November, the guard of honour was once again assembled and escorted them to the city limits. They flew the Arab flag, which Sykes had designed some time ago for the Arab Revolt, and which he now saluted to the cheers of the crowd.[30]

After similar but briefer welcomes in towns along the way, by nightfall the mission entered Aleppo in northern Syria, a place Sykes had often visited on his Eastern trips. After having driven out the Turks a month before, Maj. Gen. Henry Mcandrew and the 5th Cavalry Division now occupied this ancient trading city with its magnificent citadel. Sykes decided to make Aleppo, a favourite place of his, the mission's headquarters. Once situated in his quarters he sent a report of his trip to General Allenby and made preparations for the general's visit to the city.

On 9 December Sykes travelled south to Hama to meet Allenby and returned the following day to Aleppo with the general, who "made his official entry into the city and addressed the notables and people from the steps of the Government Serai." Immediately after Allenby's departure Sykes heard reports of Turkish atrocities against Christians in nearby Aintab. So he quickly assembled and led a contingent of troops there, where he confronted the surprised Turkish military governor still stationed there. After securing the town he returned to Aleppo, where he had Maj. Gen. MacAndrew send additional troops to guard the American mission at Aintab until Allied troops could arrive and occupy the town two weeks later. Back in Aleppo, Sykes drafted a scheme for the administration of the city, which was approved by the Commander-in-Chief.[31]

During the time he was in the Middle East, from November 1918 to January 1919 Sykes was also standing for re-election to his Central Hull parliamentary seat in a general election. His only challenger was the Liberal candidate R.M. Kedward, a Methodist minister, as the Labour Party had declined to put up a candidate against the popular Sykes. The only issue Kedward could come up with was Sykes's military record, saying he had avoided going to France with his battalion and fighting on the Western Front. Thus, he had "shirked his duty" and used the war for political gain. At Sykes's request, the prime minister wrote a letter in support of Sykes countering the charges and praising his activities on behalf of his country during the war. Sykes also cabled his wife (relayed from Syria through Cairo) disputing the scurrilous claims and explained Kitchener's summoning him to the War Office and his activities since. Lady Sykes made Lloyd George's letter and her husband's cable available to the press and shared both at Central Hull meetings. It worked, and Sykes won by a landslide with a majority of over 10,000 votes.[32]

On 20 December Sykes was driven 150 miles north to visit Adana in southern Anatolia. Adana was the site of the ancient Armenian Kingdom of Cilicia (1198–1375) and more recently of the Adana massacre of Armenians in 1909.[33] Picot came from Beirut to meet him there and the two men returned to Aleppo together. In Aleppo Sykes introduced Picot to the Arab Club, an organisation of local notables and officials, after which the two went with Maj. Gen. MacAndrew to Aintab, where final arrangements were made to occupy it. After Picot returned to Beirut, Sykes went to a nearby village to meet local Kurdish chiefs to discuss the problem of Armenian refugees in their villages. On his return to Aleppo, he "put his reform scheme for the local administration of the city into operation and interviewed the Arab leaders on Anti-Semitic propaganda and arranged for its cessation in Aleppo."[34]

The rest of Sykes's time in Aleppo was taken up trying to resolve the problems of the city. As Gladstone recorded, Sykes's time was not his own:

[The] whole of his time was devoted to promoting the welfare of the people, he allowed himself no relaxation from the work he had undertaken in his endeavour to create a more satisfactory state of affairs. All nationalities thronged to the house from early morning till late at night in pursuit of his counsel and advice, Arabs, Armenians, Priests, Archbishops, Staff Officers, Ex-Consuls, Dervishes, Kurdish Chiefs, Missionaries, Notables of the city, Merchants, Refugees and others too numerous to mention, none were allowed to be turned away and interviews often continued to the small hours of the morning.[35]

On 11 January an exhausted Sykes and his mission left Aleppo for the last time for Damascus to another grand send-off by local dignitaries. He had picked up a virus in Aintab, which prevented his keeping down any food. So for the next three weeks he lived "on nothing but three tins of condensed milk a day," Gladstone later told his wife Edith.[36] After arriving in Damascus two days later, he met members of the Arab Club and arranged a meeting on 15 January between the Arabs and the Zionists, which was also a reception for M. Picot as the French representative to the region. Leaving Damascus the day after the reception, Sykes travelled to Haifa where he reported to General Allenby on his activities since he had last seen the general in Aleppo six weeks earlier in December. On Allenby's instructions, Sykes boarded a French cruiser for Italy, where he landed five days later after a stormy crossing and hurried to Paris to join the British delegation at the Peace Conference.[37]

On his arrival in Paris for the Peace Conference, he met his wife, who was waiting there for him. Lady Sykes was alarmed at her husband's appearance and pleaded with him to go home for a rest before attending the Conference. Taking her advice, Sykes hurried home to London with his wife for a short rest and to see their daughters before returning to Paris. On 22 January, although he was still sick and tired, Sykes managed to complete the report of his trip along with recommendations and suggested guidelines for the future governance of the area after the war. Entitled "Appreciation of the Situation in Syria, Palestine and Lesser Armenia," it was a practical approach to the existing situations in the Middle East at the time as he saw them.[38] However, as he was soon to learn, there were no longer politicians willing to listen to what he had to say.

On their return to Paris Lady Sykes came down with influenza and stayed in her bed at the Hotel Lotti while her husband hurried off to the Conference. Once there, he was surprised to learn he was not to be an accredited member of the British delegation, or any of its committees or subcommittees. This man of action, so used to being at the centre of things, giving advice and being listened to on Middle East matters by his own government and other Entente government leaders, found himself shut out of the most important meeting in the world at the time. Frustrated by the inaction and believing his input was vital to the decisions of the Conference on the Middle East, Sykes tried to get himself accredited as General Allenby's representative, but his request was turned down.[39] He soon found that no one but a few friends had time for him. Despite all the work he had done in the Middle East since 1915 and coming as he did directly from postwar Palestine and Syria, few were interested in seeing him or listening to what he had to say.

In November and December 1918, while Sykes was in Palestine and Syria, the Middle Eastern Committee was convened by Lord Curzon to consider the whole question of the Middle East. During this time it had prepared the "British Middle Eastern desiderata for the Peace Conference."[40] So by the time Sykes had arrived in Paris in late January all the lines had been drawn, positions had been taken and decisions made: his input was not required.

Thus, unencumbered by meetings, deadlines for position papers or anything related to the Conference, Sykes spent his time catching up on events by meeting his old friends and those he knew who were not busy attending meetings and would spare him some time. On 10 February he dropped in on Edmund Sandars an old friend from before the war who was doing some work for the War Office. Sandars wrote later in his diary how thin Sykes had looked that day and noted that he had commented on it to Sykes, who passed it off with a joke. He told Sandars, "He had been 'poisoned in Aleppo.'" He then angrily ranted to his friend "about the Turkish Armistice terms made by 'that little man' Lloyd George, and the attitude of British officers in the Middle East."[41]

Before he had left for the Middle East in September 1918, Sykes had presented Hankey at the War Cabinet with proposed armistice terms and details of the territory needed, listing the towns to be occupied in order to establish an Armenian state in Cilicia for those who were left from the Armenian Genocide. No doubt with an eye to history, he sought to re-establish an Armenian state in the area where Armenians had lived since prehistory and where the Crusader-supported last Armenian Kingdom of Cilicia (1198–1375) existed before the arrival of the Turks. However, the armistice presented to the Ottomans by the British a month later on 30 October 1918 "cut out all mention of specific Cilician towns to be occupied [that he had listed] and only made reference to possible occupation of the eastern vilayets and Turkish withdrawal from (but not necessarily Allied occupation of) Cilicia."[42] Upset at learning this on the day of the departure of his mission to Palestine and Syria, Sykes was further frustrated by his inability to do anything about it at the Conference, where he found he no longer had privileged access to the prime minister.

With his wife confined to bed, Sykes invited Sandars and his wife to join him and Sir Arthur Hirtzel of the India Office at the opera that night. Sandars recalled his friend was in good spirits that night, having enjoyed their dinner together and the opera.[43] The next morning, however, Sykes was unable to get out of bed. His wife got up from her own sickbed to nurse him for what appeared to be a chill, but soon turned into influenza. Already in a weakened condition from his Middle East mission, Sykes was unable to fight it off and the influenza became pneumonia. After several more days his condition worsened until a priest was called and Sykes was given the Last Rites of the Catholic Church. Shortly afterwards he died in his room at the Hotel Lotti in

Paris at 6:30 pm on Sunday 16 February 1919.[44] Totally exhausted and spent from his unswerving efforts to secure a postwar Middle East peace on British terms, Sir Mark Sykes died virtually ignored, a victim of his own passionate enthusiasms.

Sykes's unexpected death a month before his fortieth birthday came at a time that should have been a celebration of all his efforts in the Middle East. However, in his last hours it may have been more a time of misgivings. Lying in his sickbed, he may have reflected that what was being done at the Peace Conference was not what he had promised the people who had put their trust in him nor what he thought was best for the region. In truth, he was merely "an establishment handyman elevated to the role of plenipotentiary."[45] Control over policy had never been his, but it was always in the hands of those who used him for their own purposes. Despite knowing this, Sykes was always eager to show off his vaunted expertise of the Middle East and had been a willing accomplice in the designs of others. However, perhaps after seeing delegates at the Conference fighting over the fresh carcass of the Ottoman Empire he came to realise the enormity of what he had done to bring this about. Perhaps in his weakened state, a revived conscience contemplating its mortality made him recall all the promises made, fabrications told and distortions of truth he had encouraged, all done "for King and Country" and the Entente.

In *Seven Pillars of Wisdom*, T.E. Lawrence's cryptic comments about Sykes at the Peace Conference mentioned a change in him. "His help did us good and harm. For this his last week in Paris he tried to atone." Elaborating on this, Lawrence noted:

> He had returned from a period of political duty in Syria, after his awful realization of the true shape of his dreams, to say gallantly, "I was wrong: here is the truth." His former friends would not see his new earnestness, and thought him fickle and in error ... It was a tragedy of tragedies, for the Arabs sake.[46]

However, Sykes's attempts to atone, if that was what they were, as Lawrence suggested, came too late. Today, his legacy in the Middle East is a region in constant turmoil due in no small part to his plans and schemes, and his involvement in those of others with total disregard for the local population.

CHAPTER 13

THE LEGACY

The evil that men do lives after them, the good is oft interred with their bones.[1]
Shakespeare, *Julius Caesar*

Thus died a man whose singular passion was the Middle East and whose contributions to the events in the region during the war have been routinely overlooked.

In 1905 Sykes reported to the Foreign Office the existence of oil deposits in Mosul,[2] which would lead to the founding of the Iraqi Petroleum Company and in 1954 become the British Petroleum Company, one of the world's largest companies. More recently, these oil fields have funded the activities of the Islamic State. In Allenby's battles against the Turks in 1917–1918, his Egyptian Expeditionary Force used maps of the region that Sykes made before the war and donated to the War Office.

While he was in Cairo in 1917, Sykes held meetings with a select group of Syrians to discuss the future of a postwar Palestine and he deliberately excluded Palestinians, whom he knew would never agree to any such arrangement. This set a precedent in which decisions about Palestine would be made without consulting the Palestinians who have had to live with the consequences. On the other hand, Sykes almost single-handedly prevented the large-scale relocation of Indians from the subcontinent to Mesopotamia, thereby preventing the Indian colonial government from making Iraq a colony of India.

At a critical point in the Arab Revolt, a disingenuous Sykes convinced King Husayn of the Hijaz, leader of the revolt, that rumors the British were reneging on their promises to him in the Husayn–McMahon correspondence were false. Thus encouraged by Sykes, the king renewed his efforts against the Turks and revived the flagging Arab revolt. This was after Sykes had negotiated the Sykes–Picot agreement the year before, dividing up the Ottoman Arab lands between Britain and France at the very same time

Britain was encouraging Husayn to believe the same land would be his postwar Arab kingdom in the Husayn–McMahon correspondence. And, at the time Sykes met Husayn, he was also actively working on a Zionist declaration that would establish a Jewish homeland in Palestine that would further diminish Husayn's postwar Arab kingdom. These documents, two with which he was intimately involved – and deliberately kept secret from Husayn – have had a profound, far-reaching and long-lasting impact on the Middle East up to the present day.

The Sykes–Picot agreement, although some of its original details were altered by the end of the war and afterwards, was used by the victors as a general blueprint to establish today's Middle Eastern countries of Syria, Lebanon, Palestine and Iraq. It resulted in lines drawn on a map in London with little regard for the region's history or its indigenous populations, all to satisfy the insatiable appetites of Britain and France in furthering their empires. After the war these former Ottoman lands became mandates and were placed under the auspices of Britain and France by the League of Nations in 1920 under the Treaty of San Remo. As a result, Western ideas of a country with borders and citizenship were imposed on what had been for a millennia open and undivided lands, where local populations roamed freely and had rarely known such restrictions. Eventually, after almost thirty years of unrest, revolts, and uprisings, Britain and France were forced to give up the mandates for which they had fought so hard.

The Balfour Declaration, which Sykes had a hand in writing, promoting and defending at home and abroad, would have even more far-reaching consequences in the region over the next 100 years. While Sykes and the British leadership saw it as a tool to help them win the war as well as ensure British control of Palestine after the war, humanitarians saw it as a solution to the so-called Jewish Problem of centuries of Jewish persecution and statelessness in Central and Eastern Europe. Others, however, saw it as a way of getting rid of the Jews in their countries. The Zionists, however, saw all these attitudes – one way or another – as helping them to achieve their goal of a Jewish state and return the Jews to their ancestral homeland. It would result in the eventual migration of hundreds, then thousands of European Jews to the Middle East, all with a sense of entitlement to a land their ancestors had left almost 2,000 years earlier.

The subsequent political and cultural clashes with the indigenous population who had no say in the matter would lead to riots, massacres and wars that continue to the present day. The Holocaust of World War II would flood the small country of Palestine with even more people intent on making a new life there free from persecution. It led to the largest displacement of people in recent history, a diaspora of an estimated five million Palestinian refugees fleeing their homeland to live in camps in surrounding countries, leave the area altogether, or eke out a meager existence in what remains of Palestine.[3]

Sykes never lived to see any of this, but as one who was at the centre of the events that led to today's turbulent Middle East he must bear some responsibility for what has happened there over the past 100 years. He interfered in the world and the lives of people he purported to know, for others whom he did not know, all the while convincing himself he was doing what was best for all. As a man of his time and class, Sir Mark Sykes was an imperialist driven more by ego and political considerations than humanitarian reasons. While he did sympathise with refugees of the Armenian genocide and wanted to help them find a home, most of his time and efforts were spent on the Zionist cause. This was for the far more practical and less altruistic reasons of what has been termed "strategic Zionism," gaining the influential support of worldwide Jewry in winning the war, while at the same time ensuring there would be support for a strong British presence in Palestine by establishing the Zionists there to counter the French in Syria.[4]

At no time did Sykes show any feeling or compassion for the Palestinian Arabs, seeing them as being of little use in his grandiose plans and otherwise intruders on the land they had occupied for millennia. Reflecting Whitehall's view,[5] he showed little interest in the Arabs in general except as a tool to be used for the war effort. In fact, he had no respect at all for what he called the "Eastern mind." As evidence of this, in an eleven-page letter to the newly appointed Secretary of State for India Edwin Montagu in August 1917, Sykes suggested that Montagu consider special education reforms for India. This was because, in his view, the Indian intellect was degenerate, like that of the Arabs with which he was most familiar and he stated manner-of-factly, due to "the following physical factors influencing his environment," which Sykes listed as:

I.   Malaria.
II. Harem life, leading to early masturbation say at the age of 6 or 7 and thence to sodomy, bestiality, etc., as tolerable things.
III. Completely disorderly domestic habits.

So, he informed Montagu, special measures should be taken when trying to educate an Eastern mind with higher Western learning entirely "absent from the East" and based on

   I. Roman Law and the theory of order and authority based on order.

  II. Medieval Christianity with the corollary of "chivalry" based on sacrifice and respect for women.

 III. Baconian science and the reformation viz., material progress and the theory of individual judgment.

 IV. The French Revolution and the development of the constitutional democratic idea.

Unless their education was handled properly, as Sykes suggested, "you produce this devastating intelligentsia of parrots, who cheat, steal, kill, bomb, peculate or shatter as the evil spirit moves them."[6] There is no evidence that Montagu responded to this letter.[7] One can only imagine what he must have thought when reading these bizarre and bigoted comments.

With unsolicited letters like this to people in positions of power as well as in official letters, papers and reports, along with more balanced work, Sykes did all he could to promote his reputation as an authority on the East. Surprisingly, although not in this case, ministers did sometimes take such letters seriously, when they fitted preconceived notions and facilitated policy decisions that would otherwise not be made.[8]

Sykes also wanted to be appreciated for his ability to get things done and straighten out bureaucratic muddles. His suggestion, resulting in the establishment of the Arab Bureau was one such instance, as was his suggestion to consolidate the many committees overseeing Eastern affairs that resulted in the Eastern Committee under Lord Curzon. However, his key role in critical events and policy making has contributed much to a region since torn apart by wars and sectarian violence.

In his defense, as Elizabeth Monroe wrote in her seminal work, *Britain's Moment in the Middle East 1914–1956*, he should not take the blame alone. While "Sykes is often held responsible for the worst of the inconsistencies [of British Middle East policy]," she wrote, "the War Cabinet discussed many of them and must take the blame for consenting to, even insisting on, courses he recommended." Moreover, "a chronology of certain events of 1917 reveals his weak points, but also those of a Cabinet that stampeded after him under a mercurial Prime Minister bent on acquiring a British Palestine."[9] The roots of the Arab Israeli conflict can be attributed directly to the Balfour Declaration and this conflict only intensified with the establishment of the state of Israel in 1948. Since then, it has led to the dispossession of a people, numerous wars, wholesale destruction, deaths of thousands and the prospect of further conflict into the unforeseeable future.

What the British and the French did not do after World War I, Britain's Tony Blair and the United States' George W. Bush completed almost a century later. They further destabilised the region with their joint Anglo-American invasion of Saddam Husayn's Iraq in 2003, sending millions more to refugee camps and also unleashing the so-called Islamic State, ISIS or Daesh on that region and the world.

On 29 June 2014, while proclaiming a new Islamic caliphate, the media-savvy militant organisation released a video entitled "The End of Sykes–Picot," to highlight the destruction of the border between Syria and Iraq. This was released "alongside its Arabic-language counterpart, *Kaser al-Hudud* ('the Breaking of the Borders') as well as a photo campaign called "Smashing the Sykes–Picot Border" and a Twitter hashtag, #SykesPicotOver."[10]

For ISIS this showed just how important and odious the Sykes–Picot agreement was to the region and its people. In the view of one analyst, not only did it represent "the fragmentation of the Islamic world as well as historical Western, Christian intervention in the region," but as

ISIS's only symbol, it is a particularly revealing one. It encompasses real and perceived Western interests and threats to the Middle East. For ISIS's potential funders, recruits, and unconvinced observers in the Arab world, it appeals to a shared aversion toward foreign intervention and fragmentation of the region.[11]

Such is the legacy of Sir Mark Sykes. He was convinced he could change the Middle East into a proverbial Eden and pro-British enclave, while simultaneously providing homelands for persecuted and disinherited Armenians and Jews. He could hardly have imagined what his efforts would eventually yield, although by doing it at the expense of the local population he should have realised there would have been repercussions.

Sykes was no Julius Caesar. He led no armies, conquered no countries and ruled no empire, nor was he assassinated for his political ambitions. He was, however, a soldier and politician, charismatic, passionate, ambitious, a traveller and writer, an aristocrat, member of his country's landed gentry and privileged upper class, a loving husband, father and friend. He also saw himself as someone in a position to right the world's wrongs, a knight in shining armor on a white horse, a crusader against evil. That is exactly how he is depicted on the Eleanor Cross War Memorial at the family estate at Sledmere in East Riding, Yorkshire. Resplendent in brass on the memorial as a Crusader knight, Sledmere's 6th Baronet Sir Mark Sykes greets visitors dressed in full armour holding a sword and standing triumphantly over the body of a fallen Saracen [Arab] with Jerusalem in the background. Above his head is the Latin inscription *Laetare Jerusalem*, which translates as Rejoice Jerusalem. As far as symbolism goes, given his activities in the Middle East from 1915 to 1919, the irony of this portrayal cannot be lost on the informed observer. It symbolises misery and tragedy for millions.

As Oscar Wilde is purported to have said, "no good deed goes unpunished." So, while Sykes may have meant well, 100 years later, his legacy in the Middle East is not what he hoped for, or could ever have imagined, and with good reason.

NOTES

Prelims

1. See "IJMES Translation and Transliteration Guide", *International Journal of Middle East Studies*, https://ijmes.chass.ncsu.edu/IJMES_Translation_and_Transliteration_Guide. htm, accessed August 21, 2017.

Introduction

1. Shane Leslie, *Mark Sykes: His Life and Letters* (London: Cassell and Company, Ltd, 1923).
2. Roger Adelson, *Mark Sykes: Portrait of an Amateur* (London: Jonathan Cape, 1975).

Chapter 1   The "Middle East Expert"

1. *Introduction*, The Right Hon. Winston Churchill, in Shane Leslie, *Mark Sykes: His Life and Letters* (London: Cassell and Company, Ltd, 1923), vii.
2. Leslie, *Mark Sykes*, 217–18.
3. Adelson, *Portrait of an Amateur* (London: Jonathan Cape, 1975), 147. See also Harold E. Gorst, *Much of Life is Laughter* (London: George Allen & Unwin Ltd., 1936), 212–15. Gorst was Sykes's brother-in-law, who also described Sykes's maiden speech in the House, not so much its content but Sykes's nervousness and impatience to speak at the time.
4. Adelson, *Portrait of an Amateur*, 146–74.
5. Ibid., 147–8.
6. Ibid., 54–9.
7. Leslie, *Mark Sykes*, 202.
8. At this time, "Kurdistan" consisted of the predominantly Kurdish-populated areas of eastern Anatolia, north-eastern Syria and much of Mesopotamia.
9. Lt. Col. Sir Mark Sykes, Bart, MP, *The Caliph's Last Heritage: A Short History of the Turkish Empire* (London: Macmillan and Co., Limited, 1915), 507–51. His previous publications on Turkey include *Through Five Turkish Provinces* (London: Bickers and Son, 1900), published while he was still a student at Cambridge, and *Dar-Ul-Islam: A Record of a Journey Through Ten of the Asiatic Provinces of Turkey* (London: Bickers & Son, 1904).
10. Sykes, *Caliph's Last Heritage*, 507–508.
11. Ibid., 508.

12. Ibid., 508–9. The Italo-Turkish War for Tripoli between Italy and Turkey began on September 29, 1911 and ended on October 18, 1912, with Italy taking over the Ottoman province from the Turks.

13. Ibid., 509.

14. Ibid., 509–11.

15. *Hansard's Parliamentary Debates*, LVI (1914), 65–9; hereafter cited as *Hansard's*.

16. H.G. Wells, *The War that Will End War* (London: Frank & Cecil Palmer, 1914).

17. Christopher Clark, *The Sleepwalkers: how Europe Went to War in 1914* (London: Penguin Books, 2012), xxi.

18. Adelson, *Portrait of an Amateur*, 170.

19. Ibid., 90–2.

20. Ibid., 164–70. Officially, the territorial 5th Green Howards Battalion was Alexandra, Princess of Wales's Own Yorkshire Regiment and later the 50th (1st Northumbrian) Division. However, all who served in it were proud to be known as the Green Howards. See Mark Marsay, *Baptism of Fire: An Account of the 5th Green Howards at the Battle of St. Julien, during the Second Battle of Ypres, April 1915; Part One of the 'Yorkshire Gurkhas' Series* (Scarborough: Great Northern Publishing, 1988), xi–xiii.

21. Adelson, *Portrait of an Amateur*, 170.

22. "War Miscellaneous Projects," Sykes Papers, quoted in Adelson, *Portrait of an Amateur*, 172–3. See Historical Inflation Rates and Calculator at http://inflation.iamkate.com, accessed 30/1/2018.

23. Ibid., 173.

24. "A History of the First World War in 100 Moments: the Day the Lights Went Out," *The Independent*, 9 September 2014 (http://www.independent.co.uk/news/world/world-history/history-of-the-first-world-war-in-100-moments/a-history-of-the-first-world-war-in-100-moments-the-day-the-lights-went-out-9239572.html, accessed 29 July 2016).

25. M. Sykes to W. Churchill, 24 August 1914, CHAR 13/45/127, *The Sir Winston Churchill Archive Trust*, hereafter cited as Churchill Papers, Churchill Archives Centre. See also Martin Gilbert, *Winston S. Churchill: Companion Volume III, Pt. I, Documents, July 1914–April 1915* (London: Heinemann, 1972), 52.

26. Adelson, *Portrait of an Amateur*, 32–3.

27. M. Sykes to W. Churchill, 24 August 1914, CHAR 13/45/127, Churchill Papers, Churchill Archives Centre. See also Gilbert, *Winston S. Churchill: Companion Volume III, Pt. I, Documents, July 1914–April 1915*, 52.

28. M. Sykes to W. Churchill, 24 August 1914, CHAR 13/45/127, Churchill Papers, Churchill Archives Centre. See also Gilbert, *Winston S. Churchill: Companion Volume III, Pt. I, Documents, July 1914–April 1915*, 52.

29. W. Churchill to M. Sykes, 26 August 1914, U DDSY2/1/28/69, The Papers of the Sykes Family of Sledmere and The Papers of Sir Mark Sykes, 1879–1919, hereafter cited as Sykes Papers.

30. M. Sykes to E. Sykes, 5 November 1914, quoted in Adelson, *Portrait of an Amateur*, 176.

31. Adelson, *Portrait of an Amateur*, 168–70.

32. Ibid., 173–4.

33. M. Sykes to W. Churchill, 27 January 1915, U DDSY 2/4/81, Sykes Papers.

34. Sykes was a great admirer of Maurice, Comte de Saxe, Marshal of France (1696–1750), who was a brilliant military leader, tactician and innovator. He wrote an article about Saxe for the battalion's journal, the *Green Howards' Gazette*. Entitled "Maréchal Saxe," it was published serially in the *Green Howards' Gazette* in 1905 and can be found along with copies of the journal (incomplete) in the Sykes Papers, U DDSY 2/6/1.

35. Frederick II, King of Prussia, also known as Frederick the Great (1712–86), was a successful commander and innovative military tactician who further organised and expanded the formidable Prussian army developed by his father, King Frederick Wilhelm I, who was known as the Soldier King.

36. M. Sykes to W. Churchill, 27 January 1915, U DDSY2/4/81, Sykes Papers. Parentheses are Sykes's.

37. Like many Europeans, Sykes persisted in calling the Ottoman capital by its Roman name Constantinople long after the Turkish conquest in 1453, after which the Turks renamed the city Istanbul, meaning the city in Turkish. Occasionally, Europeans in the area, including Sykes, would also refer to the city as Stamboul, a variant of its Turkish name.

38. M. Sykes to W. Churchill, 27 January 1915, U DDSY 2/4/8, Sykes Papers.

39. Ibid.

40. Ibid.

41. Graham T. Clews, *Churchill's Dilemma: The Real Story Behind the Origins of the 1915 Dardanelles Campaign* (Santa Barbara, CA: Praeger, 2010), xvi.

42. M. Sykes to W. Churchill, 27 January 1915, U DDSY 2/4/8, Sykes Papers.

43. Ibid.

44. Ibid., See Amanda Capern, "Winston Churchill, Mark Sykes and the Dardanelles Campaign of 1915," *Historical Research*, 71, no. 174 (February 1998): 110, where the author discusses Sykes's letter of 27 January 1915 to Churchill, and its possible importance in the planning of the Dardanelles Campaign, as well as addressing its curious omission by the biographers of both men.

45. Adelson, *Portrait of an Amateur*, 177.

46. Churchill, *World Crisis 1911–1918*, 2 Vols. First Two-Volume Edition (London: Odhams Press Limited, 1939), 595–610.

47. M. Sykes to W. Churchill, 26 February 1915, CHAR 26/2, Churchill Papers, Churchill Archives Centre. See also Gilbert, *Winston S. Churchill, Vol. III: The Challenge of War*, 317–18, and Gilbert, *The Churchill Documents Vol. 6: At the Admiralty July 1914–April 1915*, 581–3.

48. M. Sykes to W. Churchill, 26 February 1915, CHAR 26/2, Churchill Papers, Churchill Archives Centre. In an undated and unpublished series of notes, entitled "The British Soldier and the Turk," which he did not share with Churchill, Sykes wrote of his appreciation of the Turkish soldier. In it, he noted that Turkish soldiers were, like their British counterparts, "courageous . . . [and able to] go into battle after battle dogged persistent and determined, as little elated by victory as they are depressed by defeat. On the other hand, the Turk has not that capacity for grumbling at trifles . . . or love of swilling beer which [are] our soldiers worst points, nor again has our soldier that ferocity which lies at the bottom of every Turks heart." Sir Mark Sykes, "The British Soldier and the Turk," U DDSY 2/5/25, Sykes Papers.

49. M. Sykes to W. Churchill, 27 January 1915, U DDSY 2/4/8 Sykes Papers.

50. Ibid., See also Capern, "Winston Churchill, Mark Sykes and the Dardanelles Campaign of 1915," 108–18, in which she discusses Sykes's letters and the possibility of their influence on Churchill.

51. Adelson, *Portrait of an Amateur*, 177, cites a telegram from Churchill not found in either the Sykes Papers or The Churchill Papers. However, there is a telegram in the Sykes Papers U DDSY2/128/100, dated 2 October 1914 saying essentially the same thing and referring to a letter that is also apparently missing letter from the Sykes Papers.

52. W. Churchill to M. Sykes, 27 February 1915, U DDSY2/1/29/126, Sykes Papers.

53. Winston S. Churchill, *The World Crisis*.

54. Mustafa İsmet İnönü (1884–1973) was a Turkish general and statesman who served as the second president of Turkey from 1938, the day after the death of Mustafa Kemal

Atatürk, to 1950, when his Republican People's Party was defeated in Turkey's second free elections. He also served as the first chief of the general staff from 1922 to 1924 and as the first prime minister after the declaration of the Republic, serving three terms: from 1923 to 1924, from 1925 to 1937 and from 1961 to 1965.

55. Martin Gilbert, *In Search of Churchill: A Historian's Journey* (London: HarperCollins Publishers, 1995), 57–8.

Chapter 2   Kitchener's Man and Agent-at-Large

1. T.E. Lawrence, *Revolt in the Desert* (New York: Garden City Publishing Company, Inc., 1927), 2.
2. George H. Cassar, *Kitchener: Architect of Victory* (London: William Kimber, 1977), 170–7.
3. Winston S. Churchill, *The River War* (Melbourne, Australia: The Book Jungle On-Line, 2007), 85–6, describes Kitchener's early career and time in the Middle East, during which he learned Arabic, and attributes the subsequent success of his military career to his knowledge of Arabic.
4. Cassar, *Kitchener*, 23–110, 136–69.
5. Ibid., 276.
6. R. Storrs to O. Fitzgerald, 8 March 1915, PRO 30/57/45/QQ18, Kitchener Papers, The National Archives; hereafter referred to as TNA. See also George H. Cassar, *Kitchener's War: British Strategy from 1914–1916* (Washington DC: Potomac Books, Inc., 2004), 147. The presumption was that after the war, Kitchener would return to Cairo and resume his post, possibly as a viceroy of a British Middle East state.
7. Grey's note on Kitchener's letter of 14 November 1914, F.O. 800/102, Grey Papers/ Foreign Office Papers, The National Archives; hereafter cited as TNA. See also Cassar, *Kitchener's War*, 147.
8. Kitchener, "Memorandum: Alexandretta and Mesopotamia," 16 March 1915, CAB 42/2/10, Cabinet Papers, TNA.
9. Adelson, *Portrait of an Amateur*, 179. See also Elie Kedourie, *Arabic Political Memoirs and Other Studies* (London: Frank Cass, 1974), 236. Oliphant was third secretary at the British Embassy in Istanbul when Sykes was there as an honorary attaché.
10. M. Sykes to O. Fitzgerald, 4 March 1915, quoted in Adelson, *Portrait of an Amateur*, 179.
11. M. Sykes to E. Sykes, 6 April 1915, U DDSY 2/1/2f/38, Sykes Papers.
12. Aaron S. Klieman, "Britain's War Aims in the Middle East in 1915," *Journal of Contemporary History* 3 (1968): 240–41. See also Cassar, *Kitchener's War*, 147–78.
13. Cassar, *Kitchener's War*, 147. See also Kedourie, *England and the Middle East: The Destruction of the Ottoman Empire 1914–1921* (London: Mansell Publishing Limited, 1987), 30–1.
14. Cassar, *Kitchener's War*, 150. Sir Maurice de Bunsen (1852–1932) was a British diplomat, with a long career serving abroad from 1879 to 1914 in Japan, Siam (Thailand), Constantinople, Paris, Portugal, Spain and Austria. He was recalled to the Foreign Office from Vienna at the beginning of the war.
15. Ibid., 150.
16. M. Sykes to Sir George Arthur, 12 September 1916, PRO 30/57/91, Kitchener Papers, TNA. This was taken from correspondence between Sykes and Sir George Arthur, 3rd Bt., Kitchener's private secretary from 1914 to 1916, who interviewed Sykes for his book on Kitchener. His three-volume biography, *Life of Lord Kitchener*, was published in London in 1920 by Macmillan and Co. Limited, four years after Kitchener's death. See also Kedourie, *Arabic Political Memoirs*, 236–7.
17. War Council Meeting, 10 March 1915, p. 5, CAB 22 $\frac{1}{2}$, Cabinet Papers, TNA.

18. M. Sykes to G. Arthur, 12 September 1916, PRO 30/57/91, Kitchener Papers, TNA.
19. Kedourie, *Arabic Political Memoirs*, 237.
20. M. Sykes to G. Arthur, September 12, PRO 30/57/91, Kitchener Papers, TNA. See also Adelson, *Portrait of an Amateur*, 180.
21. David Fromkin, *A Peace to End all Peace: The Fall of the Ottoman Empire and the Creation of the Modern Middle East* (New York: Avon Books, 1989) 148. Report of the Committee on Asiatic Turkey, 30 June 1915, CAB 27/1, Cabinet Papers, TNA. See also Adelson, *Portrait of an Amateur*, 182.
22. Adelson, *Portrait of an Amateur*, 181–2.
23. Aaron S. Klieman, "Britain's War Aims in the Middle East in 1915," 250–1, n. 25.
24. David Fromkin, *A Peace to End all Peace*, 146–7, n. 147. In fact, both Sykes and Kitchener belonged to the Other Club, a political dining club founded by Winston Churchill and F.E. Smith (aka Lord Birkenhead) in 1911.
25. Adelson, *Portrait of an Amateur*, 180.
26. "British Desiderata in Turkey and Asia: Report, Proceedings and Appendices of a Committee Appointed by the Prime Minister, 1915," CAB 27/1, 3–29, Cabinet Papers, TNA.
27. M. Sykes, "Note by Sir Mark Sykes on the Proposed Maintenance of a Turkish Empire in Asia without Spheres of Influence," 3 May 1915, U DDSY 2/11/3, Sykes Papers. See also M. Sykes, "The Question of a Railway Connecting Haifa with the Euphrates," 3 May 1915, CAB 27/1, Cabinet Papers, TNA. Fromkin, *A Peace to End All Peace*, 250–1, n. 25.
28. According to *Encyclopædia Britannica*, "Al-Jazīrah (Arabic: 'Island'), [is] the northern reaches of Mesopotamia, now making up part of northern Iraq and extending into eastern Turkey and extreme northeastern Syria. The region lies between the Euphrates and Tigris rivers and is bounded on the south by a line running between Takrīt and Anbar. It consists of a rolling and irregular plateau 800–1,500 feet (240–460 m) above sea level."
29. M. Sykes, "The Proposed Maintenance of a Turkish Empire in Asia without Spheres of Influence," 3 May 1915, CAB 27/1, Cabinet Papers, TNA. Adelson, *Portrait of an Amateur*, 184.
30. Fromkin, *Peace to End All Peace*, 148–9.
31. Klieman, "Britain's War Aims in the Middle East in 1915," 246–7. The detailed reasons for partition are given here. For the official De Bunsen Committee report, see "British Desiderata in Turkey and Asia: Report, Proceedings and Appendices of a Committee Appointed by the Prime Minister, 1915," CAB 27/1, 3–29, Cabinet Papers, TNA.
32. M. Sykes, "The Question of a Railway Connecting Haifa with the Euphrates," 3 May 1915, CAB 27/1, Cabinet Papers, TNA.
33. Adelson, *Portrait of an Amateur*, 183.
34. Ibid., 184.
35. M. Sykes to E. Sykes, 6 April 1915, U DDSY 2/1/2f38, Sykes Papers.
36. M. Sykes to E.C. Callwell, 3 June 1915, U DDSY 2/4/97, Sykes Papers. This file contains copies of twenty more dispatches from Sykes to Maj. Gen. Callwell at the War Office. Many of the same dispatches can be found in "Policy in the Middle East II, Select Reports and Telegrams from Sir Mark Sykes. Reports (Secret). M. Sykes to E. C. Callwell, 12th June ... 30th November 1915," IOR/L/P&S/18/B218, India Office Records and Papers, British Library. The letter referred to here, however, is not included in the India Office Records and Papers in the British Library collection. See also Adelson, *Portrait of an Amateur*, 185.

37. Vladimir Borisovich Lutsky, *Modern History of the Arab Countries* (Moscow: Progress Publishers, 1969), 335–7.

38. "Policy in the Middle East II, Select Reports and Telegrams from Sir Mark Sykes. Reports, No. 4 (Secret), Athens, 12th June 1915," IOR/L/P&S/18/B218, India Office Records and Papers, British Library. See also Lutsky, *Modern History of the Arab Countries*, 336–7. The Koraysh, or Quraysh, were the tribe of the Prophet Muhammad and his descendants were leaders of the tribe and considered the rightful heirs to the Caliphate. At the time, the leading candidate for a Quraysh Caliph was the Sherif of Mecca.

39. "Policy in the Middle East II, Select Reports and Telegrams from Sir Mark Sykes. Reports, No. 4 (Secret), M. Sykes to E. C. Callwell, Athens, 12th June 1915," IOR/L/P&S/18/B218, India Office Records and Papers, British Library.

40. Sir Ignatius Valentine Chirol (1852–1929) was a British journalist and a prolific author, as well as an historian, diplomat and passionate imperialist. Raised in France and Germany by British parents and fluent in French and German, he worked at the Foreign Office from 1872 to 1876, but after he had learnt Arabic he left the Foreign Office in 1876 and moved to Cairo, then to Beirut in 1879, where he travelled through Syria to Constantinople and throughout the Balkans. In the Middle East he took up journalism, writing first for the *Levant Herald*. He then became a correspondent and editor of *The Times*, travelling across the globe for several years writing about international events until he returned to London in 1896. He became director of the foreign department of *The Times* in 1899, a post that he held until 1912. After his death in 1929, it was said of him that he was "The friend of viceroys, the intimate of ambassadors, one might almost say the counselor of ministers, he was [also] one of the noblest characters that ever adorned British journalism." Linda B. Fritzinger, *Diplomat without Portfolio: Valentine Chirol, His Life and* The Times (London: I.B.Tauris, 2006), 455 and R. Crewe to E. Grey, 14 June 1915, London, F.O. 800/95/105, Grey Papers/Foreign Office Papers, TNA.

41. For a full discussion on the Berlin to Baghdad Railway, see Sean McMeekin, *The Berlin– Baghdad Express: The Ottoman Empire and Germany's Bid for World Power 1898–1918* (London: Allen Lane, 2010).

42. "Memorandum by Sir Valentine Chirol, enclosure in Sir M. Sykes' despatch No. 9, of July 7, 1915," U DDSY 2/11/6, Sykes Papers.

43. P. J. Vatikiotis, *The Modern History of Egypt* (New York: Frederick A. Praeger, Publishers, 1969), 244.

44. "Policy in the Middle East II, Select Reports and Telegrams from Sir Mark Sykes. Reports, No. 14 (Secret), M. Sykes to E. C. Callwell, Shepherds Hotel, Cairo, 12th July 1915," IOR/L/P&S/18/B218, 2–3, India Office Records and Papers, British Library.

45. "Policy in the Middle East II, Select Reports and Telegrams from Sir Mark Sykes. Reports, No. 14 (Secret), M. Sykes to E. C. Callwell, Shepherds Hotel, Cairo, 12th July 1915," IOR/L/P&S/18/B218, India Office Records and Papers, British Library.

46. G. Clayton to R. Wingate, 27 July 1915, Strictly Private, SAD 158/6/61, Wingate Papers.

47. Said Pasha Shoucair was given a KBE in 1925 and his daughter, Celeste Leila Beatrix Shoucair (? − 1990) met and married British diplomat Sir Frank Kenyon Roberts (1907–98) when he was posted to Cairo in the early 1930s. She later came to be known as Lady Cella Roberts.

48. "Policy in the Middle East II, Select Reports and Telegrams from Sir Mark Sykes. Reports, No. 14 (Secret), M. Sykes to E. C. Callwell, Shepherds Hotel, Cairo, 12th July 1915," IOR/L/P&S/18/B218, India Office Records and Papers, British Library.

49. George Antonius, *The Arab Awakening: The Story of the Arab National Movement* (Beirut: Libraire du Liban, 1969, 81). Antonius (1891–1942) was close to the events as they unfolded in Cairo. A Lebanese Christian by birth and graduate of Cambridge University, he worked for the Egyptian government in the political intelligence department as a press censor from 1914 to 1919. After the war he worked for the Palestinian Mandate Administration Department of Education and later was assistant secretary for Arab Affairs from 1921 to 1930. In 1925, Antonius became a Palestinian citizen. For the definitive biography of Antonius, see Susan Silsby Boyle, *Betrayal of Palestine: The Story of George Antonius* (Boulder, CO: Westview Press, 2001). Faris Nimr Pasha, according to Antonius (who became his son-in-law), was one of the founders of the first organised Arab national movement in 1875 by a group of Christian students at the Syrian Protestant College in Beirut (later the American University of Beirut). Ibid., 79–81, 270 n. 1. He was also one of the founders in Cairo of *Al-Muqtalaf*, a scientific monthly, and *Al-Muquttam*, a widely published daily newspaper in Cairo.

50. "Policy in the Middle East II, Select Reports and Telegrams from Sir Mark Sykes, Reports. No. 14 (Secret). M. Sykes to E. C. Callwell, Shepherds Hotel, Cairo, 14th July 1915," IOR/L/P&S/18/B218, India Office Records and Papers, British Library.

51. Prophet Muhammad belonged to the Quraysh tribe and therefore to be a member of this tribe brings with it a certain amount of prestige. Sharif Husayn of the Hijaz was not only a member of the Quraysh, but also a *Sayyid*, a direct descendent of the Prophet, and was accorded highest honours in Muslim circles.

52. M. Sykes to E.C. Callwell, Aden, No. 16, 23 July 1915 (Secret), U DDSY 2/11/7, Sykes Papers. See also "Decipher of Telegram, Relaying Report from Mark Sykes, dated 29 July 1915, to Director of Military Operations," IOR/P&S/11/94, No. 5, India Office Records and Papers, British Library.

53. Lt. Col. Harold F. Jacob was political agent in Dala, 96 miles north of Aden, from 1904 to 1907, 1st assistant resident, JP and acting resident Aden, 1910–17, chief political officer Aden Field Force 1914–17, and 1919 (http://www.myjacobfamily.com/faversham jacobs/haroldfentonjacob.htm, accessed 1/22/2015).

54. M. Sykes to E. C. Callwell, Aden, July 23, 1915 (Secret), No. 16, U DDSY 2/11/7, Sykes Papers. See Harold F. Jacob, *Kings of Arabia: The Rise and Set of the Turkish Sovranty* [sic] *in the Arabian Peninsula* (London: Mills and Boon, 1923; reprint: Reading: Garnet Publishing Limited, 2007), 158–85, for a description of World War I in Aden.

55. The Caisse de la Dette Public (Public Debt Commission) was an international commission established in 1876 after Khedive Ismail failed to repay loans made to him by European bankers to build the Suez Canal and to pay for his lavish lifestyle. Under the so-called Dual Control of Britain and France after 1877, the Caisse supervised the repayment of these loans to the European governments taken directly from Egyptian taxes on behalf of their country's bankers and moneylenders. See David S. Landes, *Bankers and Pashas: International Finance and Economic Imperialism in Egypt* (New York: Harper Torchbooks, 1969), 317–18.

56. "Policy in the Middle East II, Select Reports and Telegrams from Sir Mark Sykes, Reports. No. 19 (Secret)," M. Sykes to E. C. Callwell, Aden, August 10, 1915," IOR/L/P&S/18/B218, 11, India Office Records and Papers, British Library.

57. M. Sykes to E.C. Callwell, Aden, 23 July 1915 (Secret), No. 16, U DDSY 2/11/7, Sykes Papers.

58. M. Sykes to E.C. Callwell, Aden, 2 August 1915 (Secret), No. 17, U DDSY 2/11/7, Sykes Papers.

Chapter 3   Islam, India, Iraq and the Arab Bureau

1. This was the ambiguous response given by W.J. Childs, a former member of the Arab Bureau, when the Public Accounts Office inquired about it to the Foreign Office in 1923. Priya Satia, *Spies in Arabia: The Great War and the Cultural Foundations of Britain's Covert Empire in the Middle East* (Oxford: Oxford University Press, 2010), 45.
2. Antonius, *Arab Awakening*, 109.
3. John L. Esposito, *Islam: The Straight Path*, 3rd edn (New York: Oxford University Press, 1998), 132–4.
4. "Policy in the Middle East II, Select Reports and Telegrams from Sir Mark Sykes, Reports. No. 14 (Secret). M. Sykes to E.C. Callwell, Shepherds Hotel, Cairo, 14th July 1915." IOR/L/P&S/18/B218, India Office Records and Papers, British Library.
5. Ibid.
6. Ibid.
7. Esposito, *Islam: The Straight Path*, 132–4.
8. "Policy in the Middle East II, Select Reports and Telegrams from Sir Mark Sykes, Reports. No. 14 (Secret). M. Sykes to E.C. Callwell, Shepherds Hotel, Cairo, 14th July 1915." IOR/L/P&S/18/B218, India Office Records and Papers, British Library.
9. Ibid.
10. Ibid.
11. Ibid.
12. Ibid.
13. Charles Hardinge, 1st Baron Hardinge of Penshurst (1858–1944) was a diplomat and British statesman who served as Viceroy of India (1910–16). After graduating from Trinity College, Cambridge, Hardinge entered the diplomatic service in 1880. He was appointed first secretary at Tehran in 1896 and first secretary at St Petersburg in 1898. After a short time as assistant under secretary for foreign affairs, he was appointed ambassador to Russia in 1904. In 1906 he was made permanent under secretary at the Foreign Office under Sir Edward Grey. In 1910 he was raised to the peerage as Baron Hardinge of Penshurst and appointed viceroy of India, a post in which he served until 1916. He returned to his former post as Permanent Under Secretary at the Foreign Office in 1916, under Arthur Balfour, and in 1920 he became ambassador to France, before retiring in 1922.
14. V. Chirol to C. Hardinge, 6 July 1915, Vol. 72, f. 234–5, Hardinge Papers, University of Cambridge Library; hereafter referred to as Hardinge Papers.
15. C. Hardinge to A. Chamberlain, 7 June 1915, IOR/L/P&S/11/93, India Office Records and Papers, British Library. See also Adelson, *Portrait of an Amateur*, 189.
16. C. Hardinge to A. Chamberlain, 6 August 1915, Hardinge Papers. See also Adelson, *Portrait of an Amateur*, 189.
17. Elizabeth Monroe, *Britain's Moment in the Middle East 1914–1956* (London: Chatto & Windus, 1963), 35–6.
18. M. Sykes to E. Sykes, 3, 9 September 1915, quoted partly in Sykes, *Portrait of an Amateur*, 190–1 and Leslie, *Mark Sykes*, 247.
19. By Anglo-Indians Sykes meant the Indians of mixed Indian and British ancestry who dominated administrative positions in the service of the British Raj. During the time of its predecessor, the British East India Company's rule in India (1600–1874) in the late eighteenth and early nineteenth centuries, it was fairly common for British officers and soldiers to take local wives and have Eurasian children due to the lack of British women in India. They eventually formed a distinct minority community apart from the rest of the

Indian population. They spoke English, dressed in Western clothes, were educated in the English schools of the Raj and maintained English customs at home. As such, they were considered valuable assets to the Indian colonial rulers. See Geoffrey Moorhouse, *India Britannica* (New York: Harper & Row, 1983), 181–93. See also Archie Baron, *An Indian Affair: From Riches to Raj* (London: Channel 4 Books, 2001), 28–33, 114–17, 193–8.

20. M. Sykes to E. Sykes, 3, 9 September 1915, quoted partly in Sykes, *Portrait of an Amateur*, 190–1, and Leslie, *Mark Sykes*, 247.

21. "Memorandum on Indian Moslems," U DDSY 2/11/8, Sykes Papers.

22. Lt. Col. Sir Mark Sykes, MP, "Memorandum," Secret, dated 28 October 1915, U DDSY 2/4/141, Sykes Papers.

23. Ian Rutledge, *Enemy on the Euphrates: The British Occupation of Iraq and the Great Arab Revolt 1914–1921* (London: Saqi Books, 2014), 73–4.

24. Lt. Col. Arnold T. Wilson, *Loyalties Mesopotamia 1914–1917: A Personal and Historical Record* (London: Oxford University Press, 1930), 152.

25. M. Sykes, "Memorandum," Secret, dated 28 October 1915, U DDSY 2/4/141, Sykes Papers.

26. Ibid.

27. Ibid. It is difficult to understand how someone like Sykes who was not familiar with Arabic and had not been trained in the complexities of Islamic scholarship could make such a comparisons and state them as fact. Without giving any evidence to support his statements other than opinion, he further described Indian Muslims as gullible and easily led to believe in the most ludicrous things.

28. Stanley Wolpert, *A New History of India, 4th ed.* (New York: Oxford University Press, 1993), 291.

29. Antonius, *Arab Awakening*, 140–1. See also Tilman Lüdke, *Jihad Made in Germany: Ottoman and German Propaganda and Intelligence Operations in the First World War* (Münster: LIT Verlag, 2005), 51, which states that jihad was declared on 14 November 1914.

30. Lüdke, *Jihad Made in Germany*, 1,12, 31–4.

31. Gary Troeller, *The Birth of Saudi Arabia: Britain and the Rise of the House of Sa'ud* (London: Frank Cass, 1987), 76.

32. M. Sykes, "Memorandum," Secret, Dated 28 October 1915, U DDSY 2/4/141, Sykes Papers.

33. Ibid.

34. Ibid.

35. Ibid.

36. Ibid.

37. Antonius, *Arab Awakening*, 111–19.

38. "Policy in the Middle East III, The Arab Movement, Memorandum from Sir Mark Sykes, Bart., M.P.," IOR/L/P&S/18/B219, India Office Records and Papers, British Library.

39. Ibid.

40. Sir Mark Sykes, "Minute," FO 371/2781/201201, no. 304720, Foreign Office Papers, TNA. See also Bruce Westrate *The Arab Bureau: British Policy in the Middle East 1916–1920* (University Park, PA: Pennsylvania State University Press, 1992), 23–4.

41. Sir Percy Cox (1864–1937) was a British Indian Army officer and colonial administrator in the Middle East. In 1904, he was appointed acting political resident in the Persian Gulf and consul general for southern Iran, residing at Bushire. In 1909, he was made resident, a post he held until 1914, when he was appointed secretary to the Government of India and promoted to the rank of lieutenant colonel. Shortly after returning to India in 1914 he was sent back to the Persian Gulf as chief political officer with the Indian Expeditionary Force to fight against the Ottoman Empire during World War I. In 1917

he was promoted honorary Major General. During this time he established strong relations with Ibn Saʿud, with whom he already had dealings when he was resident. After the war, he was appointed acting minister in Tehran until 1920, when he returned to Iraq to serve as British high commissioner in Baghdad until 1923, after which he retired and returned to England. He was one of the major figures in the creation of Iraq in 1921. See John Townsend, *Proconsul to the Middle East: Sir Percy Cox and the End of Empire* (London: I.B.Tauris, 2010).

42. Westrate, *Arab Bureau*, 24.
43. Ibid., 25–6.
44. G. Clayton to War Office, n.d., SAD 135/7, Papers of General Sir Reginald Wingate, Sudan Archive at Durham; hereafter cited as Wingate Papers.
45. Sir George Russell Clerk (1874–1951) was a Foreign Office civil servant and British diplomat. Beginning his career at the Foreign Office in 1899, he had several overseas postings before being appointed first secretary at the British Embassy in Constantinople in 1910. He returned to the Foreign Office in London as senior clerk in 1913 and became acting counsellor in 1917. In 1919, he was sent to Czechoslovakia where he served as first minister and in 1921 as counsel general. Over a long and distinguished career he was knighted (1917), made PC (1926) and served as ambassador to Turkey (1926–1933), ambassador to Belgium (1933–1934) and ambassador to France (1934–1937), after which he retired from the diplomatic service.
46. M. Sykes to R. Cecil, 4 October 1915, U DDSY 2/12/1, Sykes Papers.
47. Ibid.
48. Ibid.
49. Ibid.
50. M. Sykes to R. Cecil, 4 October 1915, U DDSY 2/12/1, Sykes Papers.
51. Secretary of State to Viceroy, 10 December 1915, IOR/L/P&S/10/576, India Office Records and Papers, British Library.
52. M. Sykes to G. Clayton, 28 December 1915, FO 882/2, Foreign Office Papers, TNA.
53. See "India Office to Foreign Office re Attitude of Government of India," CAB 42/7, Cabinet Papers, TNA, and Chamberlain to Grey, 30 December 1915, IOR/L/P&S/10/523, India Office Records and Papers, British Library. See also H. V. F. Winstone, *The Illicit Adventure: The Story of Political and Military Intelligence in the Middle East from 1898–1926* (London: Jonathan Cape, 1982), 197.

Chapter 4   The Husayn–McMahon Correspondence, the Arab Revolt and Advising the War Cabinet

1. C. Hardinge (private) to R. Rodd, 18 May 1917, *Hardinge Papers*, Cambridge University Library, 32. Briton Cooper Busch, *Hardinge of Penshurst: A Study in the Old Diplomacy* (Hamden, CT: Archon Books, 1980), 257.
2. Adelson, *Portrait of an Amateur*, 194.
3. Ibid., 194.
4. M. Sykes to E.C. Callwell, No. 18 (Telegram), 19 November 1915, IOR/L/P&S/18/B218, India Office Records and Papers, British Library.
5. Ibid.
6. Ibid.
7. Ibid.
8. Taken from Clayton's 11 October 1915 report of his interview of al-Faruqui found in F.O. 371/2486, 165761/34982, quoted in Kedourie, *Anglo-Arab Labyrinth*, 73.

9. Ibid., Al-Faruqui claimed direct descent from Omar ibn al-Khattab, the second Caliph of Islam (634–644).

10. Taken from Clayton's 11 October 1915 report of his interview of al-Faruqui found in F.O. 371/2486, 165761/34982, quoted in Kedourie, *Anglo-Arab Labyrinth*, 73–4.

11. Ibid., 74.

12. M. Sykes to E.C. Callwell, No. 19 (Telegram), 21 November 1915, IOR/L/P&S/18/B218, India Office Records and Papers, British Library.

13. Ibid.

14. Ibid.

15. Isaiah Friedman, *British Miscalculations: The Rise of Muslim Nationalism, 1918–1935* (New Brunswick, New NJ: Transaction Publishers, 2012), 16–17.

16. Andrew, Christopher M. and A.S. Kanya-Forstner, *France Overseas: The Great War and the Climax of French Imperial Expansion* (London: Thames and Hudson, 1981), 88.

17. Barr, *Setting the Desert on Fire*, T.E. Lawrence and *Britain's Secret War in Arabia, 1916–1918* (London: Bloomsbury Publishing, 2007), 19–20.

18. M. Sykes to E.C. Callwell, 21 November 1915, SAD 135/6, Wingate Papers.

19. Sir George Arthur, *Life of Lord Kitchener*, Vol. III, 154–5.

20. M. Sykes to G. Clayton, 28 December 1915, F.O. 882/2, Foreign Office Papers, TNA. See James Barr, *A Line in the Sand: Britain, France and the Struggle for the Mastery of the Middle East* (London: Simon & Schuster, Ltd, 2011), 7.

21. Lord Hankey, *The Supreme Command 1914–1918*, 2 Vols (London: George Allen and Unwin Limited, 1961).

22. "Evidence of Lieutenant-Colonel Sir Mark Sykes, Bart, M.P., on the Arab Question," (Secret) War Committee. Meeting Held at 10 Downing Street, on Thursday December 16, 1915, at 11:30 A.M. CAB 24/1/51, Cabinet Papers, TNA. Both Lloyd George and Balfour were new to their positions, having been appointed in a Cabinet reshuffle in May 1915. Churchill, who had lost his post at the Admiralty and in the Cabinet as a result of the disastrous Dardanelles campaign at Gallipoli, was succeeded by Balfour. Lloyd George was appointed to head a new wartime cabinet position at the Ministry of Munitions, created as the result of the Shell crisis of 1915. For the details of this first wartime coalition government, see the Earl of Oxford and Asquith, *Memories and Reflections 1852–1927* (Boston, MA: Little, Brown and Company, 1928), 115–26. Sykes's handwritten notes for the meeting can be found in U DDSY2/12/2, *Sykes Papers*.

23. "Evidence of Lieutenant-Colonel Sir Mark Sykes, Bart, M.P., on the Arab Question," (Secret) War Committee. Meeting Held at 10 Downing Street, on Thursday December 16, 1915, at 11:30 A.M. CAB 24/1/51, Cabinet Papers, TNA.

24. Ibid.

25. Ibid.

26. Ibid.

27. Ibid.

28. Ibid.

29. Ibid., Christopher M. Andrew and A.S. Kanya-Forstner, *France Overseas: The Great War and the Climax of French Imperial Expansion* (London: Thames and Hudson, 1981), 44, in support of Sykes's first two contentions.

30. "Evidence of Lieutenant-Colonel Sir Mark Sykes, Bart, M.P., on the Arab Question," (Secret) War Committee. Meeting held at 10 Downing Street, on Thursday December 16, 1915, at 11:30 A.M. CAB 24/1/51, Cabinet Papers, TNA.

31. Ibid., Andrew and Kanya-Forstner, *France Overseas*, 40. For an extensive discussion of France's *mission civilisatrice* see Alice L. Conklin, *A Mission to Civilize: The Republican Idea*

of Empire in France and West Africa, 1895–1930 (Stanford, CA: Stanford University Press, 1997), 11–37.

32. "Evidence of Lieutenant-Colonel Sir Mark Sykes, Bart, M.P., on the Arab Question," (Secret) War Committee. Meeting held at 10 Downing Street, on Thursday December 16, 1915, at 11:30 A.M. CAB 24/1/51, Cabinet Papers, TNA.

33. Ibid. See Eugene Rogan, *The Fall of the Ottomans: The Great War in the Middle East, 1914–1920* (London: Allan Lane, 2015), 159–84, for a detailed explanation of this complex event known today as the Armenian genocide.

34. "Evidence of Lieutenant-Colonel Sir Mark Sykes, Bart, M.P., on the Arab Question," (Secret) War Committee. Meeting Held at 10 Downing Street, on Thursday December 16, 1915, at 11:30 A.M. CAB 24/1/51, Cabinet Papers, TNA.

35. Ibid.

36. Sir Ronald Storrs, *The Memoirs of Sir Ronald Storrs* (New York: G.P. Putnam's Sons, 1937), 133–4. For a description of von Oppenheim's Cairo activities, see Lionel Gossman, *The Passion of Max von Oppenheim: Archeology and Intrigue in the Middle East from Wilhelm II to Hitler* (Cambridge: Open Book Publishers, 2014), 33–45.

37. Lüdke, *Jihad Made in Germany*, 115–16. For an expanded description of von Oppenheim's operations, see Chapter 4: "Manufacturing Support: German and Ottoman Propaganda Operations," 115–47. For a detailed description of his memorandum proposing using Islam as a weapon in the war against the Allies, see Gossman, *Passion of Max von Oppenheim*, 81–105.

38. Lüdke, *Jihad Made in Germany*, 51. Rogan, *The Fall of the Ottomans*, 51–2.

39. Barr, *Line in the Sand*, 14.

40. "Evidence of Lieutenant-Colonel Sir Mark Sykes, Bart, M.P., on the Arab Question," (Secret) War Committee. Meeting held at 10, Downing Street, on Thursday December 16, 1915, at 11:30 A.M., CAB 24/1/51, Cabinet Papers, TNA.

41. Ibid.

42. Ibid.

43. Ibid., The reference here was to Ibrahim Pasha (1789–1848), the eldest son of Muhammad Ali, the Ottoman-appointed ruler of Egypt, who was sent to conquer Syria in 1831 as a result of his father's quarrel with the Porte. Going from Egypt through the Sinai peninsula into Palestine, he took Acre in May 1832, marched north and occupied Damascus, defeated an Ottoman army at Homs in July and invaded Anatolia, where he defeated the forces of the grand vizier in December 1832. He controlled these areas until 1839, when a combined force of British, Ottoman and Austria armies forced him to evacuate and return to Egypt, thus securing the continued integrity of the Ottoman Empire.

44. "Evidence of Lieutenant-Colonel Sir Mark Sykes, Bart, M.P., on the Arab Question," (Secret) War Committee. Meeting Held at 10, Downing Street, on Thursday December 16, 1915, at 11:30 A.M., CAB 24/1/51, Cabinet Papers, TNA.

45. Ibid.

46. Francis Leveson Bertie, 1st Viscount Bertie of Thame (1844–1919), was a British diplomat. He entered the Foreign Office as a clerk in 1863. He served as private secretary to Robert Bourke, the under secretary of state and clerk in the Eastern Department from 1882 to 1885, after which he was the Eastern Department's senior clerk and assistant under secretary. He was ambassador to Italy (1903–1905) and ambassador to France (1905–1918). While in Paris, he played a key role in strengthening the *Entente Cordiale* between Britain and France, and had promoted British support of France in the Moroccan crises of 1905 and 1911.

47. Minutes of the War Committee, 16 December 1915, CAB 42/6/9, Cabinet Papers, TNA. See R. Crewe to F. Bertie, 17 December 1915, and F. Bertie to R. Crewe, December 21,1915, FO 800/58, Grey Papers/Foreign Office Papers. Adelson, *Portrait of an Amateur*, 197–8.

48. Adelson, *Portrait of an Amateur*, 196.

49. G. Clayton to M. Sykes, Cairo, 13 December 1915, U DDSY 2/11/16, Sykes Papers.

50. See Sykes's handwritten copy, The Constitution and Functions of the Arabian Bureau, 16 December 1915, U DDSY 2/4/92, *Sykes Papers*.

51. M. Sykes, The Constitution and Function of the Arabian Bureau, 28 December 1915, F.O. 882/2, Foreign Office Papers, TNA. Adelson, *Portrait of an Amateur*, 198.

52. M. Sykes to G. Clayton, 28 December 1915, F.O. 882/2, Foreign Office Papers, TNA.

53. Westrate, *Arab Bureau*, 27.

54. Ibid., 27–8.

55. Admiralty Intelligence to Clayton, December 20, 1915, FO 882/2, Foreign Office Papers, TNA. Westrate, *Arab Bureau*, 28.

56. M. Sykes to G. Clayton, 28 December 1915, F.O. 882/2, Foreign Office Papers, TNA. Westrate, *Arab Bureau*, 28.

57. G. Clayton to R. Hall, 13 January 1916, F.O. 882/2, Foreign Office Papers, TNA. Westrate, *Arab Bureau*, 28.

58. Report of the Inter-Departmental Conference, 7 January 1916, CAB 42/7, Cabinet Office Papers, TNA.

59. Adelson, *Portrait of an Amateur*, 199.

60. H.V.F. Winstone, *Gertrude Bell* (London: Barzan Publishing, 2004), 278.

Chapter 5 The Sykes–Picot Agreement

1. Michael Williams, "Sykes–Picot drew lines in the Middle East sands that blood is washing away," *Reuters, Analysis & Opinion/ The Great Debate*, 24 October 2014. http://blogs.reuters.com/great-debate/2014/10/24/sykes-picot-drew-lines-in-the-middle-easts-sand-that-blood-is-washing-away/, accessed September 2, 2016.

2. Barr, *Line in the Sand*, 13–14.

3. Robert Offley Ashburton Crewe-Milnes, Ist Marquess of Crewe (1858–1945) was a British Liberal statesman and writer. He was private secretary to Lord Granville, when he was secretary of state for Foreign Affairs (1883–1884), lord lieutenant of Ireland (1892–1895), and lord president of the Council (1905–1908). From 1908 to 1910 he served as secretary of state for the colonies and leader of the House of Lords. He served as secretary of state for India (1910–1911, 1911–1915), and lord president of the Council again from 1915 to 1916. He was president of the Board of Education (1916), but refused to serve in the Lloyd George coalition government (1916–1922), and acted as leader of the Liberal opposition in the House of Lords. After the fall of the coalition government, he was ambassador to France (1922–1928) and later secretary of state for war (1931). See James Pope Hennessy, *Lord Crewe: The Likeness of a Liberal* (London: Constable & Co. Ltd, 1955), 63–152.

4. F. Bertie to R. Crewe, 21 December 1915, in correspondence of Sir Edward Grey, Viscount Grey of Fallodon alias Falloden: Correspondence with Lord Bertie: Vol. XXX (ff. 214). December. 1915–January 1916, Add 63040, British Library.

5. Théophile Delcassé (1852–1923) was a French statesman. He was under secretary for the colonies (1883–1893) and it was largely due to his efforts that the French Colonial Office was made a separate department with a minister at its head, a position to which he was

appointed as minister of the colonies in 1894–1895, until there was a change in government. When in office, he made French colonial expansion, particularly in Africa, his main focus. From 1989–1905 he was minister of Foreign Affairs, during which time he arranged the *Entente Cordiale* in 1904, settling long standing disputes with Great Britain. He retired to private life in 1906, returning to political office as minister of Marine (1911–1913), retiring once again but returning to office as minister of war in 1914 and minister of Foreign Affairs (1914–1915), before finally retiring in 1915.

6. Barr, *Line in the Sand*, 15–16.

7. Robert de Caix, known variously as Robert Caix, and Robert de Caix de St Aymour (1869–1970) was a French journalist, politician, writer and diplomat with a lifelong interest in politics and France's image in the world. Between 1898 and 1909 he travelled extensively in Asia and between 1896–1897 he became one of the official propagandists of the colonial party. He was also a member of other several organisations, including the colonial Geographic Society of Paris, the Society for the History of the French Colonies, the French Colonial League, the France-America Committee and the Committee of French Oceana. In 1901 he became editor of the *Bulletin du Comité de L'Asie Française* and later secretary-general of the *Comité* itself. After the war, he would become involved in the Syrian Mandate and the League of Nations.

8. Barr, *Line in the Sand*, 16. See also Andrew and Kanya-Forstner, *France Overseas*, 40.

9. Andrew and Kanya-Forstner, *France Overseas*, 47.

10. Barr, *Line in the Sand*, 17.

11. Ibid., 18.

12. Ibid., 18–19.

13. Barr, *Line in the Sand*, 26–7. Paul Cambon (1843–1924) was a French diplomat, who had served as ambassador to Spain (1886–1890), ambassador to the Ottoman Empire (1890–1998), and ambassador to Britain (1898–1920). He helped negotiate the *Entente Cordiale* between Britain and France in 1904, and helped secure British involvement in World War I on France's side.

14. See Keith Robbins, *Sir Edward Grey: A Biography of Lord Grey of Fallodon* (London: Cresswell & Company Ltd., 1971), 301, 303, 321–2, 325 and 333, regarding Grey's declining health and overwork.

15. Andrew and Kanya-Forstner, *France Overseas*, 87. The Egyptian party referred to here included British officials in Cairo, ministers and politicians in London, and their supporters, who advocated a proactive British policy in the Middle East. Key figures of the Egyptian party in Cairo were Sir Milne Cheetham, acting British high commissioner (August 1914 to January 1915), Ronald Storrs, oriental secretary, General Sir Reginald Wingate, *Sirdar* (commander-in-chief) of the Egyptian Army and governor-general of the Sudan, and Brig. Gen. Gilbert Clayton, director of intelligence. In London, besides Kitchener, they included Lord Robert Cecil, parliamentary under secretary of state at the Foreign Office and Sir Mark Sykes who, while affiliated to the War Office, Foreign Office, and Cabinet Secretariat (1915–1919), either proposed, promoted or facilitated its various schemes and supported its Eastern agenda in Parliament. See Robbins, *Sir Edward Grey*, 328–9, on Grey's deferring to Kitchener on Middle Eastern matters at this time.

16. H. McMahon to Foreign Office, 28 October 1915, F.O. 371/2486, Foreign Office Papers, TNA; Barr, *Line in the Sand*, 27.

17. Andrew and Kanya-Forstner, *France Overseas*, 66. See also Adelson, *Portrait of an Amateur*, 199.

18. Andrew and Kanya-Forstner, The Climax of French Imperial Expansion: 1914–1924. Stanford: Stanford University Press, 1981, 75.

19. P. Cambon to R. Viviani, 21 October 1915, R. Viviani to P. Cambon (drafted by Picot) 25 October 1915: AE A Paix 128; F. Picot to A. Defrance, 1 November 1915, AE Defrance MSS 2, in Andrew and Kanya-Forstner, *France Overseas*, 89. René Viviani (1863–1925) was a French politician who served as prime minister during the first years of the war from 13 June 1914 to 29 October 1915.

20. Barr, *Line in the Sand*, 27.

21. F. Picot to A. Defrance, 24 December 1915 and 17 March 1916, AE Defrance MSS, 2, quoted in Andrew and Kanya-Forstner, *France Overseas*, 99.

22. Aristide Briand (1862–1932) was a French statesman who served as prime minster of France eleven times during the Third Republic. This was his fifth term as prime minister (29 October 1915–12 December 1916), succeeding René Viviani as prime minister and also replacing Declassé and serving as his own foreign minister.

23. Andrew and Kanya-Forstner, *France Overseas*, 89–90.

24. Sir Arthur Nicolson, Bart (1849–1928), later 1st Baron Carnock (1916–1928), was a British diplomat and politician. He worked at the Foreign Office (1870–1874), at the Embassy in Berlin (1874–1876), at the Embassy in Beijing (1876–1878), was chargé in Athens (1884–1885) and in Tehran (1885–1888), consul general in Budapest (1888–1893), at the Embassy in Constantinople (1893), minister at Tangiers (1895–1904), ambassador to Spain (1804–1805), ambassador to Russia (1906–1910) and permanent under secretary of Foreign Affairs (1910–1916). He was the father of Harold Nicolson (1886–1968), diplomat, author and diarist, who, after serving at the Foreign Office (1909–1911) was an attaché at Madrid (1911) and Constantinople (1912–1914) before returning to London, where he served as second secretary at the Foreign Office during the war.

25. Andrew and Kanya-Forstner, *France Overseas*, 90–1.

26. "Minutes of Anglo-French discussions, 23 November 1915," F.O. 371/2486, Foreign Office Papers, TNA. See also "Results of Second Meeting to discuss the Arab Question and Syria," 23 November 1915, F.O. 882/2, Foreign Office Papers, TNA; and Andrew and Kanya-Forstner, *France Overseas*, 91–2.

27. F. Picot to P. Margerie (Pierre de Margerie, political director at the Quai d'Orsay), 2 December 1915, AE Paix, 129; A. Briand to P. Cambon, 14 December 1915, AE Paix, 129; "Picot was in Paris while his instructions were being prepared and probably had a hand in drafting them. It is clear from his letters to Defrance (24 December 1915, 17 March 1916, AE Defrance MSS 2) that he was in agreement with them," Andrew and Kanya-Forstner, *France Overseas*, 92 and 270.

28. "Minutes of Second Round of Negotiations, 21 December 1915, P. Cambon to A. Briand, 22 December 1915: AE A Paix, 129," Andrew and Kanya-Forstner, *France Overseas*, 93.

29. Ibid., 93.

30. Minute by Clerk on H. McMahon to E. Grey, 11 December 1915, F.O. 371/2486, Foreign Office Papers, TNA.

31. "Memorandum of Third Meeting of Nicolson's Committee, 21 December 1915," FO 882/2, Foreign Office Papers, TNA. See also P. Cambon to A. Briand, 22 December 1915, AE A Paix, 129, Andrew and Kanya-Forstner, *France Overseas*, 93.

32. Adelson, *Portrait of an Amateur*, 199.

33. "Evidence of Lieutenant-Colonel Sir Mark Sykes, Bart., M.P., on the Arab Question," (Secret) War Committee Meeting held at 10 Downing Street, on Thursday, 16 December 1915, at 11:30 A.M., CAB 24/1/51, Cabinet Papers, TNA.

34. Andrew and Kanya-Forstner, *France Overseas*, 94–5.

35. Adelson, *Portrait of an Amateur*, 200.

36. Kedourie, *In the Anglo-Arab Labyrinth*, 64.

37. F. Picot to P. Cambon, 3 January 1916, AE A Paix, 129, in Andrew and Kanya-Forstner, *France Overseas*, 95.
38. Andrew and Kanya-Forstner, *France Overseas*, 95.
39. Memorandum by Sykes and Picot, 5 January 1916, accompanying draft agreement of 3 January 1916, F.O. 371//2767, Foreign Office Papers, TNA.
40. Adelson, *Portrait of an Amateur*, 201.
41. Andrew and Kanya-Forstner, *France Overseas*, 96.
42. Ibid., 96–7.
43. Peter Johnson, *A Philosopher at the Admiralty: R.G. Collingwood and the First World War: A Philosopher at War*, Volume One (Upton Pyne: Imprint Academic, 2012), 45–6. Hall would be promoted rear admiral in 1917. See *The London Gazette*, 1 May 1917. For further information on Hall and his World War I intelligence activities. See also David Ramsay, *"Blinker" Hall, Spymaster: The Man Who brought America into World War I* (Stroud: Spellmount, 2009).
44. Adelson, *Portrait of an Amateur*, 201–202.
45. "Nicolson Committee Meeting, 21 January 1916," F.O. 371/2767, TNA. See also Jacob Rosen, "Captain Reginald Hall and the Balfour Declaration," *Middle Eastern Studies* 24, no. 1 (1988), 56.
46. Comments made by Captain W. Reginald "Blinker" Hall, on 12 January 1916, in Mayir Vereté, "The Balfour Declaration and its Makers," *Middle Eastern Studies* 6, no. 1 (1970), 54.
47. Jacob Rosen, "Captain Reginald Hall and the Balfour Declaration," 56–7.
48. M. Sykes to E. Gorst, 26 and 30 April 1900, quoted in Adelson, *Portrait of an Amateur*, 72–4.
49. "Nicolson Committee Meeting, 21 January 1916," F.O. 371/2767, Foreign Office Papers, TNA.
50. Adelson, *Portrait of an Amateur*, 202.
51. Hankey, *Supreme Command*, Vol. II, 466–7, 469.
52. See Historical UK Inflation Rates and Calculator at http://inflation.iamkate.com, accessed 30 January 2018.
53. *Hansard*, LXXX, 37ff. See also "Sir Mark Sykes on the Realities," *Leeds Mercury*, 16 February 1916. See Adelson, *Portrait of an Amateur*, 203.
54. The Rt. Hon. L. S. Amery, *My Political Life, Volume Two: War and Peace, 1914–1929* (London: Hutchinson, 1953), 78.
55. *The Times, Manchester Guardian* and *Daily Telegraph* for 16 February 1916. See also Adelson, *Portrait of an Amateur, 207.*
56. George Geoffrey Robinson (1874–1944) was an English journalist and editor of *The Times* from 1912 to 1919. In 1901 he went to South Africa as private secretary to Lord Milner, then South African high commissioner, and eventually went to work for the *Johannesburg Star*, where he became correspondent for *The Times*. In 1912 the owner of the *The Times*, Lord Northcliffe, hired him as the paper's editor until 1919, at which time the two had a falling out and he was replaced by Wickham Steed. After Northcliffe's death the new proprietor, John Jacob Astor (later Lord Astor) rehired him in 1923 and he served as editor of *The Times* until his retirement in 1941. He changed his surname from Robinson to Dawson following an inheritance in 1917. See *Encyclopædia Britannica* at https://www.britannica.com/biography/George-Geoffrey-Dawson, accessed 25 January 2017.
57. Adelson, *Portrait of an Amateur*, 203–204.
58. Andrew and Kanya-Forstner, *France Overseas*, 97.
59. Adelson, *Portrait of an Amateur*, 205. A. Nicolson to G. Buchanan, 23 February 1916, F.O. 882/16.

60. Sir George William Buchanan (1854–1924) GCB GCMG GCVO PC, was the son of Ambassador Sir Andrew Buchanan, Bart. and spent his entire career in the diplomatic service. In 1899 he served on the Venezuelan Boundary Commission and in 1901 he moved to Berlin, where he was first secretary at the British embassy. In 1908 he was appointed as minister in The Hague, The Netherlands, and in 1910 was appointed as the British ambassador to Russia.

61. Sergei Dmitrievich Sazonov (1860–1927) was a Russian statesman who served as foreign minister from November 1910 to July 1916. The brother-in-law of Prime Minister Pyotr Stolypin, Sazonov had served prior to this appointment as foreign minister in the London embassy and in the diplomatic mission to the Vatican, where he became the chief in 1906. In 1909 he was recalled to Petrograd and appointed assistant foreign minister, becoming foreign minister in 1910.

62. Adelson, *Portrait of an Amateur*, 206.

63. Ibid., 205–206.

64. For the full text of Grey's letter see E. Grey to P. Cambon, 16 May 1916, quoted in Leonard Stein, *The Balfour Declaration*. (New York: Simon and Schuster, 1961), 237–9.

65. Barr, *A Line in the Sand*, 32.

66. M. Hankey to E. Grey, 4 May 1916, F.O. 371/2777, Foreign Office Papers, TNA. See also Adelson, *Portrait of an Amateur*, 209.

67. Sir Mark Sykes, "Memorandum on the Asia-Minor Agreement, 14 August 1917," U DDSY 2/11/65, Sykes Papers. Adelson, *Portrait of an Amateur*, 238–9.

68. Elie Kedourie, *Arabic Political Memoirs and Other Studies*, 236. See also Elie Kedourie, *England and the Middle East*, 66.

69. Sir Mark Sykes, "Memorandum on the Asia-Minor Agreement, Secret, 14 August 1917," F.O. 371/3059, Foreign Office Papers, TNA. See also Sykes's original copy, U DDYS 2/11/65, Sykes Papers; Adelson, *Portrait of an Amateur*, 238–9.

Chapter 6   War Cabinet Secretariat

1. Storrs, *Memoirs*, 211.

2. Adelson, *Portrait of an Amateur*, 209–10. Sykes's first *Arabian Report* was issued on 28 June 1916, F.O. 371/2781, Foreign Office Papers, TNA.

3. Adelson, *Portrait of an Amateur*, 210.

4. Sir Mark Sykes, Secret, "The Problem of the Middle East," 20 June 1916, U DDSY 2/11/13, Sykes Papers. See also, IOR/L/PS/B223, India Office Records at the British Library.

5. Ibid.

6. Ibid., See also Adelson, *Portrait of an Amateur*, 210.

7. "The Sherif of Mecca and the Arab Movement" (Secret), General Staff, War Office, 1 July 1916, U DDSY2/11/14, Sykes Papers.

8. "War Committee (Secret), Evidence of Lieut.-Col. Sir Mark Sykes, Bart., M.P., at a meeting held at 10 Downing Street, on Thursday, July 6, 1916, at 11:30," U DDSY2/11/15, Sykes Papers, also CAB 42/16, Cabinet Papers, TNA. See also Adelson, *Portrait of an Amateur*, 210.

9. Ibid., For an examination of McMahon's two-year tenure in Cairo (9 January 1915–1 January 1917) and its problems, see "Let everything slide: Surprise appointment," in C.W.R. Long, *British Pro-Consuls in Egypt, 1914–1929* (Oxford: RoutledgeCurzon, 2010), 7–18.

10. "War Committee (Secret), Evidence of Lieut.-Col. Sir Mark Sykes, Bart., M.P., at a meeting held at 10 Downing Street, on Thursday, July 6, 1916, at 11:30,"

U DDSY2/11/15, Sykes Papers, also CAB 42/16, Cabinet Papers, TNA. See also Adelson, *Portrait of an Amateur*, 210.

11. "War Committee (Secret), Evidence of Lieut.-Col. Sir Mark Sykes, Bart., M.P., at a meeting held at 10 Downing Street, on Thursday, July 6, 1916, at 11:30," U DDSY2/11/15, Sykes Papers, also CAB 42/16, Cabinet Papers, TNA.

12. Ibid.

13. Ibid.

14. Ibid.

15. Ibid.

16. Ibid.

17. Diary entry for 21 June 1916, HNKY 1/1 – Diary for 1915–17, Archives of Lord Hankey of the Chart (Maurice Hankey); hereafter cited as the Hankey Papers, Churchill Archives Centre.

18. "Minutes of meeting of the War Committee, July 6, 1916, Appendix 1, memorandum by Sykes," CAB 42/16, Cabinet Papers, TNA. Unedited draft of memorandum, dated 7 July 1916, U DDSY2/11/16, r.

19. M. Sykes to A. Chamberlain, 7 July 1916, U DDYS 2/12/4, untitled, undated and edited handwritten letter, Sykes Papers.

20. Minutes of War Committee, 11 July 1916, CAB 42/16, Cabinet Papers, TNA. See also Adelson, *Portrait of an Amateur*, 210–11.

21. Adelson, *Portrait of an Amateur*, 210. Marmaduke William Pickthall (1875–1936) was a British novelist and journalist who had travelled extensively in the East and had gained a reputation as a Middle Eastern scholar. As a result of his studies of the Orient, he published articles and novels on the subject and was a strong ally of the Ottoman Empire. In 1915, when a propaganda campaign was launched in Britain over the Armenian Massacre, Pickthall publicly challenged the anti-Turkish sentiment it caused by arguing that the blame could not be placed entirely on the Ottoman government. His controversial stand was not popular. It was considered courageous in the wartime climate, but it resulted in much criticism. In 1917 Pickthall converted to Islam and was known afterwards as Muhammad Marmaduke Pickthall. See Peter Clark, *Marmaduke Pickthall: British Muslim* (London: Quartet Books, 1986).

22. Christopher [Simon] Sykes, *Two Studies in Virtue* (London: Collins, 1953), 206 and n. 4. Christopher Hugh Sykes (1907–1986) was Sir Mark's second son, whose own colourful career included postings in Berlin, Tehran and Cairo in the Foreign Service (the latter during World War II), and in France with the French Resistance during World War II. After the war he was a BBC radio writer and producer, as well as writing for several British and American periodicals, including *The Spectator, The New Republic* and *The Observer*. He also wrote about his father, Israel and Zionism in *Crossroads to Israel* (London: Collins, 1965).

23. Adelson, *Portrait of an Amateur*, 211.

24. Clark, *Pickthall*, 31.

25. Anne Fremantle, *Loyal Enemy* (London: Hutchison & Co, 1938), 275–6. For a detailed description of the episode and copies of the letters between Pickthall and Sykes on the matter, between 25 May 1916 and 10 August 1916, see Ibid., 270–9. See also Clark, *Pickthall*, 31.

26. James Bryce, 1st Viscount Bryce, OM GCVO PC FRS FBA (1838–1922) was a British academic, jurist, historian and Liberal politician. An MP from 1880 to 1907, he also served as under secretary of state for Foreign Affairs (1886), chancellor of the Duchy of Lancaster (1882–1894), president of the Board of Trade (1894–1895), chief secretary for Ireland (1905–1907) and ambassador to the United States (1907–1913). Following the

outbreak of World War I he was commissioned by Prime Minister Asquith to give the official *Bryce Report*, describing German atrocities in Belgium, much of which was later proved to be propaganda using the famous scholar's name. Bryce also strongly condemned the Armenian genocide in the Ottoman Empire in 1915 and was the first to speak on the subject in the House of Lords. He also served as president of the British Academy from 1913 to 1917.

27. Viscount Bryce, *The Treatment of the Armenians in the Ottoman Empire, 1915–1916: Documents Presented to Viscount Grey of Fallodon, Secretary of State for Foreign Affairs, Presented to Both Houses of Parliament by Command of His Majesty, October 1916* (London: G.P. Putnam's Sons, for His Majesty's Stationery Office, London, 1916). See also Adelson, *Portrait of an Amateur*, 212.

28. *Drang nach Osten* is a German phrase, coined in the nineteenth century, meaning "push eastward," or "yearning for the east," used to designate German expansion into Slavic lands inhabited by German-speaking people.

29. Adelson, *Portrait of an Amateur*, 212.

30. "Speech by Sir Mark Sykes: 'After the War,'" *Feilding Star, Oroua & Kiwitea Counties Gazette*, Volume XII, Issue 3089, 9 November 1916, p. 4.

31. Ibid.

32. A. Mahan, "The Persian Gulf and International Relations," *National Review* (London), September 1902. See also Roger Adelson, *London and the Invention of the Middle East: Money, Power, and War, 1902–1922* (New Haven, CT: Yale University Press, 1995), 22–3. Adelson, *Portrait of an Amateur*, 212.

33. Ibid., 212. See also Johnson, *A Philosopher at the Admiralty*, Vol. 1, 46–7 for information on Professor H.N. Dickson and I.D. 32, the Geographical Intelligence section of the Admiralty, which he proposed and of which he was the director.

34. "Sir Mark Sykes to the Committee of Imperial Defence, September 10, 1916, CAB 42/20/1," Cabinet Papers, TNA, and U DDSY2/4/113, Sykes Papers. See also Adelson, *Portrait of an Amateur*, 212.

35. Eugene Rogan, *The Fall of the Ottomans*, 297. See also Eliezer Tauber, *The Arab Movements in World War I* (London: Frank Cass, 1993), 80–2.

36. Sir Kinahan Cornwallis, GCMG, CBE, DSO (1883–1959) was a British administrator and diplomat. He was in the Sudan Civil Service, 1908; the Egyptian Civil Service, 1912; Intelligence, General Headquarters, Cairo, 1914; Arab Bureau, 1915; and director of the Arab Bureau, 1916–1920. He had been deputy director of the Bureau under D.G. Hogarth.

37. Polly A. Mohs, *Military Intelligence and the Arab Revolt: The First Modern Intelligence War* (Abington: Routledge, 2008), 40.

38. Mohs, *Military Intelligence and the Arab Revolt*, 40.

39. Antonius, *Arab Awakening*, 195–200. The description of the fall of Mecca was a translation from 14–17 August issues of the newspaper *al-Qibla* by Antonius.

40. Antonius, *Arab Awakening*, 195–9. For more details of the British naval bombardment of Jeddah, see Mohs, *Military Intelligence and the Arab Revolt*, 40–1.

41. Ibid., 41.

42. Antonius, *Arab Awakening*, 201.

43. Ibid., 202–203.

44. Major-General Sir Frederick Maurice, 1st Baronet (1871–1951), was a British general, military correspondent writer, and academic. At the outbreak of war in 1914 he was posted to France where he saw action at the Battle of Mons in August 1914. In early 1915 he was posted to London as director of Military Operations for the Imperial General Staff and in 1916 promoted to major general. He worked closely with Field Marshall

Sir William Robertson, who was promoted to chief of the Imperial General Staff at the end of 1915. Maurice was forced to resign his commission in May 1918, after writing a letter to *The Times* criticising Prime Minister David Lloyd George for making misleading statements in the House of Commons about the strength of British forces in France. He founded the British Legion in 1920 and served as its president from 1932 to 1947.

45. M. Sykes to the director of Military Operations, War Office, 8 August 1916, U DDSY2/11/18, Sykes Papers.

46. M. Sykes to P. Cox, 21 August 1916, 9 Buckingham Gate, U DDYS 2/4/107, Sykes Papers.

47. Ibid., A staunch Unionist long interested in Irish affairs from his time there in 1904–1905 as private secretary to George Wyndham, chief secretary for Ireland, Sykes saw Irish nationalists Sinn Fein as the enemy and here refers to other nationalists as being "Sinn Feiners," or subject to Sinn Feinism, that is, anti-British nationalism.

48. M. Sykes, "Summary of the Arab Situation," 30 August 1916, U DDYS 2/11/20, Sykes Papers.

49. Ibid.

50. Ibid.

51. Ibid.

52. Ibid.

53. H.H. Dodwell,. ed., *The Cambridge History of India, Vol. VI: The Indian Empire 1858–1918, With Chapters on Development of Administration 1818–1858* (Cambridge: University Press, 1932), 580.

54. M. Sykes to J. Hewett, 30 September 1916, 9 Buckingham Gate, U DDYS 2/4/117, Sykes Papers. Wilhelm Wassmuss (1880–1931) was a German agent working in the Middle East during World War I, operating in the Persian Gulf area and Persia. He came to be known as "Wassmuss of Persia" and was compared with Lawrence of Arabia by Christopher [Simon] Sykes (son of Sir Mark) in his book, *Wassmuss: The German Lawrence* (Longmans, Green and Co., 1936). When he escaped, Wassmuss left his code book behind, which later was used to translate German dispatches, in particular the Zimmerman Telegram, the translation of which would bring the USA into the war against Germany in 1917. He was eventually captured and turned over to British authorities at the end of the war. See Barbara Tuchman, *The Zimmerman Telegram* (London: Phoenix Press, 1958), 19–21. See also Peter Hopkirk, *Like Hidden Fire: The Plot to Bring Down the British Empire*, 105–20, 167–78, etc. See also Winstone, *Illicit Adventure*, 94, 137ff, 167–71, 237–8.

55. M. Sykes, Untitled Memorandum, 4 October 1916, 9 Buckingham Gate, U DDSY 2/11/22, Sykes Papers.

56. Ibid.

57. M. Sykes, Untitled Memorandum, 12 October 1916, U DDYS 2/11/24, Sykes Papers.

58. Wakfs, or Waqfs, *Waqaf* (Arabic plural) are Muslim perpetual foundations to which Muslims give charity during Ramadan and at other times, out of which the upkeep of mosques and other religious institutions is paid, the poor, orphans and travellers are fed and clothed, the salaries of religious clergy are paid and many other functions sustaining the financial needs of the Muslim community, of which it is its financial lifeblood.

59. M. Sykes, Untitled Memorandum, 12 October 1916, U DDYS 2/11/24, Sykes Papers.

60. Ibid.

61. M. Sykes to C. Hardinge, 10 October 1916, 9 Buckingham Gate, U DDYS 2/11/23, Sykes Papers.

62. Diary entries for 6 and 7 November 1916, HNKY 1/1 – Diary for 1915–17, Hankey Papers, Churchill Archives Centre.

63. M. Sykes to C. Hardinge, 7 November 1916, *Hardinge Papers*, Vol. 27, f. 81; C. Hardinge to E. Grey (n.d., c. 7 November 1916), ibid., Vol. 27, f. 83 recto; and Grey's minute on Hardinge's note (n.d.), ibid., Vol. 27, f. 83 verso.

64. Frederick John Napier Thesinger, 1st Viscount Chelmsford (1868–1933) succeeded Lord Hardinge as Viceroy of India, where he served from 1916 to 1921.

65. Sir Thomas William Holderness, 1st Baronet (1849–1924) was the first former member of the Indian Civil Service to be appointed permanent under secretary of state for India. On his retirement from the Indian Civil Service in 1901 he joined the India Office as secretary of the Revenue, Statistics and Commerce Department, and in 1912 became the permanent under secretary, the professional head of the India Office, continuing in that post until his retirement in 1919.

66. "The Future Administration of Mesopotamia SECRET," U DDSY 2/4/120, Sykes Papers.

67. For Sir Percy Cox's time as high commissioner in Iraq, see Townsend, *Proconsul to the Middle East*, 139–89.

68. Lord Hankey, *The Supreme Command 1914–1918*, Vol. Two (London: George Allen and Unwin Limited, 1961), 565–73. See Robert Blake, *The Unknown Prime Minister: The Life and Times of Andrew Bonar Law 1858–1923* (London: Eyre & Spottiswoode, 1955), 338. See also Stephen Roskill, *Hankey: Man of Secrets*, Vol. 1: 1877–1918 (London: Collins, 1970), 329.

69. Hankey, *Supreme Command*, Vol. II, 582.

70. Amery was a member of a coterie of young Oxford graduates called Lord Milner's kindergarten, who served "under Milner in South Africa in the reconstruction of the two former Boer republics . . . [and] remained after restoration of self-government in 1906 to work for Milner's goal of a united South Africa loyal to the British Empire. With the creation of the Union of South Africa in 1909 . . . [they] turned their attention to a much greater goal . . . [and] organized a movement to achieve a similar sort of organic unity for the whole Empire. This movement – the round table movement – sought to establish an imperial government responsible to all the Dominions as well as to Great Britain, a government having control over matters of common interest to the otherwise independent member nations." Walter Nimocks, *Milner's Young Men: The "Kindergarten" in Edwardian Imperial Affairs* (Durham, NC: Duke University Press, 1968), vii.

71. Hankey to Lloyd George, 14 December 1916, LG/F/23/1/2, Lloyd George Papers, Parliamentary Archives. See also Adelson, *Portrait of an Amateur*, 215.

72. M. Sykes to Sir Edward? 9 Buckingham Gate, SW (private), 7 December 1916, *Lloyd George Papers*, LG/F/46/6, Parliamentary Archives.

73. Roskill, *Hankey*, Vol. 1, 344–5. The Rt. Hon. L. S. Amery, *My Political Life*, Volume Two, 92.

Chapter 7　The Zionists and a Jewish Homeland

1. Chaim Weizmann, *Trial and Error: The Autobiography of Chaim Weizmann* (New York: Harper & Brothers, 1949), 181. Actually, Sykes was not chief secretary but one of two assistant secretaries of the War Cabinet Secretariat working under War Cabinet Secretary Sir Maurice Hankey.

2. Herbert Louis Samuel, 1st Viscount Samuel GCB OM GBE PC (1870–1963), was a British Liberal politician and MP, and was the first nominally practising Jew to serve as a cabinet minister.

3. Herbert Samuel, "The Future of Palestine," 25 January 1915, CAB 37/123/43, Cabinet Papers, TNA.

4. Ibid.

5. Ibid.
6. Ibid.
7. Earl of Oxford and Asquith, KG *Memories and Reflections 1852–1927*, Vol. Two (Boston: Little, Brown and Company, 1928), 70–1. *Tancred*, or *The New Crusade* (1847), was part of a trilogy written by the future Prime Minister Benjamin Disraeli, including *Coningsby* (1844) and *Sybil* (1845). *Tancred* had a religious and almost mystical theme, revolving around how Judaism and Christianity could be reconciled and the church could become a progressive force.
8. M. Sykes to H. Samuel, 26 February 1916, quoted in Leonard Stein, *The Balfour Declaration* (New York: Simon and Schuster, 1961), 233–4. See also ibid., 234–5; Adelson, *Portrait of an Amateur*, 204; and Isaiah Friedman, *The Question of Palestine: British-Jewish-Arab Relations: 1914–1918*. Second expanded edn (New Brunswick, NJ: Transaction Publishers, 1992), 112.
9. Mark Levene, *War, Jews, and the New Europe: The Diplomacy of Lucien Wolf 1914–1919* (Oxford: Oxford University Press, 1992), 83. See also Stein, *Balfour Declaration*, 218–24, for a description of Wolf's dealings with the Foreign Office at this time and his March formula.
10. Eugene C. Black, *The Social Politics of Anglo-Jewry 1880–1920* (Oxford: Basil Blackwell Ltd, 1988), 342.
11. Edgar Suarès was a member of a prominent Jewish Alexandrian family. As shareholders, directors and managers, the Suarès family were involved in a large number of companies in all sectors of the Egyptian economy. Edgar Suarès served as president of the Jewish community from 1914 to 1917. A non-Zionist, he was anxious to see Palestine open to Jewish immigration and that "the management of its internal affairs should be in the hands of Jews ... under British protection."
12. E. Suarès to H. McMahon, 23 February 1916, quoted in Mayir Vereté, "The Balfour Declaration and its Makers," 48–76. See also Levene, *War, Jews, and the New Europe*, 94–5.
13. Hugh James O'Beirne (1866–1916) was an Irish-born and Oxford-educated senior official at the Foreign Office. He had served as former counselor and chargé d'affaires at the embassies in St Petersburg and Sofia prior to returning to London to serve in the Foreign Office after Bulgaria entered the war on the side of the Central Powers. He died with Lord Kitchener on a trip to Russia when a mine sank their ship HMS *Hampshire* on 5 June 1916.
14. Robert Crewe-Milnes, 1st Marquess of Crewe (1858–1945) was a British statesman and writer. A Liberal, Crewe became private secretary to Lord Granville when Granville was secretary of state for Foreign Affairs (1883–1884). In the Liberal administration of 1892–1895 he was lord lieutenant of Ireland. From 1905 to 1908 he was lord president of the Council in the Liberal government and in 1908 he became secretary of state for the colonies in Asquith's cabinet (1908–1910) and leader of the House of Lords. He served as secretary of state for India twice (1910–1911, 1911–1915) and as lord president of the Council again in 1915–1916. Unwilling to serve in the Lloyd George administration in 1916, he led the independent Liberal opposition in the House of Lords. He served as chairman of the governing body of Imperial College London (1907–1922), president of the Board of Education (1916) and chancellor of Sheffield University. He was also chairman of the London Council in 1917. He later served as ambassador to France (1922–1928) and secretary of state for war (1931).
15. Levene, *War, Jews, and the New Europe*, 95. See Ronald Sanders, *The High Walls of Jerusalem*, 332–44, for an in-depth description of the discussion of Wolf and the Zionists at the Foreign Office.

16. Vereté, "Balfour Declaration and its Makers."
17. Levene, *War, Jews, and the New Europe*, 95. O'Beirne drafts and letters to Bertie (in France) and Buchanan in St Petersburg, 11 March 1916, F.O. 371/2817/42608. See also Sanders, *The High Walls of Jerusalem*, 341–2.
18. F.O. 371/2767/938, G. Buchanan to Foreign Office, March 12, 17, 1916 telegram nos. 351, 370. 382, Secret, quoted in Friedman, *The Question of Palestine*, 113, n. 60. See also Sanders, *The High Walls of Jerusalem*. 344.
19. Mark Levene, *War, Jews, and the New Europe*. See H. O'Beirne to Buchanan, 11 March 1916, F.O. 371/2817/42608, Foreign Office Papers, TNA. For more on Lucien Wolf's early relationship with Zionism, see Josef Fraenkel, *Lucien Wolf and Theodor Herzl* (London: Jewish Historical Society of England, 1960).
20. M. Sykes to A. Nicholson, 18 March 1916, Nicolson Papers, F.O. 800/381, Foreign Office Papers, TNA. See Adelson, *Portrait of an Amateur*, 206–7. See also Friedman, "Jewish Palestine – A Propaganda Card," in *The Question of Palestine*, 48–64, for an in-depth discussion of the Foreign Office involvement in deliberations on Zionism and a Jewish community in Palestine in the early years of the war. See Ronald Sanders, *The High Walls of Jerusalem*, 345–51, for a detailed description of the furor at the Foreign Office caused by Sykes's letter.
21. Stein, *Balfour Declaration*, 275–6.
22. Ibid., 207.
23. Ibid., 287–8.
24. Adelson, *Portrait of an Amateur*, 207.
25. Chaim Weizmann, *Letters and Papers of Chaim Weizmann*, Vol. VII, Series A, *August 1914–November 1917*, Meyer W. Weisgal, gen. ed.; Leonard Stein, vol. ed., in collaboration with Dvorah Barzilay and Nehama A. Chalom (London and Jerusalem: Oxford University Press and Israel Universities Press, 1975), 324; hereafter referred to as the *Weizmann Papers*.
26. *Diary of Haham Moses Gaster*, 10 May 1916, *Gaster* Papers, quoted in Stein, *Balfour Declaration*, 288, n. 1.
27. *Diary*, 7 July 1916, *ibid.*, 288, n. 11.
28. *Diary*, 24 May and 3 July 1916, ibid., 288, n. 12.
29. H. Sacher, ed., *Zionism and the Jewish Future by various writers* (New York: The Macmillan Company, 1916). M. Gaster to M. Sykes, 3 July 1916, DDSY2/4/203, *Sykes Papers*.
30. Adelson, *Portrait of an Amateur*, 213. See also Anthony Verrier, ed., *Agents of Empire: Anglo-Zionist Intelligence Operations 1915–1919, Brigadier Walter Gribbon, Aaron Aaronsohn and the NILI Ring* (London: Brassey's (UK) Ltd, 1995), 101.
31. Sachar, Howard M., *A History of Israel: From the Rise of Zionism to Our Time* (New York: Alfred A. Knopf, 2002), 103.
32. Sachar, *History of Israel*, 104. See also Anita Engle, *The NILI Spies* (London: Hogarth Press, 1959; reprint, London: Routledge, 2007), 98–9.
33. Sachar, *History of Israel*, 103–104. Engle, *The NILI Spies*, 82–90.
34. Brigadier Walter Harold Gribbon, CMG, CBE (1881–1944) was a career soldier and officer, who by this time had served in Burma, India, and Mesopotamia before the outbreak of World War I. "He was posted to the War Office, Military Intelligence Directorate, to work on the Turkish position in Palestine threatening the British imperial line of communication to India, East Africa, and the Far East." Verrier, ed., *Agents of Empire*, xiv–xv.
35. Ronald Florence, *Lawrence and Aaronsohn: T.E. Lawrence, Aaron Aaronsohn, and the Seeds of the Arab-Israeli Conflict* (New York: Viking, 2007), 3–8, 21–9. For a detailed description of Aaronsohn's background, his trip to Britain, meeting with Sykes and others, see Shmuel Katz, *The Aaronsohn Saga*. (Jerusalem: Gefen Books, 2007), 95–123, See

Friedman, *The Question of Palestine*, 120–3; Verrier, *Agents of Empire*, 2–4; and Engle, *The NILI Spies*, 71–3.

36. Katz, *Aaronsohn Saga*, 112–13.

37. Ibid., 115–16.

38. Friedman, *Question of Palestine*, 122; also Entry 30 October 1916, 120–1, *Yoman Aaron Aaronsohn* [Diary of Aaron Aaronsohn] (Hebrew) (Tel Aviv, 1970), referred to hereafter as *Aaronsohn's Diary*, quoted in Friedman, *Question of Palestine*, 336, note 3.

39. *Aaronsohn's Diary*, entry 27 April 1917, 251, quoted in Friedman, *Question of Palestine*, 366, n. 7.

40. Norman and Helen Bentwich, *Mandate Memories: 1918–1948* (London: Schocken Books, 1965), 13.

41. Howard M. Sacher, *A History of the Jews in the Modern World* (New York: Alfred A. Knopf, 2005), 52–5, 61–5, and 193–206. See Britain's Ambassador to the United States Sir Cecil Spring-Rice's cable to Lord Robert Cecil describing this anti-Russian feeling among many American Jews, in C. Spring-Rice to R. Cecil, 2 February 1916, FO 371/2579/187779, Foreign Office Papers, TNA.

42. Sacher, *History of the Jews*, 255–83.

43. Alyson Pendlebury, *Portraying "the Jew" in First World War Britain* (London: Valentine Mitchell, 2006), 119.

44. Simon Schama, *Two Rothschilds and the Land of Israel* (New York: Alfred A. Knopf, 1978), 204. See also *Weizmann Papers*, xxii.

45. David Lloyd George, *War Memoirs of David Lloyd George*, Vol. I (London: Odhams Press Limited, 1938), 348–9, hereafter referred to as Lloyd George, *War Memoirs*. Lloyd George's story that as prime minister he rewarded Weizmann for his work with "a Jewish homeland" in Palestine is considered gross hyperbole and pure fiction. On this, see also Blanche E. C. Dugdale, *Arthur James Balfour* (New York: G.P. Putnam & Sons, 1937), 165, n. 1; Viscount Herbert Samuel, *Memoirs* (London: The Cresset Press, 1945), 145–6; and Regina S. Sharif, *Non-Jewish Zionism* (London: Zed Press, 1983), 79.

46. For an extensive review of this transitional period of Zionist history, see "The Interregnum" in Walter Laqueur, *A History of Zionism: From the French Revolution to the Establishment of the State of Israel* (New York: MJF Books, 1972), 136–205.

47. Sachar, *A History of Israel*, 24–35. Jewish emigration from Europe to Palestine began in the 1830s. However, there was an upsurge in numbers between 1882 and 1903, during which time 25,000 Jews entered Palestine for the purposes of settling there. This has been called the "First Aliyah – the first emigration wave." Ibid., 26.

48. Theodor Herzl, *The Jewish State*, Jacob M. Alkow, ed. and trans (New York: Dover Publications, 1988), 159. See also Laqueur, *A History of Zionism*, 84, 87–8.

49. Alex Bein, *Theodore Herzl: A Biography* (New York: Athenaeum, 1970), 179.

50. Ibid., 273–5.

51. Herzl, *The Jewish State*, 93.

52. Ibid., 95.

53. Ibid., 98.

54. See "Chapter 5: The Sykes–Picot agreement."

55. Adelson, *Portrait of an Amateur*, 217–18.

56. Shane Leslie, *Mark Sykes*, 22–4, 259. Stein, *Balfour Declaration*, 270.

57. Stein, *Balfour Declaration*, 271.

58. John Grigg, *Lloyd George: From Peace to War 1912–1916* (London: Methuen London Ltd, 1985), 489.

59. Adelson, *Portrait of an Amateur*, 218.

60. Ibid., 219. For a detailed history of the formation of the Jewish Legion during this period, see Martin Watts, *The Jewish Legion and the First World War* (London: Palgrave Macmillan, 2004), 85–116.

61. Weizmann, *Trial and Error*, 181.

62. David Cesarani, *The Jewish Chronicle and Anglo-Jewry 1841–1991* (Cambridge: Cambridge University Press, 1994), 123–4. Adelson, *Portrait of an Amateur*, 219–20. See also James A. Malcolm, *Origins of the Balfour Declaration: Dr. Weizmann's Contribution*, unpublished manuscript (1944), British Library. Malcolm wrote this twelve-page recollection of the events of 1916–1917 in 1944 and donated it to the British Library.

63. Adelson, *Portrait of an Amateur*, 219–20.

64. Ibid., 220. In *Trial and Error*, 188, Weizmann claimed the meeting was held on 17 February 1917, while Adelson in *Portrait of an Amateur*, lists the date as 7 February 1917. In view of Weizmann's letters written after the meeting in response to Sykes's request about the Jewish Legion, it seems Adelson's date is correct, so I have used it.

65. The British Palestine Committee, formed towards the end of 1916 in Manchester, began publishing the weekly *Palestine* on 26 January 1917. *Weizmann Papers*, 326, n. 5.

66. Ibid., 325. The conference referred to was the meeting Sykes requested to be held at Dr Gaster's home on 7 February.

67. Ibid., 333, n. 3.

68. Weizmann, *Trial and Error*, 188. James de Rothschild (1878–1957) was the eldest son of French Baron Edmond de Rothschild and an ardent Zionist. After attending Cambridge University he settled in England and married an English wife. He became active in the British Zionist movement. At the beginning of the war he would serve as an officer in the French Army, but ended the war as an officer in the British Army, serving as a major in the 39th (Jewish) Battalion, the Royal Fusiliers, as part of the Jewish Legion. See Simon Schama, *Two Rothschilds and the Land of Israel*). See also Dan Cohn-Sherbok, *Dictionary of Jewish Biography* (London: Continuum, 2005), 246. Joseph Cowan (1868–1932) was an early British Zionist, a friend and collaborator of Theodor Herzl and later of Weizmann. He was a member of the Inner Actions Committee of the World Zionist Organization, member of the Jewish Colonial Trust, member of the board of directors of the *Jewish Chronicle*, and president of the British Zionist Federation at the time. See David Cesarani, *The Jewish Chronicle and Anglo-Jewry 1841–1991*, 103–105. For more on Cowan and his extensive Zionist activities, see his obituary in "Death of Joseph Cowan," *Jewish Telegraphic Agency*, 26 May 1932. Herbert Bentwich (1856–1932) was a British Zionist leader and lawyer. One of Herzl's first followers, he was founder of the British Zionist Federation in 1899, and served as its vice-chairman. He was legal advisor to the Jewish Colonial Trust and from 1916 to 1918 served on the Zionist political advisory committee under Chaim Weizmann. See Cohn-Sherbok, *Dictionary of Jewish Biography*, 30. Harry Sacher (1882–1971) was a prominent British lawyer and Zionist leader. He wrote for the *Manchester Guardian* as a political analyst and was a close associate of editor C.P. Snow and Weizmann. He was elected to the executive of the World Zionist Organization and worked closely with Weizmann in that organisation during the 1920s and 1930s. He contributed to early drafts of the Balfour Declaration. Later, he was involved in the establishment of the Hebrew University.

69. Fromkin, *Peace*, 286. Weizmann, *Trial and Error*, 188–9.

70. Adelson, *Portrait of an Amateur*, 220.

71. Simcha Kling, *Nachum Sokolow: Servant of his People* (New York: Herzl Press, 1960), 103. Weizmann, *Trial and Error*, 190.

72. Adelson, *Portrait of an Amateur*, 220.

73. Schama, *Two Rothschilds*, 206.
74. Adelson, *Portrait of an Amateur*, 220–1. Shortly afterwards, the Zionists took Sykes up on his offer and, by the end of 1917 over 200 letters and telegrams from or to the Zionist leaders had been transmitted by the Military Intelligence Directorate of the War Office. See Stein, *Balfour Declaration*, 377.
75. Adelson, *Portrait of an Amateur*, 221.
76. Stuart A. Cohen, *English Zionists and British Jews: The Communal Politics of Anglo-Jewry, 1895–1920* (Princeton, NJ: Princeton University Press, 1982), 222, n. 13.
77. Kling, *Sokolow*, 103–104.
78. Ibid., 103–104.
79. Ibid., 105.
80. Memorandum by N. Sokolow, undated (c. February 1917), UDDYS 2/4/203, *Sykes Papers*. See also N. Sokolow to M. Sykes, Regent Palace Hotel, Piccadilly Circus, London, 17 February 1917, UDDSY 2/4/203, Sykes Papers.
81. Adelson, *Portrait of an Amateur*, 222. See also *Weizmann Papers*, 330, n. 1 and 2.
82. Adelson, *Portrait of an Amateur*, 222.
83. Weizmann, *Trial and Error*, 188.
84. M. Sykes to F. Georges-Picot, 28 February 1917, Sykes Papers; F. Georges-Picot to M. Sykes, 26 March 1917, Sykes Papers, quoted in Adelson, *Portrait of an Amateur*, 222.
85. Adelson, *Portrait of an Amateur*, 222.
86. Alyson Pendlebury, *Portraying "the Jew" in First World War Britain*, 120.
87. *Weizmann Papers*, 344–5. Well-connected, active in politics and a former Liberal MP in addition to being editor of the *Manchester Guardian* (1872–1929) and owner (1905–1932), C.P. Scott was an early supporter of Zionism. He met Weizmann at the beginning of the war and the two become close friends. As a result of their friendship Scott introduced Weizmann to many important people, including Lloyd George. See Trevor Wilson, ed., *The Political Diaries of C.P. Scott 1911–1928: A Unique Record of the Lloyd George Years by the Great Editor of the Manchester Guardian* (Ithaca, NY: Cornell University Press, 1970), 113, 128, 159ff. See also Weizmann, *Trial and Error*, 148–50, 179.
88. Ibid., 345, n. 9.
89. William George Arthur Ormsby-Gore, 4th Baron Harlech (1885–1964), was known as William Ormsby-Gore until 1938, when he inherited his title on his father's death. In 1913 he married Lady Beatrice Gascoyne-Cecil, sister of Lord Robert Cecil and first cousin of Arthur Balfour. He was a British Conservative politician, serving in the House of Commons as an MP (1910–1938), when he succeeded to his title as Lord Harlech on his father's death and entered the House of Lords. In a long and illustrious career, he served as under-secretary of state for the colonies (1922–1924, 1924–1929), and as the British representative to the Permanent Mandates Commission of the League of Nations (1921–1922). He also served as postmaster-general (1931), the first commissioner of works (1931–1936), and colonial secretary (1936–1938), until he succeeded to his title and entered the House of Lords. He served as British High Commissioner to South Africa (1941–1944). After retiring from politics, Lord Harlech entered the family's Midland Bank and authored several books on art and architecture. See http://www.thepeerage.com/p4596.htm#i45952, accessed 29 January 2017.
90. Westrate, *Arab Bureau*, 41. See Adelson, *Portrait of an Amateur*, 145; Katz, *Aaronsohn Saga*, 139; and also Scott Anderson, *Lawrence in Arabia: War, Deceit, Imperial Folly and the Making of the Modern Middle East* (London: Atlantic Books, 2013), 254.
91. W. Ormsby-Gore to M. Sykes, 26 March 1917, UDDYS 2/4/203, Sykes Papers.
92. Ibid.

93. Sykes was appointed chief political officer to the general commanding, Egypt on 7 March 1917. F.O. 371/3045, Foreign Office Papers, TNA. Adelson, *Portrait of an Amateur*, 223.

94. J. Greenburg to M. Sykes, 21 March 1917, 2 Finsbury Square, London, UDDYS2/11/34, Sykes Papers.

95. *Weizmann Papers*, 349, n. 1.

96. Nahum Sokolow, *History of Zionism, 1600–1918* (New York: KTVA Publishing House, 1969), Vol. II, xxx.

97. Sanders, *The High Walls of Jerusalem*, 498.

98. Alexandre-Félix-Joseph Ribot (1842–1923) was a French politician who served four times as prime minister (1892–1893, 1895, 1914, 1917) and also served as minister of foreign affairs (1890–1893, 1917), minister of finance (1895, 1914–1917), minister of justice (1914) and minister of the interior (1893).

99. Sanders, The High *Walls of Jerusalem*, 499.

100. M. Sykes to A. Balfour, 8 April 1917, Hotel Lotti, Paris, FO 800/310, Foreign Office Papers, TNA. See also Jon Kimche, *The Second Arab Awakening* (London: Thames and Hudson, 1970), 60–1.

101. Martin Sicker, *Between Hashemites and Zionists: The Struggle for Palestine 1908–1988* (New York: Holmes & Meier), 39–40.

102. M. Sykes to A. Balfour, 8 April 1917, Hotel Lotti, Paris, F.O. 800/310, Foreign Office Papers, TNA.

103. Sanders, *The High Walls of Jerusalem*, 501.

104. Ibid., 501.

105. Sokolow, *Zionism*, 52. For a summary of Sokolow's visit to Paris, based on Sokolow's letters to Weizmann, see also Weizmann's letter to Harry Sacher, dated 22 April 1917, *Weizmann Papers*, 368.

106. Stein, *Balfour Declaration*, 404. See *Weizmann Papers*, 360, n. 2 and 3. See also M. Sykes to Balfour, 9 April 1917, Hotel Lotti, Paris, UDDYS 2/12/7, Sykes Papers.

107. M. Sykes to R. Graham, 9 April 1917, Hotel Lotti, Paris, UDDYS 2/12/7, Sykes Papers.

108. Sir John Francis Charles de Salis, 7th Count de Salis-Soglio (1864–1939), was a long-time British diplomat. In June 1888 he served as an attaché at the British Embassy in Brussels and was promoted third secretary the following year. He was posted to Madrid in 1892 and promoted to second secretary in 1893. In August 1894 he served in Cairo under Lord Cromer until 1897, when he was posted to Berlin. He was posted to Brussels in 1899 and Athens in 1901. He returned to London to work in the Foreign Office from 1901 to 1906, and was promoted to first secretary in 1904. He returned to Berlin as chargé d'affaires and counsellor at the Embassy from 1906 to 1911. From 1911 to 1916 he was envoy extraordinary and minister plenipotentiary to the King of Montenegro and then envoy extraordinary and minister plenipotentiary on a special mission to the Holy See from 1916 to 1923.

109. Eugenio Maria Giuseppe Giovanni Pacelli (1876–1958) was to become Pope Pius XII (r. 1939–1958) and remained in office until his death. Prior to that Pacelli served the Catholic Church in numerous capacities, including as secretary of the Department of Extraordinary Ecclesiastical Affairs (1914–1917), papal nuncio to Germany (1917–1929) and cardinal secretary of state (1930–1939). In the latter capacity he worked to conclude treaties with European and Latin American nations, most notably the *Reichskonkordat* with Nazi Germany, by which most historians believe the Vatican sought to protect the Church in Germany. A prewar critic of Nazism, Pius XII lobbied world leaders to avoid war, as had his predecessor Pope Benedict XV (r. 1914–1922) during World War I. However, as pope, at the outbreak of World War II Pius XII issued the *Summi Pontificatus*, expressing dismay at the invasion of Poland, reiterating Church teaching against racial persecution and calling

for love, compassion and charity to prevail over war. For a look at the extensive Church career of Eugenio Pacelli, later Pope Pius XII, see John Julius Norwich, *The Popes: A History* (London: Vintage Books, 2012), 415–36.

110. M. Sykes to R. Graham, 15 April 1917, Hotel Excelsior, Rome, UDDYS2/12/7, Sykes Papers. Sanders, *The High Walls of Jerusalem*, 501.

111. Ibid..

112. Ibid.

113. M. Sykes to N. Sokolow, 14 April 1917, Hotel Excelsior, Rome, UDDYS2/12/7, Sykes Papers. Sanders, *The High Walls of Jerusalem*, 503.

114. *Weizmann Papers*, 365, n. 1.

115. Sanders, *The High Walls of Jerusalem*, 503. Sanders wrote that Gasguet was "the French Envoy to the Vatican," which is incorrect because France had no diplomatic relations with the Vatican at this time and Cardinal Gasquet was English. After a dispute with Pope Pius X, France cut off diplomatic relations with the Holy See in 1904 and did not renew them until 1920–1921. See Andrew and Kanya-Forstner, *France Overseas*, 43, 155. Francis Aiden Cardinal Gasquet, OSB was a prominent English Benedictine monk and Church historian. At the time of this meeting he was archivist of the Vatican Secret Archives and in 1918 became Librarian of the Vatican Library. See Peter Guilday,"Francis Aiden Cardinal Gasquet," *Catholic World (Apr–Sep 1922)*, 210–16.

116. *Weizmann Papers*, 369, n. 4 and 5.

117. Kling, *Sokolow*, 107–108.

118. Sanders, *The High Walls of Jerusalem*, 504.

119. The agreement between France, Italy and Britain, was signed at Saint Jean de Maurienne on 26 April 1917 and endorsed from 18 August to 26 September 1917. It was drafted by the Italian foreign ministry as a tentative agreement to settle its Middle Eastern interests, and was mainly negotiated and signed by the Italian Foreign Minister Baron Sidney Sonnino, along with the Italian, British and French prime ministers. Russia was not represented in this agreement as the tsarist regime was in a state of collapse. The agreement was needed by the Allies to secure the position of Italian forces in the Middle East and to balance the loss of military power in the Middle Eastern theatre of World War I as Russian (tsarist) forces were pulling out of the Caucasus Campaign. See C.J. Lowe and M.L. Dockrill, *Mirage of Power: British Foreign Policy 1914–1922*, Vol. 2 (London: Routledge & Keegan Paul, 1972), 226–7; Sanders, *The High Walls of Jerusalem*, 504. See also David Lloyd George, *Memoirs of the Peace Conference* (New Haven, CT: Yale University Press, 1939), Vol. II, 507–11.

120. Stein, *Balfour Declaration*, 414.

121. Sokolow, *Zionism*, xxx.

122. Pietro Gasparri (1852–1934) was a Roman Catholic cardinal, diplomat and politician in the Roman Curia (1914–1930) and signatory of the 1929 Lateran Pacts, by which the Kingdom of Italy recognised the Vatican as an independent state. He served also as the cardinal secretary of state under Pope Benedict XV and Pope Pius XI.

123. Andrej Kreutz, *Vatican Policy on the Palestinian-Israeli Conflict: The Struggle for the Holy Land* (Westport, CT: Greenwood Press, 1990), 35. See also Sanders, *The High Walls of Jerusalem*, 508.

124. Sanders, *The High Walls of Jerusalem*, 508.

125. Ibid., 509.

126. Ibid., 509. See also Kreutz, *Vatican Policy*, 34.

127. Sanders, *The High Walls of Jerusalem*, 509–10.

128. Sokolow, *Zionism*, 53.

Chapter 8   Mesopotamia, Arabia and King Husayn

1. C. Hardinge to R. Rodd (Private) 18 May 1917, Hardinge Papers, Cambridge University Library.

2. See Lt. Col. Sir Arnold T. Wilson, *Loyalties Mesopotamia 1914–1917, From the Outbreak of War to the Death of General Maude: A Personal and Historical Record* (London: Oxford University Press, 1930), 206–37, for a detailed first-hand account of Lt. Gen. Maude's leaving Basra to retake Kut and then taking Bagdad. At the time Wilson was attached to the Mesopotamian Expeditionary Force as Sir Percy Cox's deputy political officer.

3. John Fisher, *Curzon and British Imperialism in the Middle East 1916–1919* (London: Frank Cass, 1999), 42. Italics are mine.

4. Ibid., 46.

5. Ibid., 62–3, n. 2. Regular attendees included Sir Austen Chamberlain, secretary of state for India and Sir Arthur Hirtzel, secretary of the Political Department, representing the India Office; Maj. Gen. Sir George Macdonogh, director of military intelligence, representing the War Office; Arthur Balfour, the foreign secretary, Lord Hardinge, permanent under secretary for Foreign Affairs, and Sir Ronald Graham, assistant under secretary of state for Foreign Affairs, representing the Foreign Office; and Lord Milner, minister without portfolio, representing the War Cabinet.

6. Wilson, *Loyalties*, 237–8. See Sir Mark Sykes, "To the People of Baghdad and Iraq" (unedited draft), U DDSY 2/11/32, Sykes Papers. See also Charles Townshend, *When God Made Hell: The British Invasion of Mesopotamia and the Creation of Iraq, 1914–1921* (London: Faber and Faber, 2010), 373.

7. Wilson, *Loyalties*, 237–9.

8. Ibid., 239.

9. Political Officer, Basra to Foreign Department, 8 March 1917, no. 1795, P1079/17, 189–190; minute by Hirtzel, 9 March 1917, 168ff, quoted in Fisher, *Curzon and British Imperialism in the Middle East*, 46, n. 16.

10. See Wilson, *Loyalties*, 237–41 for the final proclamation given by Lt. Gen. Maude and subsequent controversy surrounding it, along with various interpretations at the time. Adelson, *Portrait of an Amateur*, 223. Fisher, *Curzon and British Imperialism in the Middle East*, 46.

11. Adelson, *Portrait of an Amateur*, 224. Fisher, *Curzon and British Imperialism in the Middle East*, 46–7. See also Townshend, *When God Made Hell*, 373–4, for the story of Sykes's version.

12. "To the People of Baghdad and Irak," U DSSY 2/11/32, Sykes Papers, which includes Sykes's original and the subsequent additions and corrections. Also, see "Proclamation of Lieutenant-General Sir Stanley Maude at Baghdad," 19 March 1917, taken from *Curzon Papers* F112/256 and quoted in Fisher, *Curzon and British Imperialism in the Middle East*, Appendix 2, 305–306. See Wilson, *Mesopotamian Loyalties*, 237–9. In introducing Lt. Gen. Maude's proclamation, Wilson commented, "This document, for which the Cabinet accepted full responsibility, was drafted by Sir Mark Sykes and bears in every line the mark of his ebullient orientalism." Apparently Wilson was unaware that the proclamation was an edited version of Sykes's draft by Curzon, Milner, Hardinge, and Chamberlain. See Adelson, *Portrait of an Amateur*, 223–4, which discusses the background of Sykes's draft of the proclamation. See also Fisher, *Curzon and British Imperialism in the Middle East*, 45–7, for a detailed description of the discussion and arguments over the proclamation.

13. Adelson, *Portrait of an Amateur*, 224.

14. Wilson, *Loyalties*, 241.

15. Minutes of meeting of the War Committee, 6 July 1916, Appendix 1, memorandum by Sykes, CAB 42/16, Cabinet Papers, TNA. Again, see Sykes's unedited draft of the memorandum, dated 7 July 1916, U DDSY 2/11/16, Sykes Papers. See also Fisher, *Curzon and British Imperialism in the Middle East*, 48.

16. Mesopotamian Administrative Committee. Report of Sub-Committee appointed under paragraph 17 of the Minutes of 21st March 1917, Secret. IOR/L/PS/18/B254, India Office Records and Papers, British Library. See also Secretary of State to Viceroy, Foreign Department, 29 March 1917, CAB 24/9, Cabinet Papers, TNA. Fisher, *Curzon and British Imperialism in the Middle East*, 50–1.

17. Fisher, *Curzon and British Imperialism in the Middle East*, 49–50.

18. Adelson, *Portrait of An Amateur*, 226.

19. Notes of a Conference held at 10 Downing Street, at 3:30 p.m. on 3 April 1917, to Consider the Instructions to Lieutenant-Colonel Sir Mark Sykes, the Head of the Political Mission to the General Officer Commanding-in-Chief, Egyptian Expeditionary Force, with appendix, Status and Function of Chief Political Officer and French Commission, Secret, CAB 24/9/75, Cabinet Papers, TNA. See also Status and Functions of Chief Political Officer and French Commissioner, F.O. 882/16, Foreign Office Papers, TNA.

20. Notes of a Conference ... 3 April 1917 ... Secret, CAB 24/9, Cabinet Papers, TNA, Cabinet Papers, TNA. See also "Status and Functions of Chief Political Officer and French Commissioner, F.O. 882/16, Foreign Office Papers, TNA.

21. Adelson, *Portrait of An Amateur*, 226.

22. Notes of a Conference ... 3 April 1917 ... Secret, CAB 24/9, Cabinet Papers, TNA.

23. Anderson, *Lawrence in Arabia*, 197–9, 260–5, 406–407.

24. Timothy J. Paris, *In Defence of Britain's Middle Eastern Empire*, 243–44. See James Barr, *A Line in the Sand*, 40–1. See also Ali A. Allawi, *Faisal I of Iraq* (London: Yale University Press, 2014), 90.

25. A. Hirtzel to P. Cox, 29 December 1920, Cox Papers, Middle East Centre Archive, quoted in Barr, *Line in the Sand*, 41. See also Anderson, *Lawrence in Arabia*, 197–99.

26. Notes of a Conference ... 3 April 1917 ... Secret, CAB 24/9, Cabinet Papers, TNA. Adelson, *Portrait of An Amateur*, 226. See also Stein, *Balfour Declaration*, 384.

27. Notes of a Conference ... 3 April 1917 ... Secret, CAB 24/9, Cabinet Papers, TNA.

28. Ibid., Fisher, *Curzon and British Imperialism*, 82–3.

29. Adelson, *Portrait of An Amateur*, 184, 226. By including Armenia, the reference is to Sykes's postwar Ottoman Empire devolution scheme for the Ottoman Empire, which he proposed to the De Bunsen Committee in 1915 and remained a dream he held throughout the war. In his "The Proposed Maintenance of a Turkish Empire in Asia without Spheres of Influence," 3 May 1915, CAB 27/1, Cabinet Papers, TNA, Sykes proposed the Ottoman Empire be divided up into five spheres of influence, "five historical and ethnological provinces: Anatolia, Armenia, Syria, Palestine and Iraq-Jazirah ... so that they would have the 'chance to foster and develop their own resources.'"

30. M. Sykes to A. Balfour, Hotel Lotti, Paris, 8 April 1917, U DDSY 2/12/7, Sykes Papers.

31. Ibid.

32. Adelson, *Portrait of an Amateur*, 227.

33. M. Sykes to A. Balfour, Hotel Lotti, Paris, 9 April 1917, U DDSY 2/12/7, Sykes Papers. See also M. Sykes to M. Hankey and War Cabinet, Paris, FO 371/3045/73659, Foreign Office Papers, TNA, reporting on Sokolow's meeting with Cambon and Picot, 9 April 1917.

34. M. Sykes to A. Hirtzel, Hotel Lotti, Paris, 9 April 1917, U DDSY 2/12/7, Sykes Papers. Storrs, *Memoirs*, 219.

35. Storrs, *Memoirs*, 232ff. See also John Townsend, *Proconsul to the Middle East*, 101–102.

36. Wilson, *Loyalties*, 304–305.
37. M. Sykes to R. Graham, Hotel Lotti, Paris, 9 April 1917, U DDSY 2/12/7, Sykes Papers.
38. Adelson, *Portrait of an Amateur*, 227.
39. Ibid., 228. The trip courtesy of the French Navy proved quite torturous and caused Sykes to complain to Hankey. See M. Sykes to M. Hankey, Shepheard's Hotel, Cairo, 13 May 1917, U DDSY 2/12/7, Sykes Papers.
40. M. Sykes to E. Sykes, 23 April 1917, Sykes Papers, quoted in Adelson, *Portrait of an Amateur*, 229.
41. M. Sykes to R. Graham, 24 April 1917, Arab Affairs Papers, Durham, Box 145/4, quoted in Adelson, *Portrait of an Amateur*, 229.
42. Ibid., 229.
43. M. Sykes to Prodrome London [F.O.], 29 April 1917, Cairo, U DSSY 2/11/35, Sykes Papers. Prodrome was a Foreign Office term for a dispatch sent by messenger, or regular post, designed to fill in the supporting details, or arguments, of a brief telegram already sent. Adelson, *Portrait of an Amateur*, 229. See also Tauber, *Arab Movements*, 173.
44. Adelson, *Portrait of an Amateur*, 230.
45. Sharif Husayn received his position as Amir (prince or ruler) of Mecca from Turkish Sultan Abdul Hamid II in 1908. After proclaiming his revolt against the Turks, Husayn "declared himself King of the Arab Lands in an elaborate ceremony in Great Mosque in Mecca" on 2 November 1916. The British and French objected to this, fearing it might lead to misunderstandings and conflicts with other Arab rulers. On January 3, 1917, Britain, France, and Italy officially recognised Husayn as King of the Hijaz. See A. Hirtzel to R. Cecil, 17 February 1917, on the matter, confirming that the India Office would refer to him only as King of the Hejaz, F.O. 371/3044/37634, Foreign Office Papers, TNA. See also R. Wingate to P. Cox, on the issue, F.O. 371/3044/3474, Foreign Office Papers, TNA. See Randall Baker, *King Husain and the Kingdom of the Hijaz* (Cambridge: The Oleander Press, 1979), 17, 114–15. See also Kedourie, *Anglo-Arab Labyrinth*, 145–8.
46. Baker, *King Husain*, 141.
47. Ibid., 141.
48. Wingate's telegrams on 28 April and no. 446 on 30 April 1917, FO 371/3054, Foreign Office Papers, TNA. See also Kedourie, *Anglo-Arab Labyrinth*, 162–3.
49. Wingate's telegram on 28 April and no. 446, on 30 April 1917, FO 371/3054, Foreign Office Papers, TNA. See also Kedourie, *Anglo-Arab Labyrinth*, 162–3.
50. R. Wingate to Foreign Office, Confidential, Cairo, 27 April 1917 and R. Wingate to Prodrome London, Cairo, 27 April 1917, U DDSY 2/11/35, Sykes Papers.
51. G. Clayton to C.E. Wilson, Cairo, 28 April 1917, FO 880/12/241–4, Foreign Office Papers, TNA. Kedourie, *Anglo-Arab Labyrinth*, 162.
52. Westrate, *Arab Bureau*, 153.
53. Commander-in-Chief, East Indies & Egypt to M. Sykes, 27 April 1917, U DDSY 2/11/33, Sykes Papers. Handwritten note giving Sykes his itinerary for his trip to the Hijaz. See also Diary of M. Sykes's Trip East, April–June 1917, U DDSY 2/2/7, Sykes Papers.
54. M. Sykes to High Commissioner, Cairo, Red Sea, 5 May 1917, U DDSY 2/11/35, Sykes Papers. A.L. Tibawi, *Anglo-Arab Relations and the Question of Palestine, 1914–1921* (London: Luzac & Company Ltd, 1978), 181.
55. M. Sykes to R. Wingate, Red Sea, 5 May 1917, U DDSY 2/11/35, also F.O. 882/16, and F.O. 371/3054 (paper 93335), Foreign Office Papers, TNA. Kedourie, *Anglo-Arab Labyrinth*, 164. Tibawi, *Anglo-Arab Relations*, 181.

56. M. Sykes to R. Wingate, Jeddah, 6 May 1917, F.O. 371/3054, 93334/86256, Foreign Office Papers, TNA, repeated in Wingate's telegram no. 496, Cairo, 7 May 1917. See Westrate, *Arab Bureau*, 154. According to the records of the French mission on Sykes's movements while in Mecca, given the king's busy schedule that day the meeting could not have lasted more than an hour, not the three-and-a-half hours Sykes reported. Tibawi, *Anglo-Arab Relations*, 180.

57. Service historique de l'armée, Paris, section d'outremer, papers of the military mission in the Hijaz, 17 N 498, Brémond's telegram to Defrance no. 1440, Jeddah, 7 May 1917, quoted in Kedourie, *Anglo-Arab Labyrinth*, 164. After this purported meeting, Sykes wrote to Wingate saying, "after a careful study of the situation in Jeddah and Wejh I am convinced that the sooner the French military mission is removed from the Hejaz the better." M. Sykes to High Commissioner, Cairo, ... [to be sent to] Sir R. Graham, Red Sea, 5 May 1917, U DDSY 2/11/35, Sykes Papers. Note: Although Britain and France had agreed to refer to Husayn as King of the Hijaz, he was often still referred to – as in this quote – as Sharif the traditional title given descendants of the Prophet Muhammad.

58. Diary of M. Sykes's Trip East, April–June 1917, U DDSY 2/2/7, Sykes Papers.

59. Fu'ad al-Khatib (1882–1957) was a Syrian nationalist from Damascus, a poet and journalist who was educated at the American University of Beirut and taught Arabic at Gordon College in Khartoum, where he also served as inspector of education in Sudan before joining the Arab revolt in 1916. He was editor of the newspaper of the Arab revolt, *al-Qitab*, in 1916 and was appointed minister of foreign affairs for the Hijaz in 1917. See Sir Gilbert Falkingham Clayton, *An Arabian Diary*, Robert O. Collins, ed. (Berkeley and Los Angeles, CA: University of California Press, 1969), Appendix IX, 320.

60. M. Sykes to High Commissioner, Cairo, Aden, 23 May 1917, No. 28, U DDSY 2/11/35, Sykes Papers.

61. Lt. Col.Stewart Francis Newcombe (1878–1956) was a British army officer and surveyor. He fought in the Boer War (1899–1900) and was transferred to the Egyptian Army in the Sudan where he did survey work for the Sudan Railways (1901–1911). In 1912 he surveyed the desert region south of the Gaza–Beersheba line in southern Palestine to measure and map a strategic triangle of southern Palestine – today's Negev Desert – as part of a secret survey carried out on behalf of the British War Office for the War Office of Southern Palestine – the Negev Desert – under the auspices of the Palestine Exploration Fund. In 1913–1914 he surveyed the are from the Negev Desert to the Gulf of Aqaba, an area considered to be of military importance in any future conflict with Turkey, and was accompanied by C.L. Woolley and T.E. Lawrence (later Lawrence of Arabia), both of whom served under Newcombe in the Arab revolt. He served in France and Gallipoli before being transferred to the Hijaz, where he was put in command of the British Military Mission in the Hijaz (1916–1917), and with Lawrence and Arab tribal forces took part in mine-laying operations on the Hijaz Railway, a vital supply line for the Turks. See Kerry Webber, "In the Shadow of the Crescent: The Life & Times of Colonel Stewart Francis Newcombe, R.E., D.S.O – Soldier, Explorer, Surveyor, Adventurer and loyal friend to Lawrence of Arabia," *In The Shadow of the Crescent*, shadowofthecrescent.blogspot.co.uk/p/normal-0-false-false-false-en-gb-x-none.html, accessed 5 October 1917.

62. Note by Sheikh Fu'ad El Khatib taken down by Lt. Col. Newcombe, F.O. 882/16, folios 131–6, cited in Kedourie, *Anglo-Arab Labyrinth*, 176.

63. Note by Sheikh Fu'ad El-Khatib taken down by Lt. Col. Newcombe, F.O. 882/16/131–36, Foreign Office Papers, TNA. Wilson to Clayton, May 24, 1917, FO 882/16/102–14, Foreign Office Papers, TNA. See also Kedourie, *Anglo-Arab Labyrinth*, 176.

64. M. Sykes to High Commissioner, Cairo, Aden, 23 May 1917, No. 28, U DDSY 2/11/35, Sykes Papers. See also C. Wilson to Arab Bureau, Jeddah, 24 May 1917, F.O. 882/16, Foreign Office Papers, TNA. The declaration was translated from Arabic into French for Picot by Louis Massignon (1883–1962), who later became a renowned Catholic scholar of Islam and its history. He was attached to the Sykes–Picot mission as translator on loan from the Deuxième Bureau (Intelligence) of the Colonial Department. Tibawi, *Anglo-Arab Relations*, 180. See also William E. Watson, *Tricolor and Crescent: France and the Islamic World* (Westport, CT: Praeger, 2005), 43–4.

65. Kedourie, *Anglo-Arab Labyrinth*, 177.

66. C.E. Wilson to G. Clayton, 24 May 1917, F.O. 882/16/102–114, Foreign Office Papers, TNA. Mohs, *Military Intelligence and the Arab Revolt*, 138–9. Kedourie, *Anglo-Arab Labyrinth*, 178–9.

67. C.E. Wilson to Clayton, May 21 [24], 1917 SAD Wingate Papers 145/6, and C.E. Wilson to Clayton, 24 May 1917, F.O. 882/16/102–14, Foreign Office Papers, TNA. See also Kedourie, *Anglo-Arab Labyrinth*, 176. Westrate, *Arab Bureau*, 154–5. See also Wilson, *Loyalties*, 237–42, for the full text and analysis of the controversial Baghdad Proclamation.

68. C.E. Wilson to G. Clayton, 24 May 1917, F.O. 882/16/102–14, Foreign Office Papers, TNA. Mohs, *Military Intelligence and the Arab Revolt*, 138–9. Kedourie, *Anglo-Arab Labyrinth*, 178–9.

69. Wilson, *Loyalties*, 241.

70. Kedourie, *Anglo-Arab Labyrinth*, 174.

71. Westrate, *Arab Bureau*, 155.

72. A.E. Guerre, 1914–1918, vol. 877, Picot's telegrams nos. 13–18 via Aden 24 May 1917; telegram to Picot no. 127–8, Paris May 29; Picot's telegram no. 24–7, Cairo 8 June; and telegram to Picot no 138, Paris 11 June 1917, quoted in Kedourie, *Anglo-Arab Labyrinth*, 179.

73. E. Grey to H. McMahon, 26 July 1915, F.O. 371/2353, Foreign Office Papers, TNA; G. Clayton to R. Wingate, 4 August 1915, Wingate Papers, SAD 158/71. Timothy J. Paris, *The Defence of Britain's Middle Eastern Empire: A Life of Sir Gilbert Clayton* (Brighton: Sussex Academic Press, 2016), 182.

74. Sir Mark Sykes, Memorandum, 14 August 1917, F.O. 371/3059, 159558, Foreign Office Papers, TNA.

75. M. Sykes to P. Cox, Aden, 23 May 1917, U DDSY2/12/7, Sykes Papers.

76. Ibid.

77. Ibid.

78. M. Sykes, "Observations on Arabian Policy as Result of Visit to Red Sea Ports, Jeddah, Yenbo, Wejh, Kamaran, and Aden," Cairo, 6 June 1917, F.O. 371/3044/12049, Foreign Office Papers, TNA; also, U DDSY 2/11/45, Sykes Papers.

Chapter 9   The Arab Legion and the French Difficulty

1. Christopher Sykes, *Two Studies of Virtue*, 179.

2. Sykes's Arab Legion is not to be confused with the later army of Jordan established in 1921 under the command of Captain Frederick Gerard Peake. Despite its official Arabic name of *Al Jeish al Arabi*, or Arab Army, the British referred to it as the Arab Legion and the name stuck. While this could be with Sykes's Arab Legion in mind, I could find no evidence for this. "Peake Pasha" commanded the Arab Legion until 1939, when it was taken over by Sir John Bagot Glubb. "Glubb Pasha" was to make it famous. He modernised the Legion and commanded it until 1956. See Brig. John Bagot Glubb,

The Story of the Arab Legion (London: Hodder & Stoughton, 1848), 59ff. For more details about its origins and history, see P. J. Vatikiotis, *Politics and the Military in Jordan*, 57ff.

3. M. Sykes to the Chief of General Staff, Egypt Forces, Cairo, 28 April 1917, U DDSY 2/11/39, Sykes Papers.

4. *Indian Army List*, Vol. 2 (Army Headquarters, India, 1919; reprint: Uckfield, 2001), 909.

5. The Directorate of Military Operations was established in 1904 after the Second Boer War with a director of military operations, a position under the chief of the Imperial General Staff in the War Office. It was responsible for strategic considerations in military operations, records of armed strength, and the fighting efficiency of the British and Allied forces, liaison with the Allied Armies, defence policy and the collection, collation and dissemination of information on British Overseas Dominions and Colonies. Meanwhile the Directorate of Military Intelligence collected information on enemy forces. The two were combined into the Directorate of Military Operations and Intelligence in 1922. "Great Britain. War Office. Director of Military Operations," Archives Association of Ontario, http://www.archeion.ca/great-britain-war-office-director-of-military-operations-2, accessed 31 March 2017.

6. Major N.N.E. Bray, *Shifting Sands* (London: Unicorn Press, 1934), 20. See John Fisher, "The Rabegh Crisis, 1916–1917: 'A Comparatively Trivial Question' or a 'Self-Willed Disaster,'" *Middle East Studies* 38, no. 3 (2002), 75. See also John Fisher, *Gentlemen Spies: Intelligence Agents in the British Empire and Beyond* (Thrupp, Stroud: Sutton Publishing, Limited, 2002), 107–109.

7. Priya Satia, *The Great War and the Cultural Foundations of Britain's Covert Empire in the Middle East* (New York: Oxford University Press, 2008), 31, 49.

8. Sir Robert Vansittart (1881–1957) was permanent under secretary of state for Foreign Affairs from 1930 to 1938, during which time he was a major figure among British officials and politicians opposed to appeasement with Germany. His strongly anti-German views would eventually lead to a clash with Neville Chamberlain and after 1938 he was replaced as permanent under secretary and a new post was created for him, chief diplomatic adviser to His Majesty's Government, in which he would serve from 1938 to 1941 when he retired from public office. Given Vansittart's notoriety at the time his book *Shifting Sands* was published (1934), perhaps Bray did not spell out his name because he felt it may have distracted from his story, but left enough of a hint to show who his Foreign Office contact was. For the controversy surrounding Vansittart at this time see Michael L. Roi's *Alternative to Appeasement: Sir Robert Vansittart and Alliance Diplomacy, 1934–1937* (Westport, CT: Praeger, 1997). For personal insights into his anti-German beliefs, see Vansittart's own works: *Black Record: Germans Past and Present* (London: Hamish Hamilton, 1941), *Lessons of My Life* (London: Hutchinson, 1943) and *The Mist Procession: The Autobiography of Lord Vansittart* (London: Hutchinson, 1958). On page 161 in *The Mist Procession* Vansittart also talks about his cousin T.E. Lawrence, whose paternal grandmother was Louisa Vansittart Chapman (1813–1877).

9. Bray, *Shifting Sands*, 66.

10. Ibid., 67.

11. Ibid., 68–9, 70–7. See Paris, *In Defence of Britain's Middle East Empire*, 227. See also H.V.F. Winstone, ed., *The Diaries of Parker Pasha* (London: Quartet Books, 1983), 131.

12. Paris, *In Defence of Britain's Middle East Empire*, 193.

13. Lawrence, *Seven Pillars of Wisdom*, 70.

14. Fisher, "The Rabegh Crisis, 1916–1917," 74. For the biography of Sharif Ali Haidar and the story of his attempt to replace Husayn as Amir of Mecca, see George Stitt, *A Prince of Arabia: The Emir Shereef Ali Haider* (London: George Allen & Unwin, 1948), 155–78.

See also his daughter's recollection of the events in her autobiography, Her Royal Highness Musbah Haidar, *Arabesque* (Hutchinson & Co., Publishers, 1944), 87–105.

15. Fisher, "The Rabegh Crisis, 1916–1917," 74.
16. Paris, *In Defence of Britain's Middle East Empire*, 193–200.
17. Note by Colonel Lawrence, November 18, 1916, SAD 694/4/44–45, Clayton Papers.
18. Bray, *Shifting Sands*, 100.
19. Note by Colonel Lawrence, 18 November 1916, SAD 694/4/44–45, Clayton Papers.
20. Appendix A. Note by Lieut.-Colonel Sir Mark Sykes, 18 September 1916, U DDSY2/11/25, Sykes Papers.
21. Paris, *In Defence of Britain's Middle East Empire*, 200–201.
22. Bray, *Shifting Sands*, 102. Winstone, ed., *The Diaries of Parker Pasha*, 172.
23. Bray, *Shifting Sands*, 66–7.
24. Ibid., 103–104.
25. Ibid., 104–106, 109–12.
26. Report on the Situation in the Hejaz by Captain N.N.E. Bray, 18th King George's Own Lancers, 8 November 1916, printed in the Arabian Report, No, XVIII, 23 November 1916, F.O. 371/2781/236383/5–8, Foreign Office Papers, TNA.
27. Mons, *Military Intelligence and the Arab Revolt*, 88–9.
28. Ibid., 85.
29. Paris, *In Defence of Britain's Middle East Empire*, 192–3. Paris cites extensive correspondence on the matter from Clayton to Wingate, the Arab Bureau and Gen. Robertson, Chief of the Imperial General Staff, between June and October 1916, found in the Sudan Archives at Durham University.
30. Sir Mark Sykes, Memorandum, 19 June 1916, CAB 17/17, Cabinet Papers, TNA, and W.O. to F.O., 20 July 1916, FO 371/2772, Foreign Office Papers, TNA, and Bray to D.M.I. (Sir Percy Lake), 19 October 1916, L/P&S/10, 2100/16. No. 4657/16, India Office Records and Papers, British Library, quoted in Briton Cooper Busch, *Britain, India, and the Arabs 1914–1921* (Berkeley, CA: University of California Press, 1971), 174–5 n.
31. Bray, *Shifting Sands*, 146.
32. M. Sykes to the Chief of General Staff, Egypt Force, Cairo, 28 April 1917, U DDSY 2/11/39, Sykes Papers.
33. Minute by Cecil (n.d.), F.O. 371/3043/109892, 96, Foreign Office Papers, TNA; ibid., Minute by Graham, 114. Draft telegram by Curzon, no. 717, 115283, quoted in Fisher, *Curzon and British Imperialism*, 83.
34. For more on Lt. Col. Gerard Leachman (1880–1920) and his intelligence career in Mesopotamia and Arabia, see H. V. F. Winstone, *Leachman: O.C. Desert* (London: Quartet Books, 1982).
35. M. Sykes to High Commissioner, Egypt, Jeddah, 18 May 1917, U DDSY 2/11/39, Sykes Papers.
36. M. Sykes to S. of S., India Office, Aden, 23 May 1917, U DDSY 2/11/39, Sykes Papers.
37. ARBUR to Wilson, "Following for Sykes," Cairo, 18 May 1917, U DDSY 2/11/39, Sykes Papers.
38. ARBUR Cairo to Resident Aden, for Sir M. Sykes, 20 May 1917, U DDSY 2/11/39, Sykes Papers.
39. EGYPTFORCE to Resident Aden, "Following for Sir Mark Sykes," 21 May 1917, U DDSY 2/11/39, Sykes Papers.
40. Ibid.
41. Sir Mark Sykes, "The Arab Legion, War Cabinet, Secret, n.d., CAB 24/18/29." See Baker, *King Husain*, 122–3; Westrate, *Arab Bureau*, 69; Lawrence, *Pillars of Wisdom*, 389; Tauber, *Arab Movements*, 127; and H.V.F. Winstone, *The Illicit Adventure: The Story of*

Political and Military Intelligence in the Middle East from 1898–1926. (London: Jonathan Cape, 1982), 258, 281–2, 298, 352.

42. M. Sykes, *Personal Diary on Trip East*, April–June 1917, U DDSY 2/2/7, Sykes Papers.

43. M. Sykes to Prodrome (War Office) London, Cairo, 2 June 1917, No. 46, DDSY 2/11/39, Sykes Papers. Sir Mark Sykes, "The Arab Legion," War Cabinet, Secret, n.d., CAB 24/18/29.

44. M. Sykes, *Personal Diary on Trip East*, April–June 1917, U DDSY 2/2/7, Sykes Papers.

45. Ibid.

46. Wingate to War Cabinet, 10 June 1917, Nos. 605 and 606, U DDSY 2/11/46 and U DDSY2/11/47, Sykes Papers. See also nos. 606 and 609, F.O. 371/3043/109892. Reference is to M. Sykes, "Recommendations," Cairo, 17 May 1917, U DSSY2/11/45, Sykes Papers.

47. Notes by Sir Mark Sykes to Sir Reginald Wingate's telegram No. 609, Mesopotamian Administration Committee, Secret, CAB 24/17/46.

48. Wingate to War Cabinet, 10 June 1917, nos. 605 and 606, U DDSY 2/11/46 and U DDSY2/11/47, Sykes Papers. See also nos. 606 and 609, F.O. 371/3043/109892.

49. Notes by Sir Mark Sykes to Sir Reginald Wingate's telegram No. 609, Mesopotamian Administration Committee, Secret, CAB 24/17/46.

50. Ibid.

51. Lawrence, *Seven Pillars of Wisdom*, 58.

52. Sir Mark Sykes, "The Arab Legion," War Cabinet, Secret, n.d., CAB 24/18/29, Cabinet Papers, TNA. Though it is undated, given its contents and the subsequent course of events it was probably written in early July 1917.

53. Ibid.

54. Ibid.

55. Ibid.

56. Sir Mark Sykes, "The Arab Legion, War Cabinet, Secret, n.d., CAB 24/18/29," Cabinet Papers, TNA.

57. Ibid.

58. The Arab Situation, 17 June 1917, U DDSY2/4/144 and U DDSY2/11/48 (duplicate), Sykes Papers.

59. Ibid.

60. Ibid.

61. Ibid.

62. Ibid.

63. M. Sykes to G. Curzon, London, 2 July 1917, U DDSY 2/4/47, Sykes Papers.

64. Ibid.

65. Ibid.

66. D.G. Hogarth, "Note on The Anglo-Franco-Russian Agreement About the Near East," Secret, 10 July 1917, U DDSY 2/4/150, Sykes Papers.

67. M. Sykes to D.G. Hogarth, dated 11 July 1917, handwritten note on a letter from Hogarth to Sykes and returned to the sender. U DDSY2/4/150, Sykes Papers.

68. C. Hardinge (private) to R. Rodd, 18 May 1917, Hardinge Papers, Cambridge University Library, 32.

69. D.G. Hogarth to R. Clayton, London, 11 July 1917 David George Hogarth Collection, 2/16, St. Antony's College, Middle East Centre, Oxford; hereafter referred to as Hogarth Papers. See Fisher, *Curzon and British Imperialism*, 85. See also Westrate, *Arab Bureau*, 156. Sykes nickname of "Alabaster" for Curzon was based on an incident that took place when he visited Curzon at the family home Kedleston Hall. Sykes "asked Curzon if Kedleston's columns were imitation marble like those at Sledmere. 'Purest Alabaster!' [was] Curzon's

emphatic rejoinder." Adelson, *Portrait of an Amateur*, 236. According to "Compendium of the History of Derbyshire," *The Gentlemen's Magazine*, Vol. 126, August 1819, 108, "In the Entrance-hall [of Kedleston], 67 feet by 42 feet, are 20 Corinthian columns of veined alabaster, 25 feet high, brought from Lord Curzon's quarries at Red-hill in Nottinghamshire." Curzon's pompous, stiff-necked manner and natural pallor only added to the wide popularity of this nickname. In his autobiography of Curzon, Leonard Mosley dedicated an entire chapter to "Lord Alabaster". See, Leonard Mosley, *The Glorious Fault: The Life of Lord Curzon* (New York: Harcourt, Brace & Co., 1960), 230–52.

70. D.G. Hogarth to R. Clayton, London, 11 July 1917 Hogarth Papers, 2/16.

71. D.G. Hogarth to R. Clayton, 20 July 1917, Hogarth Papers, 2/17. The Fashoda Incident of 18 September 1898 took place at Fashoda in southern Sudan. A French expedition under Major Jean-Baptiste Marchand left Brazzaville in the Congo with 120 men for Fashoda on the Nile. The plan was to extend French control of sub-Saharan Africa from modern Senegal on the Atlantic coast and establish an uninterrupted link between the Niger River and the Nile. This would extend French control from Senegal through Mali, Niger and Chad and give them control of Saharan caravan routes. They also wanted to connect to their colony in French Somaliland (Djibouti) on the Red Sea, thus assuring them of an east–west belt of territory across Africa. At the time, Sir Herbert Kitchener, Commander of the Anglo-Egyptian Army, had just defeated the forced of the Mahdi at the Battle of Omdurman on 2 September and had reconquered the Sudan, which had been lost to Egypt at the Battle of Khartoum in 1885 and had been in the hands of Mahdist forces since then. Hearing of Marchand's arrival in Fashoda, Kitchener confronted the French force there on 18 September. The French and British almost went to war over the incident, but the calm and respectful approach of the two men – Marchand and Kitchener – who waited for orders from Paris and London, ended with an agreement that the source of the Nile and the Congo rivers would mark the boundary between their two areas of influence. This agreement is said to be the basis of the Entente Cordiale between the two nations in 1904, which subsequently led to the Triple Entente in 1909 of Britain, France, and Russia, allies in World War I. Curiously enough, while it actually was an example of positive cooperation between Britain and France it came to have a negative aspect and signify French imperialism in its highest form; hence, the use of the term here. Fashoda has since been renamed Kodok. See David Levering Lewis, *The Race for Fashoda: European Colonialism and African Resistance in the Scramble for Africa* (New York: Weidenfeld and Nicolson, 1988).

72. Westrate, *Arab Bureau*, 156–7.

73. G. Clayton to M. Sykes, 22 July 1917, Clayton Papers, SAD 693/12/33–34.

74. Ibid.

75. Briton Cooper Busch, *Britain, India and the Arabs 1914* (Berkeley, CA: University of California Press, 1971), 177, with reference to "Note on subcommittee, July 5, 1917," L/P&S/11, vol. 123, no. 2981/17, India Office Records and Papers, British Library.

76. A. Balfour to R. Wingate (from Sykes to Arab Legion), quoted, 30 July 1917, F.O. 371/3034, Foreign Office Papers, TNA. Busch, *Britain, India and the Arabs*, 177.

77. Busch, *Britain, India and the Arabs*, 177–8.

Chapter 10   The Balfour Declaration

1. The Rt. Hon. L.S. Amery, *My Political Life, Vol. Two: War and Peace 1914–1929* (London: Hutchinson & Co, Ltd., 1953), 115. As noted in Chapter 6, Amery worked with Sykes as one of the two assistant secretaries in the War Cabinet Secretariat during this period and,

as noted later in this chapter, would have a hand himself in the final editing of the Balfour Declaration.

2. Henry Morgenthau, Sr (1856–1946), was a lawyer, businessman and the US ambassador to the Ottoman Empire from 1913 to 1916. He was famous for exposing the Turkish Armenian genocide of 1915, about which he wrote in *Ambassador Morgenthau's Story* (Garden City, NJ: Doubleday, Page & Company, 1919). Also an anti-Zionist, Morgenthau was one of thirty prominent American Jewish signatories to a petition against Zionism given to President Wilson to take to the Paris Peace Conference, rejecting "the Zionist project of a 'national home for the Jewish people in Palestine.'" Alfred M. Lilienthal, *The Zionist Connection II: What Price Peace?* (New Brunswick, NJ: North American, 1982), 268–9.

3. Adelson, *Portrait of an Amateur*, 234–5.

4. H. Rumbold to A. Balfour, Berne, 20 June 1917, U DDSY2/11/50, Sykes Papers. The attached Constitution is in French. See Lloyd George, *War Memoirs of David Lloyd George*, Vol. Two, 1490. See also George H. Cassar, *Lloyd George at War, 1916–1918* (London: Anthem Press, 2011), 221.

5. M. Sykes to R. Cecil, London, 29 July 1917, U DDSY2/12/9, Sykes Papers.

6. *Weizmann Papers*, 548. Frankfurter, a Harvard law professor, Zionist and Brandeis protégé, was confidential assistant to the US secretary of war, and had been sent to represent Zionist interests. Ibid., 546–7.

7. *Weizmann Papers, Vol. VII, Series A, August 1914–November 1917*, 475.

8. Ibid., 449, see n. 2; 475.

9. Ibid., 460, Letter 451, n. 3 and 4. Weyl was a former head of the Turkish tobacco monopoly and Thomas was the French minister of munitions.

10. Ibid., 460, Letter 452, note 2. See also Letter 453 and Document 454: Gibraltar Conference Report, ibid., 461–7 and Letter 461, 473.

11. Sanders, *The High Walls of Jerusalem*, 557. See also *Weizmann Papers, Vol. VII, Series A, August 1914-November 1917*, 477 and n. 2, for the formal set-up of the committee within the Zionist organisation and a list of members. Originally it was led by former English Zionist Federation president Joseph Cowen and Ahad Ha'am (pen name of Asher Zvi Ginsberg), a leading figure in Jewish cultural thought and one of Weizmann's most intimate friends. Ibid., 1, n. 1.

12. Harold Nicolson (1886–1968) was the youngest son of Sir Arthur Nicolson, 1st Baron Carnock, permanent under secretary of state for Foreign Affairs under Grey (1910–1916) and a junior clerk at the Foreign Office at the time. He joined the diplomatic service in 1909 and served as attaché in Madrid (1911–1912) and then was posted to Constantinople, where he served as third secretary (1912–1914). He served at the Foreign Office during World War I, during which time he was promoted to second secretary. As the Foreign Office's most junior employee, in August 1914 he was assigned to hand Britain's declaration of war to the German ambassador in London. He would go on to have a long and distinguished career, as a diplomat, writer and politician. He was married to writer Vita Sackville-West. Nicolson would later affectionately write of Sykes that he was "exuberant, energetic, impulsive, humorous, and slightly irresponsible." Harold Nicolson, *Curzon: The Last Phase, 1919–1925: A Study in Post War Diplomacy* (London: Faber and Faber Ltd., 1934; reprint, 1988), 86, n.1.

13. Stein, *Balfour Declaration*, 467.

14. Adelson, *Portrait of an Amateur*, 235.

15. Stein, *Balfour Declaration*, 469–70.

16. W. Rothschild to A.J. Balfour, London, 18 July 1917, AMEL 1/3/52, Leopold Amery Papers, Churchill Archives Centre; hereafter referred to as Amery Papers. This document,

which is a copy of the original Rothschild formula from the Secret War Cabinet Papers, G.T. 1803, Zionist Movement, was found among the Amery Papers at the Churchill Archives. It proposed, *"that Palestine should be reconstituted as the National Home of the Jewish people"* (italics mine). This document shows an obvious attempt by its authors to have all Palestine given to the Zionists as their homeland. However, Balfour gave it to Milner to revise, who turned it over to Amery and the final version approved – the so-called Milner–Amery version – was issued by the government on 2 November 1917. See also Stein, *Balfour Declaration*, 470–1.

17. Stein, *Balfour Declaration*, 465–70.
18. James Renton, *The Zionist Masquerade: The Birth of the Anglo-Zionist Alliance, 1914–1918* (Basingstoke: Palgrave Macmillan, 2007), 3. See Mark Levene, "The Balfour Declaration: A Case of Mistaken Identity," *The English Historical Review*, 107, no. 422 (1992), 54–77. Idem, *War, Jews, and the New Europe: The Diplomacy of Lucien Wolf 1914–1919*, 77–107, 128–44. See also Tom Segev, *One Palestine Complete: Jews and Arabs under the British Mandate*, Haim Watzman, trans (New York: Metropolitan Books, 2000), 33–49.
19. Sanders, *The High Walls of Jerusalem*, 559–60.
20. See Chapter 7, "The Zionists and a Jewish Homeland." See also Adelson, *Portrait of an Amateur*, 219–20.
21. Isaiah Friedman, *The Question of Palestine: British-Jewish-Arab Relations: 1914–1918*, 184.
22. Adelson, *Portrait of an Amateur*, 237. See also Stein, *Balfour Declaration*, 491.
23. Samuel Montagu, 1st Baron Swaythling (1832–1911) was a British banker who founded the bank of Samuel Montagu & Co. He was a philanthropist and Liberal politician who sat in the House of Commons from 1885 to 1900, and was later raised to the peerage. A pious Orthodox Jew, he devoted himself to social services and advancing Jewish institutions. Born Montagu Samuel, the second son of Louis Samuel, a watchmaker of Liverpool, he would later reverse his name. He had four sons and six daughters. His daughter Lily would help establish Liberal Judaism and his second son Edwin Samuel Montagu would follow his father into politics. See Chaim Bermant, *The Cousinhood: The Anglo-Jewish Gentry* (London: Eyre & Spottiswoode, 1971), 200–15.
24. Stein, *Balfour Declaration*, 484, and n. 1.
25. E. Montagu, Memorandum to Cabinet, 23 August 1917, CAB 21/58, Cabinet Papers, TNA. See Martin Watts, *The Jewish Legion*, 110. See also Adelson, *Portrait of an Amateur*, 242.
26. Watts, *The Jewish Legion*, 110–15.
27. Sanders, *The High Walls of Jerusalem*, 561–2.
28. Richard P. Stevens, "Weizmann and Smuts", *Journal of Palestine Studies*, 3, 1, 1973. See also Stein, *Balfour Declaration*, 482.
29. Sanders, *The High Walls of Jerusalem*, 562. See also *Weizmann Papers, Vol. VII, Series A, August 1914-November 1917*, 531, n. 5. Lord Curzon, in a carefully worded statement titled "The Future of Palestine," dated 26 October 1917 outlined his position on the matter. See the complete text in Lloyd George, *Memoirs of the Peace Conference*, 727–32. See also Fisher, *Curzon and British Imperialism in the Middle East*, 209–11.
30. Sanders, *The High Walls of Jerusalem*, 564. Herbert Samuel, "The Future of Palestine," CAB 37/123/43, Cabinet Papers, TNA.
31. Edwin S. Montagu, The Anti-Semitism of the Present Government, Secret, 23 August 1917, CAB 24/24/71, Cabinet Papers, TNA.
32. Ibid.
33. Adelson, *Portrait of an Amateur*, 240.
34. "Draft Declaration, War Cabinet, G.T. 1803A, The Zionist Movement. G.T. 1803A. Alternative, by Lord Milner, to Draft Declaration, see II of Paper G.T. 1803,

AMEL 1/3/52, Amery Papers, Churchill Archives. See also Sanders, *The High Walls of Jerusalem*, 575.

35. Ibid., 576–7.

36. Ibid., 579. For Weizmann's comments on Montagu, the cabinet meeting and his astonished reaction to Wilson's response, see his letter to Harry Sacher, *Weizmann Papers, Vol. VII, Series A, August 1914-November 1917*, 513–15. Also, see Weizmann's letter to Brandeis, ibid., 506, n. 5, which goes into more detail on the cabinet meeting, Montagu and the decision to seek Wilson's approval.

37. *Weizmann Papers, Vol. VII, Series A, August 1914–November 1917*, 505–506.

38. Sanders, *The High Walls of Jerusalem*, 576, 580. Apparently, in sending his request for support of the Zionist declaration to Wilson, Cecil had omitted enclosing a copy of the declaration draft under consideration – the Milner version. The only one Weizmann had and was aware of was the one submitted to Balfour by Lord Rothschild, which he enclosed with his communication to Brandeis. See also Terrence O'Brien, *Milner: Viscount Milner of St James's and Cape Town* (London: Constable and Company Ltd, 1959), 288.

39. Edward Mandell House (1858–1928) was known as "Colonel" House, although he had never served in the military. He was a highly influential Texas politician before becoming a key supporter in President Woodrow Wilson's 1912 presidential campaign. Although he no public office, he was the president's agent and chief advisor on European politics and diplomacy during World War I (1914–1918) and afterwards at the Paris Peace Conference in 1919. See Charles E. Neu, *Colonel House: A Biography of Woodrow Wilson's Silent Partner* (London: Oxford University Press, 2014) for an excellent biography of this fascinating yet little-known man.

40. Ibid., 580–1.

41. War Cabinet (Secret) Zionist Movement, CAB24/27/98, Cabinet Papers, TNA.

42. Edwin Montagu, Secretary of State for India, to Lord Robert Cecil, Undersecretary of State for Foreign Affairs (Secret), India Office, 14 September 1916, CAB 24/27/93, Cabinet Papers, TNA.

43. M. Sykes, Note on Zionism, Sledmere Papers #80, quoted in Adelson, *Portrait of an Amateur*, 242–3. Interestingly enough, shortly before this and just after Montagu's appointment as secretary of state for India Sykes took it upon himself to write Montagu a detailed eleven-page letter of suggested reforms for India. See M. Sykes to E. Montagu, 14 August 1917, London, U DDSY 2/11/66, Sykes Papers.

44. Adelson, *Portrait of an Amateur*, 243. See Weizmann's letter to a Foreign Office official on German interest in sponsoring Zionist aspirations in Palestine, *Weizmann Papers, Vol. VII, Series A, August 1914–November 1917*, 438–42. For the German side of this issue, see Isaiah Friedman, *Germany, Turkey, and Zionism 1897–1918* (Oxford: Oxford University Press, 1977), 327–8.

45. Adelson, *Portrait of an Amateur*, 95, 185, 278–82, 288.

46. A. Albina to M. Sykes, Cairo, 10 August 1917, U DDSY 2/11/67b, Sykes Papers.

47. The small, neat handwriting is similar to that found on another document in the Sykes Papers with Christopher Hugh Sykes's initials after it. See "Moslem Interest in Palestine," a page taken from *The Near East* of an article about Marmaduke Pickthall's speech at the Central Islamic Society on 13 July 1917. U DDSY 2/11/67a.

48. Sir Mark Sykes, Note on Palestine and Zionism, 21 September 1917, U DDSY 2/12/10, Sykes Papers. This includes all the copies through the various stages from the original handwritten draft to the final typed version.

49. Ibid.

50. Ibid.

51. Ibid.

52. Ibid.

53. Ibid.

54. Ibid.
55. Ibid.
56. Ibid.
57. Sokolow, *History of Zionism*, I, 111, 215–16, 226 and 256.
58. Sir Mark Sykes, Note on Palestine and Zionism, 21 September 1917, U DSSY 2/12/10, Sykes Papers.
59. Ibid.
60. Ibid.
61. This version has been referred to as the Milner–Amery formula, "which with certain amendments, was to be finally approved by the War Cabinet on 31 October and to emerge as the Balfour Declaration." Stein, *Balfour Declaration*, 520–1.
62. War Cabinet Proceedings, 4 October 1917, CAB/23/4, quoted in Adelson, *Sykes*, 243. See also, Amery, *My Political Life*, 114–17.
63. S.D. Waley, *Edwin Montagu: A Memoir and an Account of his Visits to India* (London: Asia Publishing House, 1964), 140.
64. Edwin Montagu, Zionism (Secret), CAB 24/28/63, Cabinet Papers, TNA. See also, *Weizmann Papers, Vol. VII, Series A, August 1914-November 1917*, 521 note n. 3, and 524 note n. 5.
65. *Weizmann Papers, Vol. VII, Series A, August 1914-November 1917*, 531 n. 1. See, Edwin Montagu, "Zionism (Secret), CAB 24/28/63," TNA. See also Chaim Bermant, *The Cousinhood: The Anglo-Jewish Gentry*, 259–62, 263. Among these British Jewish leaders, "Claude Montefiore was a prominent anti-Zionist, who lectured, gave sermons, wrote letters and books railing against Zionism and was [the most prominent of] . . . the "anti-Zionists" interviewed by the War Cabinet . . . on 31 October 1917." He was also a close friend of Lord Milner. See O'Brien, *Milner*, 288–9.
66. Waley, *Montagu*, 141.
67. M. Sykes anonymously noted this on in E. Drummond to A. Balfour, 30 October 1917, FO/371/3083, Foreign Office Papers, TNA. Adelson, *Portrait of an Amateur*, 243.
68. Justin McCarthy, *The Population of Palestine: Population History and Statistics of the Late Ottoman Period and the Mandate* (New York: Columbia University Press, 1990).
69. War Cabinet Proceedings, 31 October 1917, CAB/23/4, Cabinet Papers, TNA, quoted in Adelson, *Sykes*, 243.
70. Draft Declaration, War Cabinet, G.T. 1803A, The Zionist Movement. G.T. 1803A. Alternative, by Lord Milner, to the draft declaration, See II of Paper G.T. 1803, AMEL 1/3/52, Amery Papers, Churchill Archives. See also O'Brien, *Milner*, 288.
71. *Weizmann Papers, Vol. VII, Series A, August 1914-November 1917*, 540–1. For a listing of the drafts leading up to the final Declaration, see Appendix: "Successive Drafts and Final Text of the Balfour Declaration," in Edwin Samuel Montague, *Edwin Montagu and the Balfour Declaration* (London: Arab League Office, n.d.), 23.
72. Weizmann, *Trial and Error*, 208.
73. A. Balfour to W. Rothschild, 2 November 1917, Addl. Ms. 41178, folios 1 and 3, British Museum. See also Stein, *Balfour Declaration*, 549 and John Norton Moore, ed., *The Arab-Israeli Conflict, Vol. III: Documents*, American Society of International Law (Princeton, NJ: Princeton University Press, 1974), 31–2.
74. Hansard CCCXLVII, 2129–97. See Christopher [Hugh] Sykes, *Crossroads to Israel*, 235–7, 239–41, for "the main propositions of the White Paper" and Jewish/Zionist and Arab reaction to it and overview. See also Weizmann, *Trial and Error*, 411–14, on his meeting with Churchill, who spoke against the government-sponsored White Paper in the Commons and the reaction among Zionists with its passage.
75. George Antonius, *The Arab Awakening*, 267–8.
76. Boyle, *Betrayal of Palestine*, 215.
77. Antonius, *The Arab Awakening*, 268.

78. John Marlowe, *The Seat of Pilate: An Account of the Palestine Mandate* (London: Cresset Press, 1959), 3.

79. Paris, *In Defence of Britain's Middle Eastern Empire*, 260–1.

80. G. Clayton to M. Herbert, 10 July 1917, F.O. 141/805/1, Foreign Office Papers, TNA. Paris, *In Defence of Britain's Middle Eastern Empire*, 261.

81. G. Clayton to S. Symes, 16 August 1917, and R. Wingate (for Clayton) to Foreign Office (for R. Graham), 18 August 1917 (quoted), F.O. 141/805/1, Foreign Office Papers, TNA. Paris, *In Defence of Britain's Middle Eastern Empire*, 261.

82. G. Clayton to M. Sykes, 20 August 1917, SAD 693/12/34, Clayton Papers.

83. Paris, *In Defence of Britain's Middle Eastern Empire*, 261.

84. T.E. Lawrence to M. Sykes (unsent) H.M.S. Hardinge, Aqaba, 9 September 1917, SAD 693/11/4, Clayton Papers.

85. Ibid.

86. Ibid.

87. Ibid.

88. Ibid.

89. Clayton to T.E. Lawrence (Strictly Private) Cairo, 20 September 1917, SAD 693/12/36–39, Clayton Papers.

90. Ibid.

91. Sir Mark Sykes, Memorandum on the Asia-Minor Agreement, F.O. 371/3059, Foreign Office Papers, TNA.

92. A. Milner's memorandum, 12 November, M. Sykes to M. Hankey, 14 November 1917, F.O. 371/3057, Foreign Office Papers, TNA, quoted in Adelson, *Portrait of an Amateur*, 244.

93. War Cabinet Proceedings, 5 November, 7 and 14 December 1917, CAB 23/4, Cabinet Papers, TNA. Adelson, *Portrait of an Amateur*, 244–5.

Chapter 11   Palestine

1. Sir Walter Scott, *Marmion: A Tale of Flodden Field in Six Cantos*, Canto VI, Stanza 17.

2. Adelson, *Portrait of an Amateur*, 245. War Cabinet Proceedings, 14 December 1917, Cabinet Papers 23/4, TNA. Sykes had earlier written Allenby's "Proclamation of Martial Law in Jerusalem," on 19 November 1917, U DDSY 2/4/162, Sykes Papers. It was posted on 9 December 1917 two days before his official entry into the city, at which time Allenby made the Proclamation prepared by Curzon and Sykes to the People of Jerusalem.

3. John D. Grainger, *The Battle for Palestine 1917* (Woodbridge: Boydell Press, 2006), 215.

4. M. Sykes, "Appreciation," *Eastern Report*, 13 December 1917, CAB 24/144, Cabinet Papers, TNA, cited in Adelson, *Portrait of an Amateur*, 245.

5. Ibid., 245.

6. *The Times*, 2 December; *Manchester Guardian*, 10 December 1917. Copies of Reuters News Agency's accounts of the speeches, which were used by the papers and sent telegraphically to Cairo at Sykes's request, can be found in the Sykes Papers as U DDSY 2/11/121 (London speech of 2 December 1917) and U DDSY 2/11/122 (Manchester speech of 9 December 1917).

7. "Sees Great Future for Jew and Arab," *New York Times*, 12 December 1917.

8. Ottoman Sultan Selim I (r. 1512–1520), whose nickname the Grim is more accurately translated from the Turkish as stern, or ruthless, in view of his apparently merciless personality. Between 1516 and 1517 Selim I conquered all what became modern Syria, Lebanon, Palestine, Jordan, Egypt and Western Arabia. This included the Muslim holy cities of Mecca and Medina, as well as Jerusalem (which Muslims also considered to be holy).

9. "Jerusalem and Its Future . . . Interview with Sir Mark Sykes," *The Observer*, 16 December 1917.

10. Ibid.
11. G. Clayton to M. Sykes, Cairo, 15 December 1917, U DDSY 2/11/83, Sykes Papers.
12. Ibid.
13. Lawrence, *Seven Pillars of Wisdom*, 57.
14. G. Clayton to M. Sykes, Cairo, 15 December 1917, U DDSY 2/11/83, Sykes Papers.
15. The fundamentalist Wahhabi "revivalist movement" Clayton referred to was a Saudi phenomenon, which the British were watching very carefully, even though Sykes had dismissed them on 16 December 1915 in a meeting with the War Cabinet. See "Evidence of Lieutenant-Colonel Sir Mark Sykes, Bart., M.P., on the Arab Question," (Secret) War Committee. Meeting held at 10 Downing Street, on Thursday 16 December 1915, at 11:30 A.M. CAB 24/1/51, Cabinet Papers, TNA. As mentioned previously, Sir Percy Cox was responsible for dealings with Ibn Sa'ud (1876–1953), sending first political agent William Shakespear and then Harry Philby as his "minders" during the war, while Britain paid him not to attack King Husayn and jeopardise the Arab revolt. While there are many books on Saudi Arabia and Ibn Sa'ud, those that give special insight into the history of the Wahhabis and Ibn Sa'ud are George S. Rentz, *The Birth of the Islamic Reform Movement in Saudi Arabia: Muhammad b. 'Abd al-Wahhab (1703–1792) and the Beginnings of the Unitarian Empire in Arabia* (London: Arabian Publishing, 2004); Alexei Vassiliev, *The History of Saudi Arabia* (London: Saqi Books, 1998); H. St J. B. Philby, *Arabia of the Wahhabis* (London: Frank Cass, 1977); Robert Lacey, *The Kingdom: Arabia and the House of Saud* (New York: Avon Books, 1981); and John S. Habib, *Ibn Sa'ud's Warriors of Islam: The Ikhwan of Najd and Their Role in the Creation of the Sa'udi Kingdom, 1910–1930* (Leiden: E.J. Brill, 1978). For more on Wahhabi influence elsewhere, particularly in modern India, Pakistan and Afghanistan, see Charles Allen, *God's Terrorists: The Wahhabi Cult and Hidden Roots of Modern Jihad* (London: Little, Brown, 2006).
16. G. Clayton to M. Sykes, Cairo, 15 December 1917, U DDSY 2/11/83, Sykes Papers.
17. For a detailed description of the overthrow of King Husayn by Ibn Sa'ud and his Ikhwan, see Michael Darlow and Barbara Bray, *Ibn Saud: The Desert Warrior and his Legacy* (London: Quartet Books Limited, 2010), 253–92. For the definitive work on the Ikhwan, see Habib, *Ibn Sa'ud's Warriors of Islam*.
18. G. Clayton to M. Sykes, Cairo, 15 December 1917, U DDSY 2/11/83, Sykes Papers.
19. R. Storrs to M. Sykes, Cairo, 29 November 1917, U DDSY 2/4/165, Sykes Papers. Storrs, *Memoirs*, 276; and Storrs, *Orientations: Definitive Edition* (London: Nicolson & Watson, 1945), 261. *Memoirs* is the USA version of *Orientations*. Orientation is an updated version of the earlier book(s).
20. R. Storrs to R. Graham, Cairo, 29 November 1917, U DDSY 2/4/165. Sykes Papers.
21. Explanatory note on instructions to Sir Mark Sykes, n.d., FO 371/3056/239988, Foreign Office Papers, TNA. Fisher, *Curzon and British Imperialism in the Middle East*, 99 and 108, n. 159–61.
22. Fisher, *Curzon and British Imperialism*, 98–9.
23. Mesopotamian Administration Committee, Observations by the Director of Military Intelligence on Sir Reginald Wingate's dispatch, no. 127 (of 11 June), 6 July 1917, secret, War Office, CAB 27/22, Cabinet Papers, TNA. Fisher, *Curzon and British Imperialism in the Middle East*, 101–102.
24. Wingate to Foreign Office, Cairo, 28 November 1917, No. 1281 (forwarded by Clayton to Sir Mark Sykes), U DDSY2/11/76, Sykes Papers, and R. Wingate to Foreign Office, Cairo, 29 November 1917, No. 1286, U DDSY 2/11/77, Sykes Papers.
25. Fisher, *Curzon and British Imperialism in the Middle East*, 100.
26. D. Lloyd George to M. Sykes, 27 June 1917; M. Sykes to E. Sykes, c. Dec. 1918, Sykes Papers, quoted in Adelson, *Portrait of an Amateur*, 236.

27. Adelson, *Portrait of an Amateur*, 249.

28. Ibid., 249–50. See Viscount Cecil [Lord Robert Cecil], *A Great Experiment: An Autobiography* (London: Jonathan Cape, 1941) for Cecil's autobiography as it relates to the League of Nations. See also George W. Egerton, "Cecil's War" in his *Great Britain and the Creation of the League of Nations: Strategy, Politics, and International Organization, 1914–1919* (London: Scolar Press, 1979), 63–80, for a detailed description of Cecil's struggle to get Britain to support President Woodrow Wilson's proposal for a League of Nations.

29. Adelson, *Portrait of an Amateur*, 250. See also Gaynor Johnson, *Lord Robert Cecil: Politician and Internationalist* (Farnham: Ashgate Publishing Limited, 2013), 68–9.

30. G. Curzon to R. Cecil, 6 January 1918, 1 Carlton House Terrace, private, Cecil Papers, F.O. 800/198/186, Foreign Office, TNA. Sir George William Buchanan (1854–1924) was the British ambassador to Russia (1910–1918). He left Russia after the Revolution in January 1918.

31. Johnson, *Lord Robert Cecil*, 69.

32. G. Curzon to R. Cecil, 6 January 1918, 1 Carlton House Terrace, private, Cecil Papers, F.O. 800/198/186, Foreign Office, TNA. Fisher, *Curzon and British Imperialism in the Middle East*, 111–12.

33. Pan-Turanian Movement, Arab Bureau Papers, F.O. 882/18, Document TU/17/17, Foreign Office Papers, TNA. Fromkin, *Peace to End All Peace*, 352.

34. G. Curzon to R. Cecil, 6 January 1918, 1 Carlton House Terrace, private, Cecil Papers, F.O. 800/198/186, Foreign Office Papers, TNA. Fisher, *Curzon and British Imperialism in the Middle East*, 111–12.

35. R. Cecil to A. Balfour, 8 January 1918, F 6/5/13, Lloyd George Papers, House of Lords Records. Office/Parliamentary Archives. See also Johnson, *Lord Robert Cecil*, 71.

36. Minute by Curzon, 13 March 1918, F.O. 371/3380/38817, Foreign Office Papers, TNA. Fisher, *Curzon and British Imperialism in the Middle East*, 102–103.

37. Minutes of Middle East Committee, 12 January 1918, CAB 27/23, Cabinet Papers, TNA. For the involved story behind Cecil's proposal including Sykes's involvement, see John Fisher, "Lord Robert Cecil and the Formation of a Middle East Department of the Foreign Office," *Middle Eastern Studies* 42, no. 3, 365–80. See also Adelson, *Portrait of an Amateur*, 251.

38. Nahum Sokolow, Sir Mark Sykes, Bart., M.P (A Tribute), *History of Zionism 1600–1918*, Vol. II (New York: Ktav Publishing House, 1969), xxxii–xxxiii.

39. Ibid., xxxiii.

40. M. Sykes to M. Hankey, London, 14 November 1917, AMEL 1/3/7, Amery Papers, Churchill Archives Centre.

41. Joseph "Jack" N. Mosseri (1884–1934) was a member of the prominent Egyptian Sephardic Jewish Mosseri family and leader of the new Zionist Federation of Egypt, founded in June 1917. See Gudrun Krämer, *The Jews in Modern Egypt 1914–1952* (London: I.B.Tauris, 1989), 184.

42. R. Wingate (Cairo), 20 December 1917, No. 1374, to Foreign Office, "Following from Clayton," Files Only, U DDSY 2/22/84, Sykes Papers. According to Scott Anderson in *Lawrence of Arabia: War, Deceit, Imperial Folly and the Making of the Modern Middle East*, William "Billy" Ormsby-Gore (1885–1964), later 4th Baron Harlech, was a recent convert to Judaism (254). A member of the Arab Bureau in Cairo (1916–1917), he had recently returned to London as Lord Milner's parliamentary private secretary and then replaced Sykes's as assistant secretary in the War Cabinet Secretariat with Amery. He was a close friend of Sykes and Aaron Aaronsohn and would become an avid Zionist.

43. R. Graham to R. Wingate, London, 11 January 1918, SAD 167/1/128, Wingate Papers.

44. R. Graham to R. Wingate, London, 18 January 1918, SAD 167/1/237, Wingate Papers.

45. R. Wingate to G. Clayton, Cairo, 1 February 1918, SAD 470/8/1–2, Clayton Papers.
46. G. Clayton to M. Sykes, General Headquarters Egyptian Expeditionary Force (Jerusalem), January 26, 1918, SAD 693/12/31–36, Clayton Papers.
47. M. Sykes to R. Wingate, London, 3 March 1918, SAD 168/1/12, Wingate Papers.
48. M. Sykes to G. Curzon, London, 2 July 1917, U DDSY 2/4/47, Sykes Papers.
49. Adelson, *Portrait of an Amateur*, 255.
50. Weizmann, *Trial and Error*, 212.
51. Sokolow, *History of Zionism*, Vol. II, 140.
52. Weizmann, *Trial and Error*, 214.
53. Ibid., 214–15.
54. R. Wingate to Lord Hardinge, Cairo, 23 March 1918, SAD 168/1/137, Wingate Papers.
55. G. Clayton to M. Sykes, General Headquarters, Egyptian Expeditionary Force (Palestine) (Private), 4 April 1918, SAD 693/13/47–48, Clayton Papers.
56. Dr. Faris Nimr was a wealthy Syrian Christian originally from Beirut, who moved to Egypt in 1883 and was founder and editor of the popular Cairo newspaper *El Mokattam*. See Chapter 2, "Kitchener's Man," describing Sykes's meeting with Dr Nimr during his visit to Cairo in the summer of 1915.
57. G. Clayton to M. Sykes, General Headquarters, Egyptian Expeditionary Force (Palestine) (Private), April 4, 1918, SAD 693/13/47–48, Clayton Papers.
58. Ibid.
59. Ibid.
60. Ibid.
61. W. Ormsby-Gore to M. Sykes, Tel Aviv/Jaffa (Confidential and Private), 9 April 1918, U DDSY 2/11/96, Sykes Papers.
62. W. Ormsby-Gore to M. Sykes, Tel Aviv/Jaffa (Confidential and Private), 9 April 1918, U DDSY 2/11/96, Sykes Papers.
63. "Jewish Charitable Institutions & Organizations," Political Intelligence Officer Jerusalem, Weekly Summary for the period 1–7 July inclusive, U DSSY 2/11/103, Sykes Papers; "In the Jewish Community," Political Intelligence Officer Jerusalem, Weekly Summary for the period 12–18 August inclusive, U DSSY 3/11/104, Sykes Papers; "Moslems, Jews," Political Intelligence Officer Jerusalem, Weekly Summary for the period 19–25 August inclusive, U DSSY 2/11/105, Sykes Papers; and "Jews," Political Intelligence Officer Jerusalem, Weekly Summary for the period August 26 to September 1 inclusive, U DSSY 2/11/106, Sykes Papers.
64. Rough Notes and Final Notes for Speech at Manchester on Friday 22 June 1918 to the Syrians, U DDSY 2/11/124, Sykes Papers.
65. Zeine N. Zeine, *The Struggle for Arab Independence* (Beirut: Khayat's, 1960), 22–3. For the petition and discussion between Wingate and Balfour, see Wingate to A. Balfour, May 7, 1918, "enclosing 'Syrian Moslem Opinion in Egypt; and Balfour to Wingate, 11 June 1918, FO/371/3380." For the "Declaration to the Seven [Translation]," see Antonius, *Arab Awakening*, Appendix D, 433. See also Adelson, *Portrait of an Amateur*, 267 and Kedourie, *England and the Middle East*, 113–17.
66. T.E. Lawrence, "To the Editor of *The Times*," 11 September 1919, T.E. Lawrence, *The Letters of T.E. Lawrence*, David Garnett, ed. (London: Jonathan Cape, 1938), 282.
67. Rough Notes and Final Notes for Speech at Manchester on Friday 22 June 1918 to the Syrians, U DDSY 2/11/124, Sykes Papers.
68. Ibid.
69. Ibid.
70. Sir Mark Sykes, "Note on Palestine and Zionism, September 21, 1917," U DSSY 2/12/10, Sykes Papers.

71. "Notes used for Speech to the Zionists at Hull Sunday July 7th 1918," U DDYS 2/11/125, Sykes Papers. Along with Sykes's notes are a programme of the event in English and Hebrew and the "Hatikvah" in English. The "Hatikvah" was a poem whose theme reflected the Jews' hope of returning to the Land of Israel, later put to music and chosen as the anthem of the first Zionist Congress in 1897. In 1948 it would become the Israeli national anthem.

72. Ibid., The brackets include words that seemed appropriate to fill out Sykes's notes.

73. Ibid.

74. Ibid.

75. Ibid., Sykes must have confused his audience by singling out the Mosque of Omar as the most important mosque in Jerusalem, instead of Al-Aqsa Mosque, or *Masjid al-Aqsa*. As its name translates, *Masjid al-Aqsa* is the "Farthest Mosque" referred to in the Qur'an in Chapter 17, *Sura Al Isra'* ("The Night Journey"), verse 1. It is the third most important mosque for Muslims after (in order) the Grand Mosque in Mecca and the Prophet's Mosque in Medina. The Mosque of Omar is not on the Temple Mount, but nearby opposite the Church of the Holy Sepulchre. It is important to Muslims, but not nearly as important as *Al-Aqsa*.

Chapter 12   Final Days

1. Leslie, *Mark Sykes*, 300. Everard Feilding (1867–1936) was the second son of William Everard, the 8th Earl of Denbigh, a prominent Catholic peer. He was also the hon. secretary of the Society of Psychical Research.

2. M. Sykes to D. Lloyd George (Personal and Private), 2 September 1918, U DDSY 2/11/107, Sykes Papers.

3. Ibid.

4. R. Cecil to M. Sykes, Foreign Office, 7 September 1918, Add MS 51094, f. 66, Cecil of Chelwood Papers, Vol. XXIV, British Library; hereafter referred to as Cecil Papers.

5. M. Sykes to R. Cecil, Foreign Office, 9 September 1918, Add MS 51094, f. 70, Cecil Papers, Vol. XXIV, British Library.

6. R. Cecil to M. Sykes, Foreign Office, 7 September 1918, Add MS 51094, f. 66, Cecil Papers, Vol. XXIV, British Library.

7. R. Cecil to M. Sykes, Foreign Office, 7 September 1918, Add MS 51094, f. 66, Cecil Papers, Vol. XXIV, British Library, and M. Sykes to R. Cecil, Foreign Office, September 9, 1918, Add MS 51094, f. 70, Cecil Papers, Vol. XXIV, British Library.

8. "XII. The Turkish portion of the present Ottoman Empire should be assured a secure sovereignty, but the other nationalities which are now under Turkish rule should be assured an undoubted security of life and an absolutely unmolested opportunity of autonomous development, and the Dardanelles should be permanently opened as a free passage to the ships and commerce of all nations under international guarantees." *President Woodrow Wilson's Fourteen Points, 8 January 1918*, The Avalon Project, Documents in Law, History and Diplomacy, Yale Law School, Lillian Goldman Law Library, http://avalon.law. yale.edu/20th_century/wilson14.asp, accessed 21 August 2016.

9. M. Sykes to R. Cecil, Foreign Office, 9 September 1918, Add MS 51094, f. 70, Cecil Papers, Vol. XXIV, British Library.

10. Sir Eyre Crowe (1864–1925), a career British diplomat, entered the Foreign Office in 1885, where he was resident clerk until 1895. He served in the Consular Department and then as assistant to Clement Lloyd Hill, supervisor of the African Protectorates Department from 1896 until 1905. In 1906 he was appointed senior clerk in the Western Department, where he became chief authority on German problems in the

Foreign Office. In 1912 he became assistant under secretary and then supervising under secretary in 1913. During much of World War I he served in the Contraband Department until he was moved to the Eastern Department, which lasted only a few months. In January 1919 he was appointed assistant secretary of state for Foreign Affairs and was to play a significant role on the British delegation at the Paris Peace Conference, and by mid-June he was made head of the political section of the delegation. In 1920 until his death in 1925, Crowe served as permanent under secretary at the Foreign Office. See Zara Steiner, *The Foreign Office and Foreign Policy, 1898–1914* (Cambridge: At the University Press, 1969), 108–18, 140–7 and numerous other references. For more on Crowe, see Richard A. Cosgrove, "The Career of Sir Eyre Crowe: A Reassessment," *Albion: A Quarterly Journal Concerned with British Studies*, 4, no. 4 (1972), 193–205. See also J.S. Dunn, *The Crowe Memorandum: Sir Eyre Crowe and Foreign Office Perceptions of Germany, 1918–1925* (Newcastle-upon-Tyne: Cambridge Scholars, 2013).

11. M. Sykes to R. Cecil, Sledmere, 12 October 1918, U DDSY 2/4/179, Sykes Papers.
12. Ibid.
13. Ibid.
14. Ibid.
15. Lt. Gen. Sir William R. Marshall (1865–1939) succeeded Lt. Gen. Sir Frederick Stanley Maude on the latter's death as commander in chief of the British forces in Mesopotamia. He kept that position until the end of World War I.
16. Ibid.
17. Ibid.
18. Briton Cooper Busch, *Mudros to Lausanne: Britain's Frontier in West Asia 1918–1923* (Albany, NY: SUNY Press, 1976), 9.
19. R. Gladstone, "Sir Mark Sykes Last Mission," 10 March 1919, U DDSY 2/4/188, Sykes Papers. Major Ronald Gladstone accompanied Sykes on this mission and after Sykes's death wrote a detailed description of the mission for Lady Sykes and the family.
20. Ibid.
21. Adelson, *Portrait of an Amateur*, 280. See Busch, *Mudros to Lausanne*, 15, for a description of Townshend's involvement in the signing of the Mudros Armistice.
22. R. Gladstone, "Sir Mark Sykes Last Mission," 10 March 1919, U DDSY 2/4/188, Sykes Papers.
23. Kamil al-Husayni (1882–1921) was the mufti of Jerusalem from 1908 to 1921. Unlike his younger brother, Amin al-Husayni, who would succeed him in the post of grand mufti on his death in 1921, Kamil sought to work with the British and the Jews. The British would appoint him as grand mufti and put him in charge of the sharia courts and waqaf (pl. of waqf, religious charitable foundations). Philip Mattar, *The Mufti of Jerusalem: Al-Hajj Amin Al-Husayni and the Palestinian National Movement* (New York: Columbia University Press, 1988), 22.
24. Rabbi Moshe Tzvi Hirsch Segal (1876–1968) was a specialist in Jewish studies and a Biblical scholar. Born in Lithuania, he moved to Britain, where for a time he worked as a journalist in London before moving to Oxford in 1901 to serve the Jewish community there. In 1918 he joined Dr Weizmann's Zionist Commission and went to Palestine, where he settled in Jerusalem and became a professor of the Bible and Semitic languages at the Hebrew University. Dan Cohn-Sherbok, *Dictionary of Jewish Biography*, 260.
25. R. Gladstone, "Sir Mark Sykes Last Mission," 10 March 1919, U DDSY 2/4/188, Sykes Papers.
26. Dr Montague David Eder (1865–1936) was an English psychoanalyst and physician. Born in London, he founded the Psychoanalytical Association, established a children's

clinic and edited the journal *School Hygiene*. Active in Zionist affairs, he served on the Zionist Commission to Palestine from 1918 to 1921. Later he would serve on the World Executive Committee of the Zionist Organization in Jerusalem and London. Dan Cohn-Sherbok, *Dictionary of Jewish Biography*, 68.

27. R. Gladstone, "Sir Mark Sykes Last Mission," 10 March 1919, U DDSY 2/4/188, Sykes Papers.
28. Lt. Gen. Sir Edward Bulfin (1862–1939) commanded the XXI Corps of the Egyptian Expeditionary Force in the Sinai and Palestine Campaign (1917–1918), successfully leading his Corps against Turkish forces at the Third Battle of Gaza, opening the way for the capture of Jerusalem and the last battle of the war at the Battle of Megiddo. See John D. Granger, *The Battle for Palestine 1917* (Woodbridge: Boydell Press: 2006), 93ff. See also Sir Edward Bulfin (1862–1939), *Oxford Dictionary of National Biography*, http://www.oxforddnb.com/view/article/32162, accessed 6 September 2016.
29. R. Gladstone, "Sir Mark Sykes Last Mission," 10 March 1919, U DDSY 2/4/188, Sykes Papers.
30. Ibid.
31. Ibid.
32. M. Sykes to Prodrome (Foreign Office), Cairo, 8 December 1918, U DSSY 2/4/185, Sykes Papers. See also Adelson, *Portrait of an Amateur*, 290–1.
33. See Bedross Der Matossian, "From Bloodless Revolution to Bloody Counterrevolution: The Adana Massacres of 1909" (2011), Faculty Publications, Department of History, Paper 124, University of Nebraska-Lincoln, accessed 5 February 2017.
34. R. Gladstone, "Sir Mark Sykes Last Mission," 10 March 1919, U DDSY 2/4/188, Sykes Papers.
35. Ibid.
36. R. Gladstone to E. Sykes, 10 March 1919, Sykes Papers, quoted in Adelson, *Portrait of an Amateur*, 287.
37. R. Gladstone, "Sir Mark Sykes Last Mission," 10 March 1919, U DDSY 2/4/188, Sykes Papers.
38. M. Sykes, "Appreciation of the Situation in Syria, Palestine and Lesser Armenia," January 22, 1919, F.O. 608/105, Foreign Office Papers, TNA.
39. Adelson, *Portrait of an Amateur*, 290–4.
40. Paris, *Britain, the Hashemites and Arab Rule* (London: Frank Cass Publishers, 2003), 111–12. H.V.F. Winstone, *Gertrude Bell*, 321.
41. Diary of E. Sandars, as quoted in Adelson, *Portrait of an Amateur*, 294.
42. Busch, *Mudros to Lausanne*, 75–6.
43. Diary of E. Sandars, as quoted in Adelson, *Portrait of an Amateur*, 294.
44. Adelson, *Portrait of an Amateur*, 294–5.
45. Verrier, ed., *Agents of Empire*, 11.
46. Lawrence, *Seven Pillars of Wisdom*, 58.

Chapter 13   The Legacy

1. William Shakespeare, *Julius Caesar* (1599).
2. M. Sykes, "A Report on the Petroliferous Districts of Mesopotamia," 12 October 1905, F.O. 78/5398, Foreign Office Papers, TNA. Adelson, *Portrait of an Amateur*, 112.
3. See details at United Nations Relief and Works Agency for Palestine Refugees in the Near East https://www.unrwa.org/sites/default/files/content/resources/unrwa_in_figures_2017_english.pdf (accessed 30 January 2018).

4. Verrier, ed., *Agents of Empire*, 7, 92–3.
5. Ibid., 92.
6. M. Sykes to E. Montagu, 14 August 1917, U DDSY 2/11/66, Sykes Papers.
7. Robert H. Liesthout, *Britain and the Arab Middle East: World War I and its Aftermath* (London: I.B.Tauris, 2016), 1–2.
8. Ibid., 2.
9. Monroe, *Britain's Moment in the Middle East 1914–1956*, 40–1.
10. Meghan Tinsley, "ISIS's Aversion to Sykes–Picot Tells us Much About The Group's Future Plans," *Muftah*, http://muftah.org/the-sykes-picot-agreement-isis/#.V4dKc1eCy-8, accessed 14 July 2016.
11. Ibid.

BIBLIOGRAPHY

Primary

Private Papers

Aubrey Herbert Papers	Somerset History Centre
Cabinet Papers	National Archives
Cecil of Chelwood Papers	The British Library
Churchill Papers	Churchill Archives Centre
Clayton Papers	Sudan Archive at Durham University
Correspondence of Sir Edward Grey	The British Library
Dardanelles Commission Report	The National Archives
Foreign Office Papers	The National Archives
George Lloyd Papers	Churchill Archives Centre
Grey Papers/Foreign Office Papers	The National Archives
Hankey Papers	Churchill Archives Centre
Hardinge Papers	Cambridge University Library
India Office Records and Papers	The British Library
Kitchener Papers	The National Archives
Lady Jessica Sykes Papers	East Riding of Yorkshire Council Archives
Leopold Amery Papers	Churchill Archives Centre
Lloyd George Papers	Parliamentary Archives/House of Lords Record Office
Papers of Sir Mark Sykes	British Online Archives
Sir Ronald Storrs Papers	Pembroke College, Cambridge
Sykes Family of Sledmere Papers	Hull History Centre
Wingate Papers	Sudan Archives at Durham University

Government Documents

Hansard's Parliamentary Debates

Newspapers and Magazines

"A History of the First World War in 100 Moments: The Day the Lights Went Out," *The Independent*, 9 September 2014, http://www.independent.co.uk/news/world/world-history/history-of-the-first-world-war-in-100-moments/a-history-of-the-first-world-war-in-100-moments-the-day-the-lights-went-out-9239572.html, accessed 29 July 2016).

"Compendium of the History of Derbyshire," *The Gentlemen's Magazine*, Vol. 126, August 1819, 108.

"Death of Joseph Cowan," *Jewish Telegraphic Agency*, 26 May 1932.

"Interview of Mark Sykes," *Daily Mail*, 25 January 1907.

"Jerusalem and Its Future... Interview with Sir Mark Sykes," *The Observer*, 16 December 1917.

"Journeys in North Mesopotamia," by Sir Mark Sykes, *The Geographical Journal*, 30, no. 3 (2007): 237–54, 284–98. The Royal Geographical Society (with the Institute of British Geographers).

"Maréchal Saxe," by Sir Mark Sykes, *Green Howards Gazette* (1908), U DDSY 2/6/1, Sykes Papers.

"Mark Sykes and Zionism," *Hull Daily Mail*, 3 December 1917.

"Sees Great Future for Jew and Arab," *New York Times*, 12 December 1917.

"Sir Mark Sykes on the Realities," *Leeds Mercury*, 16 February 1916.

"Sir Mark Sykes, M.P., On the War: Address to Wounded Soldiers," *Daily Mail*, 18 September 1916.

"Speech by Sir Mark Sykes: 'After the War,'" *Feilding Star, Oroua & Kiwitea Counties Gazette*, Volume XII, Issue 3089, 9 November 1916.

"The Italian Adventure," by Sir Mark Sykes, *Saturday Review* (London). 28 October 1911.

Memoirs, Diaries, Private Papers and Document Collections

Amery, The Rt. Hon. Leopold S., C.H. *In the Rain and the Sun: A Sequel to Days of Fresh Air.* Hutchinson & Co. (Publishers) Ltd., 1946.

—— *My Political Life.* Two volumes. Volume One: *England Before the Storm 1895–1914.* Volume Two: *War and Peace, 1914–1929.* London: Hutchinson, 1953.

Asquith, the Earl of Oxford and K.G. *Memories and Reflections 1852–1927.* Two volumes. Boston: Little, Brown and Company, 1928.

Asquith, H.H. *Letters to Venetia Stanley.* Selected and edited by Michael and Eleanor Brock. Oxford: Oxford University Press, 1985.

Bray, Major N.N.E. *Shifting Sands.* Forward by Sir Austen Chamberlain, K.G., M.P. London: Unicorn Press, 1934.

Bryce, Viscount. *The Treatment of the Armenians in the Ottoman Empire, 1915–1916: Documents Presented to Viscount Grey of Fallodon, Secretary of State for Foreign Affairs, Presented to Both Houses of Parliament by Command of His Majesty, October 1916.* London: G.P. Putnam's Sons, London, 1916.

Cecil, Viscount (Lord Robert Cecil). *A Great Experiment: An Autobiography.* London: Jonathan Cape, 1941.

Churchill, Winston S. *The River War.* Melbourne, Australia: The Book Jungle Online, 2007.

Clayton, Sir Gilbert. *An Arabian Diary: Sir Gilbert Falkingham Clayton, K.C.M.G., K.B.E., C.B., C.M.G,* ed. Robert O. Collins. Berkeley and Los Angeles, CA: University of California Press, 1969.

Gladstone, Major Ronald. "Sir Mark Sykes Last Mission," 10 March 1919, U DDSY 2/4/188, Sykes Papers.

Gilbert, Martin, ed. *The Churchill Documents, Volume 6: At the Admiralty July 1914–April 1915.* Hillsdale, MI: Hillsdale College Press, 1972.

—— *Winston S. Churchill, Companion Volume III, Pt. 1, Documents, July 1914–April 1915.* London: Heinemann, 1972.

Gooch, G.P., and H.W.V. Temperley, eds. *British Documents on the Origins of the War,* vol. X, Part II, *The Last Years of Peace.* London: H.M. Stationery Office, 1926.

Grey, Viscount of Fallodon. *Twenty-Five Years 1892–1916.* Two volumes. London: Hodder and Stoughton, 1925.

Haidar, H.R.H. Musbah. *Arabesque.* London: Hutchinson & Co., Publishers, 1944.

Herbert, Aubrey. *Mons, Anzac and Kut: By an MP* (Lieutenant Colonel The Hon. Aubrey Herbert). London: Edward Arnold, 1919. Reprint, ed. Edward Melotte. Barnsley: Pen & Sword Military, 2009.

Herzl, Theodor. *The Jewish State*, ed. and trans. Jacob M. Alkow. New York: Dover Publications, 1988.

Hurewitz, J.C. *Diplomacy in the Near and Middle East, A Documentary Record: 1914–1956*. New York: Octagon Books, 1972.

Indian Army List, January 1919, Vol. 2. Army Headquarters, India, 1919. Reprint: Uckfield: Naval and Military Press, 2001.

Lawrence, T.E. *Oriental Assembly*. London: Williams and Northgate Ltd., 1939.

—— *Revolt in the Desert*. New York: Garden City Publishing Company, Inc., 1927.

—— *Seven Pillars of Wisdom a Triumph*. Garden City, New York: Doubleday, Doran & Company, Inc., 1935.

—— *The Letters of T.E. Lawrence of Arabia*. Ed. David Garnett. London: Jonathan Cape, 1938.

Lloyd George, David. *Memoirs of the Peace Conference*. New Haven, CT: Yale University Press, 1939.

—— *War Memoirs of David Lloyd George*. Two volumes. London: Odhams Press Limited, 1936.

Malcolm, James A. "Origins of the Balfour Declaration: Dr. Weizmann's Contribution" (unpublished manuscript, 1944). The British Library.

Morganthau, Henry. *Ambassador Morganthau's Story*. New York: Doubleday, Page and Co., 1918. Reprint. Memphis: General Books, LCC, 2012.

Parker, Alfred Chevalier. *The Diaries of Parker Pasha*, ed. H.V.F. Winstone. London: Quartet Books, 1983.

President Woodrow Wilson's Fourteen Points, 8 January 1918, The Avalon Project, Documents in Law, History and Diplomacy, Yale Law School, Lillian Goldman Law Library, http://avalon.law.yale.edu/20th_century/wilson14.asp, accessed 21 August 2016.

Samuel, The Rt. Hon. Viscount. *Memoirs*. London: Cresset Press, 1945.

Storrs, Sir Ronald. *The Memoirs of Sir Ronald Storrs*. New York: G.P. Putnam's Sons, 1937.

—— *Orientations. Definitive Edition*. London: Nicolson & Watson, 1945.

Sykes, Mark. "The British Soldier and the Turk" (undated and unpublished notes), U DDSY 2/5/25, Sykes Papers.

—— *The Caliph's Last Heritage: A Short History of the Turkish Empire*. London: Macmillan and Co., Limited, 1915.

—— *Dar-Ul-Islam: A Record of a Journey Through Ten of the Asiatic Provinces of Turkey*. London: Bickers & Son, 1904. Reprint: Charleston, SC: Forgotten Books, 2012.

—— "Notes used for Speech to the Zionists at Hull Sunday July 7th 1918," U DDYS 2/11/125, Sykes Papers. p. 322.

—— "Note on Palestine and Zionism," 21 September 1917, U DSSY 2/12/10, Sykes Papers.

—— "A Report on the Petroliferous Districts of Mesopotamia," 12 October 1905, F.O. 78/5398, Foreign Office Papers, TNA.

—— "Rough Notes and Final Notes for Speech at Manchester on Friday June 22 1918 to the Syrians," U DDSY 2/11/124, Sykes Papers.

—— "Socialism and Education," *Things Political, No. 2*, September–December, (1907): 26–32 (unpublished speech). U DDYS 2/8/2, Sykes Papers.

—— *Through Five Turkish Provinces*. London: Bickers and Son, 1900.

—— "Unionist Ideals," *Things Political, No.4*, October–December *1908*, 30 (unpublished speech). U DDY S2/8/4, Sykes Papers.

Von Sanders, General Liman. *Five Years in Turkey*. Originally published in Berlin as *Fünf Jahre Türkei* by Scherl Publishers in 1919. Trans. Colonel Carl Reichmann. 1929; reprint, Uckfield, East Sussex: The Naval and Military Press, 2015.

Weizmann, Chaim. *Letters and Papers of Chaim Weizmann*, Vol. VII, Series A, *August 1914– November 1917*, Meyer W. Weisgal, gen. ed.; Leonard Stein, vol. ed., in collaboration with Dvorah Barzilay and Nehama A. Chalom. London and Jerusalem: Oxford University Press and Israel Universities Press, 1975.

—— *Trial and Error: The Autobiography of Chaim Weizmann*. New York: Harper and Brothers, 1949.

Wilson, Lt.-Col. Sir Arnold T. *Loyalties Mesopotamia 1914–1917, From the Outbreak of War to the Death of General Maude: A Personal and Historical Record.* London: Oxford University Press, 1930.

Secondary Sources

Adelson, Roger. *London and the Invention of the Middle East: Money, Power and War, 1902–1922.* New Haven, CT: Yale University Press, 1995.
—— *Mark Sykes: Portrait of an Amateur.* London: Jonathan Cape, 1975.
Adler, Cyrus. *Jacob Henry Schiff: A Biographical Sketch.* New York: The American Jewish Committee, 1921.
Allawi, Ali A. *Faisal I of Iraq.* London: Yale University Press, 2014.
Allen, Charles. God's Terrorists: The Wahhabi Cult and the Hidden Roots of Modern Jihad. London: Little, Brown, 2006. (Italicise the complete name of the book, God's Terrorists . . . Modern Jihad).
Anderson, Scott. *Lawrence in Arabia: War, Deceit, Imperial Folly and the Making of the Modern Middle East.* London: Atlantic Books, 2013.
Andrew, Christopher M. and A.S. Kanya-Forstner. *France Overseas: The Great War and the Climax of French Imperial Expansion.* London: Thames and Hudson, 1981.
—— *The Climax of French Imperial Expansion: 1914–1924.* Stanford, CA: Stanford University Press, 1981.
Antonius, George. *The Arab Awakening: The Story of the Arab National Movement.* Beirut: Librairie du Liban, 1969.
Arthur, Sir George. *Life of Lord Kitchener.* Three volumes. London: Macmillan and Co., Limited, 1920.
Baker, Randall. *King Husain and the Kingdom of the Hijaz.* Cambridge: The Oleander Press, 1979.
Baron, Archie. *An Indian Affair: From Riches to Raj.* London: Channel 4 Books, 2001.
Barr, James. *A Line in the Sand: Britain, France and the Struggle that Shaped the Middle East.* London: Simon & Schuster UK Ltd., 2011.
—— *Setting the Desert on Fire: T.E. Lawrence and Britain's Secret War in Arabia, 1916–1918.* London: Bloomsbury Publishing, 2007.
Bein, Alex. *Theodore Herzl: A Biography.* New York: Athenaeum, 1970.
Bentwich, Norman and Helen Bentwich. *Mandate Memories: 1918–1948.* London: Schocken Books, 1965.
Bermant, Chaim. *The Cousinhood: The Anglo-Jewish Gentry.* London: Eyre & Spottiswoode, 1971.
Black, Edwin. *British Petroleum and the Redline Agreement: The West's Secret Pact to Get Mideast Oil.* Washington, DC: Dialog Press, 2011.
Black, Eugene C. *The Social Politics of Anglo-Jewry 1880–1920.* Oxford: Basil Blackwell Ltd, 1988.
Blake, Robert. *The Unknown Prime Minister: The Life and Times of Andrew Bonar Law 1858–1923.* London: Eyre & Spottiswoode, 1955.
Boyle, Susan. *Betrayal of Palestine: The Story of George Antonius.* Boulder, CO: Westview Press, 2001.
Browne, Horace B. *The Story of The East Riding of Yorkshire.* Hull: publisher unknown, 1912; reprint, Charleston, SC: Forgotten Books, 2012.
Busch, Briton Cooper. *Britain, India, and the Arabs 1914–1921.* Berkeley, CA: University of California Press, 1971.
—— *Hardinge of Penshurst: A Study in the Old Diplomacy.* Hamden, CT: Archon Books, 1980.
—— *Mudros to Lausanne: Britain's Frontier in West Asia 1918–1923.* Albany, NY: SUNY Press, 1976.

Capern, Amanda. "Winston Churchill, Mark Sykes and the Dardanelles Campaign of 1915," *Historical Research*, vol. 71, no. 174 (1998): 108–18.

Cassar, George H. *Kitchener: Architect of Victory*. London: William Kimber, 1977.

—— *Lloyd George at War 1916–1918*. London: Anthem Press, 2011.

Cesarani, David. *The Jewish Chronicle and Anglo-Jewry 1841–1991*. Cambridge: Cambridge University Press, 1994.

Charmley, John. *Lord Lloyd and the Decline of the British Empire*. New York: St Martin's Press, 1987.

Churchill, Winston S. *The World Crisis 1911–1918, Vol. 1*. London: Odhams Press Limited, 1938.

Clark, Christopher. *The Sleepwalkers: How Europe Went to War in 1914*. London: Penguin Books, 2012.

Clark, Peter. *Marmaduke Pickthall: British Muslim*. London: Quartet Books, 1986.

Clews, Graham T. *Churchill's Dilemma: The Real Story Behind the Origins of the 1915 Dardanelles Campaign*. Santa Barbara, CA: Praeger, 2010.

Cloarec, Vincent. *La France et la question de Syrie, 1914–1918*. Paris: CNRS, 2010.

Cohen, Stuart A. *English Zionists and British Jews: The Communal Politics of Anglo-Jewry, 1895–1920*. Princeton, NJ: Princeton University Press, 1982.

Cohn-Sherbok, Dan. *Dictionary of Jewish Biography*. London: Continuum, 2005.

Collier, Peter. "Covert Mapping the Ottoman Empire: The Career of Francis Maunsell," http://icaci.org/files/documents/ICC_proceedings/ICC2013/_extendedAbstract/14_proceeding.pdf, accessed 25 August 2016.

Conklin, Alice L. *A Mission to Civilize: The Republican Idea of Empire in France and West Africa, 1895–1930*. Stanford, CA: Stanford University Press, 1997.

Cooper, Duff. *Old Men Forget*. New York: Carroll & Graf Publishers, 1953.

Cosgrove, Richard. "The Career of Sir Eyre Crowe: A Reassessment," *Albion*, 4, no. 4 (1972): 193–205.

Daftary, Farhad. *The Ismailis: Their History and Doctrines*. Cambridge: The Cambridge University Press, 2000.

Darlow, Michael and Barbara Bray. *Ibn Saud: The Desert Warrior and his Legacy*. London: Quartet Books Limited, 2010.

De Gaury, Gerald. *The Rulers of Mecca*. New York: Dorsett Press, 1991.

Dodwell, H.H., M.A., ed. *The Cambridge History of India, Vol. VI: The Indian Empire 1858–1918*. With Chapters on Development of Administration 1818–1858. Cambridge: Cambridge University Press, 1932.

Dugdale, Blanche E.C. *Arthur James Balfour: First Earl of Balfour, K.G., O.M., F.R.S., Etc.* Volume One: 1848–1905. Volume Two: 1906–1930. New York: G.P. Putnam & Sons, 1937.

Dunn, J.S. *The Crowe Memorandum: Sir Eyre Crowe and Foreign Office Perceptions of Germany, 1918–1925*. Newcastle-upon-Tyne: Cambridge Scholars Publishing, 2013.

Egerton, George W. *Great Britain and the Creation of the League of Nations: Strategy, Politics, and International Organization, 1914–1919*. London: Scolar Press, 1979.

Egremont, Max. *Balfour: A Life of Arthur James Balfour*. London: Phoenix Giant, 1980.

Engle, Anita. *The NILI Spies*. London: Hogarth Press, 1959; reprint, London: Routledge, 2007.

Erickson, Edward J. *Ottomans and Armenians: A Study in Counterinsurgency*. New York: Palgrave Macmillan, 2013.

Esposito, John L. *Islam: The Straight Path*, 3rd edn. New York: Oxford University Press, 1998.

Faulkner, Neil. *Lawrence of Arabia's War: The Arabs, the British and the Remaking of the Middle East in WWI*. New Haven, CT: Yale University Press, 2016.

Ferguson, Niall. *The House of Rothschild: The World's Banker 1849–1999*. New York: Viking, 1999.

Fisher, John. *Curzon and British Imperialism in the Middle East 1916–1919*. London: Frank Cass, 1999.

—— *Gentleman Spies: Intelligence Agents in the British Empire and Beyond*. Stroud: Sutton Publishing Limited, 2002.

—— "Lord Robert Cecil and the Formation of a Middle East Department of the Foreign Office," *Middle Eastern Studies* 42, no. 3 (2006): 365–80.

—— "Major Norman Bray and Eastern unrest in the aftermath of World War I," *Asian Affairs* 31, no. 2 (2010): 189–97.

—— "The Rabegh Crisis, 1916–1918: 'A Comparatively Trivial Question' or 'A Self-willed Disaster,'" *Middle Eastern Studies* 38, no. 3 (2010): 73–92.

Fitzherbert, Margaret. *The Man Who Was Greenmantle: A Biography of Aubrey Herbert.* Oxford: Oxford University Press, 1985.

Florence, Ronald. *Lawrence and Aaronsohn: T.E. Lawrence, Aaron Aaronsohn, and the Seeds of the Arab-Israeli Conflict.* New York: Viking, 2007.

Fraenkel, Josef. *Lucien Wolf and Theodor Herzl.* London: The Jewish Historical Society of England, 1960.

Freemantle, Anne. *Loyal Enemy: The Life of Marmaduke Pickthall.* London: Hutchinson & Co., 1938.

Friedman, Isaiah. *British Miscalculations: The Rise of Muslim Nationalism, 1918–1935* New Brunswick, NJ: Transaction Publishers, 2012.

—— *Germany, Turkey, and Zionism 1897–1918.* Abingdon: Oxford University Press, 1977.

—— *Palestine: A Twice Promised Land?* Volume 1: *The British, the Arabs and Zionism 1915–1920.* New Brunswick, NJ: Transaction Publishers, 2000.

—— *The Question of Palestine: British-Jewish-Arab Relations: 1914–1918.* Second Expanded Edition. New Brunswick, NJ: Transaction Publishers, 1992.

Fritzinger, Linda B. *Diplomat Without Portfolio: Valentine Chirol, His Life and* The Times. London: I.B.Tauris, 2006.

Fromkin, David. *A Peace to End All Peace: The Fall of the Ottoman Empire and the Creation of the Modern Middle East.* New York: Avon Books, 1989.

Gilbert, Martin. *In Search of Churchill: A Historian's Journey.* London: HarperCollins Publishers, 1995.

—— *Winston S. Churchill, Volume III: The Challenge of War 1914–1916.* Hillsdale: Hillsdale College Press, 2008.

Glubb, Brigadier John Bagot. *The Story of the Arab Legion.* London: Hodder & Stoughton Limited, 1948.

Gossman, Lionel. *The Passion of Max von Oppenheim: Archaeology and Intrigue in the Middle East from Wilhelm II to Hitler.* Cambridge: Open Book Publishers, 2014.

Gorst, Harold E. *Much of Life is Laughter.* London: George Allen & Unwin Ltd., 1936.

Grainger, John D. *The Battle for Palestine 1917.* Woodbridge: Boydell Press, 2006.

"Great Britain. War Office. Director of Military Operations," Archives Association of Ontario, http://www.archeion.ca/great-britain-war-office-director-of-military-operations-2, accessed 4 September 2016.

Grigg, John. *Lloyd George: From Peace to War 1912–1916.* London: Methuen Ltd, 1985.

Guilday, Peter. "Francis Aiden Cardinal Gasquet," *Catholic World (Apr–Sep 1922)*, 210–16.

Gurnham, Richard. *The Story of Hull.* Andover: Phillimore & Co. Ltd, 2011.

Habib, John S. *Ibn Sa'ud's Warriors of Islam: The Ikhwan of Najd and Their Role in the Creation of the Sa'udi Kingdom, 1910–1930.* Leiden: E.J. Brill, 1978.

Hankey, Lord. *The Supreme Command 1914–1918.* Two volumes. London: George Allen and Unwin Limited, 1961.

Hazelhurst, Cameron. *Politicians at War, July 1914–May 1915: A Prologue to the Triumph of Lloyd George.* London: Jonathan Cape Ltd, 1971.

Hennessy, James Pope. *Lord Crewe: The Likeness of a Liberal.* London: Constable & Co Ltd, 1955.

Historical UK Inflation Rates and Calculator, http://inflation.stephenmorley.org, accessed 3 August 2016.

Historical Society of the New York Courts: Abram Elkus (1867–1847), http://www.nycourts.gov/history/legal-history-new-york/luminaries-court-appeals/elkus-abram.html">http://www.nycourts.gov/history/legal-history-new-york/history-legal-bench-court-appeals.html? http://www.nycourts.gov/history/legal-history-new-york/luminaries-court-appeals/elkus-abram.html, accessed 6 September 2016.

Hopkirk, Peter. *Like Hidden Fire: The Plot to Bring Down the British Empire*. London: Kodansha International, 1994.

Hughes, Matthew. *Allenby and British Strategy in the Middle East 1917–1919*. Reprint. London: Frank Cass, 2005.

Hunter, Archie. *Power and Passion in Egypt: A Life of Sir Eldon Gorst, 1861–1911*. London: I.B.Tauris, 2007.

Hyde, H. Montgomery. *Carson: The life of Sir Edward Carson, Lord Carson of Duncairn*. London: Heinemann, 1953.

Ingrams, Doreen. *Palestine Papers 1917–1822: Seeds of Conflict*. London: John Murray, 1972; reprint, London: Eland Publishing Limited, 2009.

Jacob, Harold F. *Kings of Arabia: The Rise and Set of the Turkish Sovranty {sic} in the Arabian Peninsula*. London: Mills and Boon, 1923, reprint, Reading: Garnet Publishing Limited, 2007.

James, Lawrence. *Aristocrats, Power, Grace & Decadence: Britain's Ruling Classes from 1066 to the Present*. London: Abacus, 2009.

—— "Sykes, Sir Mark, sixth baronet (1879–1919)," *Oxford Dictionary of National Biography*. Oxford: Oxford University Press, 2004; online edition, May 2011, accessed 7 June 2014.

Jenkins, Roy. *Asquith: Portrait of a Man and his Era*. New York: E.P. Dutton & Co, 1966.

Johnson, Gaynor. *Lord Robert Cecil: Politician and Internationalist*. Farnham: Ashgate Publishing Limited, 2013.

Johnson, Peter. *A Philosopher at the Admiralty: R.G. Collingwood and the First World War: A Philosopher at War*, Volume One. Upton Pyne: Imprint Academic, 2012.

Karsh, Ephraim & Inari Karsh. *Empires of the Sand: The Struggle for Mastery in the Middle East 1789–1923*. Cambridge, MA: Harvard University Press, 1999.

Katz, Shmuel. *Lone Wolf: A Biography of Vladimir (Ze'ev) Jabotinsky*. Two volumes. New York: Barricade Books, Inc., 1996.

—— *The Aaronsohn Saga*. Jerusalem: Gefen Books, 2007.

Kedourie, Elie. *Arabic Political Memoirs and Other Studies*. London: Frank Cass, 1974.

—— *England and the Middle East: The Destruction of the Ottoman Empire 1914–1921*. London: Mansell Publishing Limited, 1987.

—— *In the Anglo-Arab Labyrinth: The McMahon–Husayn Correspondence and its Interpretations 1914–1939*. London: Frank Cass, 2000.

—— "Sir Mark Sykes and Palestine 1915–1916," *Middle Eastern Studies* 6, no. 3 (1970): 340–45.

Kimche, Jon. *The Second Arab Awakening*. London: Thames and Hudson, 1970.

Klieman, Aaron S. "Britain's War Aims in the Middle East in 1915," *Journal of Contemporary History* 3, no. 3 (1968): 237–51.

Kling, Simcha. *Nachum Sokolow: Servant of his People*. New York: Herzl Press, 1960.

Krämer, Gudrun. *The Jews in Modern Egypt 1914–1952*. London: I.B.Tauris, 1989.

Kreutz, Andrej. *Vatican Policy on the Palestinian-Israeli Conflict: The Struggle for the Holy Land*. Westport, CT: Greenwood Press, 1990.

Lacey, Robert. *The Kingdom: Arabia and the House of Saud*. New York: Avon Books, 1981.

Landes, David S. *Bankers and Pashas: International Finance and Economic Imperialism in Egypt*. New York: Harper Torchbooks, 1969.

Laqueur, Walter. *A History of Zionism: From the French Revolution to the Establishment of the State of Israel*. New York: MJF Books, 1972.

Leslie, Shane. *Mark Sykes: His Life and Letters*. With an Introduction by The Right Hon. Winston Churchill. London: Cassell and Company Ltd, 1923.

—— *Salutation to Five*. London: Hollis & Carter, 1951; reprint, Freeport, NY: Books for Libraries Press, 1970.

Levene, Mark. "The Balfour Declaration: A Case of Mistaken Identity," *English Historical Review* 107. no. 422 (1992): 54–77.

—— *War, Jews, and the New Europe: The Diplomacy of Lucien Wolf 1914–1919*. Oxford: Oxford University Press, 1992.

Levine, Naomi B. *Politics, Religion & Love: The Story of H.H. Asquith, Venetia and Edwin Montagu, Based on the Life and Letters of Edwin Samuel Montagu.* New York: New York University Press, 1991.

Lewis, Bernard. *The Assassins: A Radical Sect in Islam.* London: Orion Publishing Group, 1967.

Lewis, David Levering. *The Race to Fashoda: European Colonialism and African Resistance in the Scramble for Africa.* New York: Weidenfeld & Nicolson, 1987.

Lewis, Geoffrey. *Balfour and Weizmann: The Zionist, the Zealot and the Emergence of Israel.* London: Continuum, 2009.

—— *Carson: The Man Who Divided Ireland.* London: Hambledon Continuum, 2006.

Lief, Alfred. *Brandeis: The Personal History of an American Ideal.* Harrisburg, PA: Stackpole Sons, 1936; fourth reprint 1937.

Liesthout, Robert H. *Britain and the Arab Middle East: World War I and its Aftermath.* London: I.B.Tauris, 2016.

Lilienthal, Alfred M. *The Zionist Connection II: What Price Peace?* New Brunswick, NJ: North American, 1982.

Long, C.W.R. *British Pro-Consuls in Egypt, 1914–1929: The Challenge of Nationalism.* Abington: RoutledgeCurzon, 2010.

Lowe, C.J. and M.L. Dockrill. *Mirage of Power: British Foreign Policy 1914–1922.* Volume Two. London: Routledge & Kegan Paul, 1972.

Lüdke, Tilman. *Jihad Made in Germany: Ottoman and German Propaganda and Intelligence Operations in the First World War.* Münster: LIT Verlag, 2005.

Lukit, Liora. *A Quest in the Middle East: Gertrude Bell and the Making of Modern Iraq.* London: I.B.Tauris, 2013.

Lutsky, Vladimir Borisovich. *Modern History of the Arab Countries.* Moscow: Progress Publishers, 1969.

McCarthy, Justin. *The Population of Palestine: Population History and Statistics of the Late Ottoman Period and the Mandate.* New York: Columbia University Press, 1990.

McKale, Donald M. "'The Kaiser's Spy': Max von Oppenheim and the Anglo-German Rivalry Before and During the First World War," *European History Quarterly* 27, no. 2 (1997): 199–219.

McMeekin, Sean. *The Berlin–Baghdad Express: The Ottoman Empire and Germany's Bid for World Power 1898–1918.* London: Allen Lane, 2010.

MacMillan, Margaret. *Paris 1919: Six Months that Changed the World.* New York: Random House Trade Paperbacks, 2002.

Mahan, A.T. "The Persian Gulf and International Relations," *The National Review,* September 1902, 38–9.

Marlowe, John. *The Seat of Pilate: An Account of the Palestine Mandate.* London: Cresset Press, 1959.

Marsay Mark. *Baptism of Fire: An Account of the 5th Green Howards at the Battle of St. Julien, during the Second Battle of Ypres, April 1915; Part One of the 'Yorkshire Gurkhas' Series.* Scarborough: Great Northern Publishing, 1988.

Mattar, Philip. *The Mufti of Jerusalem: Al-Hajj Amin Al-Husayni and the Palestinian National Movement.* New York: Columbia University Press, 1988.

Minerbi, Sergio I. *The Vatican and Zionism: Conflict in the Holy Land 1895–1925.* New York: Oxford University Press, 1990.

Mohs, Polly A. *Military Intelligence and the Arab Revolt: The First Modern Intelligence War.* London: Routledge, 2008.

Momen, Moojan. *An Introduction to Shi'i Islam: The History and Doctrines of Twelver Shi'ism.* Oxford: George Roland Publisher, 1985.

Monger, David. *Patriotism and Propaganda in First World War Britain: The National War Aims Committee and Civilian Morale.* Liverpool: Liverpool University Press, 2012.

Monroe, Elizabeth. *Britain's Moment in the Middle East 1914–1956.* London: Chatto & Windus, 1963.

Montague, Edwin Samuel. *Edwin Montagu and the Balfour Declaration.* London: Arab League Office, n.d.

Moorhouse, Geoffrey. *India Britannica*. New York: Harper & Row, Publishers, 1983.

Moore, John Norton, ed. *The Arab-Israeli Conflict, Vol. III: Documents*, American Society of International Law. Princeton, NJ: Princeton University Press, 1974.

Morgan, Kenneth O. *Consensus and Disunity: The Lloyd George Coalition Government 1918–1922*. Oxford: Clarendon Press, 1979; reprint, 2001.

Mosley, Leonard. *The Glorious Fault: The Life of Lord Curzon*. New York: Harcourt, Brace and Company, 1960.

Motassian, Bedross Der. "From Bloodless Revolution to Bloody Counterrevolution: The Adana Massacres of 1909," Faculty Publications, Department of History, Paper 124 (2011), University of Nebraska-Lincoln, http://digitalcommons.unl.edu/historyfacpub/124, accessed 5 February 2017.

Murphy, David. *The Arab Revolt 1916–1918: Lawrence Sets Arabia ablaze*. Oxford: Osprey Publishing, 2008.

Neu, Charles E. *Colonel House: A Biography of Woodrow Wilson's Silent Partner*. New York: Oxford University Press, 2014.

Nevakivi, Jukka. *Britain, France and the Arab Middle East 1914–1920*. London: University of London, The Athlone Press, 1969.

Nicolson, Harold. *Curzon: The Last Phase, 1919–1925: A Study in Post War Diplomacy*. London: Faber and Faber Ltd., 1934; reprint, 1988.

Nimocks, Walter. *Milner's Young Men: the "Kindergarten" in Edwardian Imperial Affairs*. Durham, NC: Duke University Press, 1968.

Norwich, John Julius. *The Popes: A History*. London: Vintage Books, 2012.

O'Brien, Terrence. *Milner: Viscount Milner of St James's and Cape Town*. London: Constable and Company Ltd, 1959.

O'Gorman, Frank. *Voters, Patrons, and Parties: The Unreformed Electoral System of Hanoverian England, 1734–1832*. Oxford: Oxford University Press, 1989.

Oliver, Neil. "Was World War One propaganda the birth of spin?" World War One/BBC, http://www.bbc.co.uk/guides/zq8c7ty, accessed 6 September 2016.

Paris, Timothy J. *Britain, the Hashemites and Arab Rule 1920–1925*. London: Frank Cass Publishers, 2003.

—— *In Defence of Britain's Middle Eastern Empire: A Life of Sir Gilbert Clayton*. Brighton, UK: Sutton Academic Press, 2016.

Pendlebury, Alyson. *Portraying "the Jew" in First World War Britain*. London: Valentine Mitchell, 2006.

Philby, H. St J.B. *Arabia of the Wahhabis*. London: Frank Cass, 1977.

Pickthall, Marmaduke. *With the Turk in Wartime*. London: J.M. Dent & Sons, Ltd., 1914.

Presland, John. *Deedes Bey: A Study of Sir Wyndham Deedes 1883–1923*. London: Macmillan & Co. Ltd., 1942.

Rabinowitz, Ezekiel. *Justice Louis D. Brandeis: The Zionist Chapter in His Life*. New York: Philosophical Library, 1968.

Ramsay, David. *"Blinker" Hall, Spymaster: The Man Who Brought America into World War I*. Stroud: Spellmount, 2009.

Reinharz, Jehuda. *Chaim Weizmann: The Making of a Statesman*. New York: Oxford University Press, 1993.

Renton, James. *The Zionist Masquerade: The Birth of the Anglo-Zionist Alliance, 1914–1918*. Basingstoke: Palgrave Macmillan, 2007.

Rentz, George S. *The Birth of the Islamic Reform Movement in Saudi Arabia: Muhammad b. 'Abd al-Wahhab (1703–1792) and the Beginnings of the Unitarian Empire in Arabia*. London: Arabian Publishing, 2004.

Robbins, Keith. *Sir Edward Grey: A Biography of Lord Grey of Fallodon*. London: Cresswell & Company Ltd., 1971.

Rogan, Eugene. *The Fall of the Ottomans: The Great War in the Middle East, 1914–1920*. London: Penguin Books, 2015.

Roi, Michael L. *Alternative to Appeasement: Sir Robert Vansittart and Alliance Diplomacy, 1934–1937*. Westport, CT: Praeger, 1997.

Rosen, Jacob. "Captain Reginald Hall and the Balfour Declaration," *Middle Eastern Studies* 24, no. 1 (1988): 56–67.

Roskill, Stephen. *Hankey: Man of Secrets, Vol. 1: 1877–1918*. London: Collins, 1970.

Rutledge, Ian. *Enemy on the Euphrates: The British Occupation of Iraq and the Great Arab Revolt 1914–1921*. London: Saqi Books, 2014.

Sabini, John. *Armies in the Sand: The Struggle for Mecca and Medina*. London: Thames and Hudson Ltd, 1981.

Sachar, Howard M. *A History of Israel: From the Rise of Zionism to Our Time*. New York: Alfred A. Knopf, 2002.

—— *A History of the Jews in the Modern World*. New York: Alfred A. Knopf, 2005.

Sanders, Ronald. *The High Walls of Jerusalem: A History of the Balfour Declaration and the Birth of the British Mandate for Palestine*. New York: Holt, Rinehart and Winston, 1983.

Satia, Priya. *Spies in Arabia: The Great War and the Cultural Foundations of Britain's Covert Empire in the Middle East*. New York: Oxford University Press, 2008.

Sattin, Anthony. *Young Lawrence: A Portrait of the Legend as a Young Man*. London: John Murray Publishers, 2014.

Schama, Simon. *Two Rothschilds and the Land of Israel*. New York: Alfred A. Knopf, 1978.

Schneer, Jonathan. *The Balfour Declaration: The Origins of the Arab-Israeli Conflict*. London: Bloomsbury Publishing, 2011.

Segev, Tom. *One Palestine Complete: Jews and Arabs under the British Mandate*. Trans. Haim Watzman. New York: Metropolitan Books, 2000.

Sharif, Regina S. *Non-Jewish Zionism*. London: Zed Press, 1983.

Sicker, Martin. *Between Hashemites and Zionists: The Struggle for Palestine 1908–1988*. New York: Holmes & Meier, 1989.

Sokolow, Nahum. *History of Zionism 1600–1918*. With an Introduction by the Rt. Hon. A.J. Balfour, M.P. New York: KTAV Publishing House, Inc., 1969.

—— "Sir Mark Sykes, Bart., M.P. (A Tribute)," *History of Zionism 1600–1918*, Vol. II. New York: Ktav Publishing House, 1969.

Spender, J.A. and Cyril Asquith, *Life of Lord Oxford & Asquith*, Vol. II. London: Hutchison, 1932.

Stein, Leonard. *The Balfour Declaration*. New York: Simon and Schuster, 1961.

Steiner, Zara. *The Foreign Office and Foreign Policy 1898–1914*. Cambridge: Cambridge University Press, 1969.

Stevens, Richard P. *Weizmann and Smuts. Journal of Palestine Studies* 3, no. 1 (1973), published by University of California Press on behalf of the Institute for Palestine Studies.

Stitt, George. *A Prince of Arabia: The Emir Ali Haider*. George Allen & Unwin, Ltd., 1948.

Sykes, Christopher Simon. *The Big House: The Story of a Country House and its Family*. London: Harper Perennial, 2005.

Sykes, Christopher (Hugh). *Crossroads to Israel: Palestine from Balfour to Bevin*. London: Collins, 1965.

—— *Two Studies in Virtue*. London: Collins, 1953.

—— *Wassmuss: The German Lawrence*. London: Longmans, Green, and Co., 1936.

Tauber, Eliezer. *The Arab Movements in World War I*. London: Frank Cass and Co., Ltd. 1993.

Tibawi, A.L. *Anglo-Arab Relations and the Question of Palestine 1914–1921*. London: Luzac & Company Ltd, 1978.

Tinsley, Meghan. "ISIS's Aversion to Sykes–Picot Tells Us Much about the Group's Future Plans," *Muftah*, http://muftah.org/the-sykes-picot-agreement-isis/#.V4dKc1eCy-8, accessed 14 July 2016.

Townsend, John. *Proconsul to the Middle East: Sir Percy Cox and the End of Empire*. London: I.B.Tauris, 2010.

Townshend, Charles. *When God Made Hell: The British Invasion of Mesopotamia and the Creation of Iraq, 1914–1921*. London: Faber and Faber, 2010.

Troeller, Gary. *The Birth of Saudi Arabia: Britain and the Rise of the House of Sa'ud*. London: Frank Cass, 1987.

Tuchman, Barbara. *The Zimmerman Telegram*. London: Phoenix Press, 1958.

United Nations Relief and Works Agency for Palestinian Refugees in the Near East. https://www.unrwa.org/sites/default/files/content/resources/unrwa_in_figures_2017_english.pdf

Unknown. "Lieutenant-Colonel Harold Fenton Jacob, CSI, Officer of the Legion of Honour," myjacobfamily.com/favershamjacobs/haroldfentonjacob.htm, accessed 25 August 2016.

Vanssitart, Sir Robert. *Black Record: Germans Past and Present*. London: Hamish Hamilton, 1941.

—— *Lessons of My Life*. London: Hutchinson, 1943.

—— *The Mist Procession: The Autobiography of Lord Vansittart*. London: Hutchinson, 1958.

Vassiliev, Alexei. *The History of Saudi Arabia*. London: Saqi Books, 1998.

Vatikiotis, P.J. *The Modern History of Egypt*. New York: Frederick A. Praeger Publishers, 1969.

—— *Politics and the Military in Jordan: A Study of the Arab Legion 1921–1957*. London: Frank Cass & Co. Ltd., 1967.

Vereté, Mayir. "The Balfour Declaration and its Makers," *Middle Eastern Studies* 6, no. 1 (2006): 48–76.

Verrier, Anthony, ed. *Agents of Empire: Anglo-Zionist Intelligence Operations 1915–1919, Brigadier Walter Gribbon, Aaron Aaronsohn and the NILI Ring*. London: Brassey's (UK) Ltd, 1995.

Waley, S.D. *Edwin Montagu: A Memoir and an Account of his Visits to India*. London: Asia Publishing House, 1964.

Wasson, Ellis. *Born to Rule: British Political Elites*. Stroud: Sutton Publishing Ltd., 2000.

Watson, William E. *Tricolor and Crescent: France and the Islamic World*. Westport, CT: Praeger, 2005.

Watts, Martin. *The Jewish Legion and the First World War*. London: Palgrave Macmillan, 2004.

Wavell, Archibald. *Allenby, A Study in Greatness: The Biography of Field-Marshall Viscount Allenby of Megiddo and Felixstowe*. New York: Oxford University Press, 1941; reprint, Whitefish, MT: Kessinger Publishing, LLC, 2010.

Webb, R.K. *Modern England: From the Eighteenth Century to the Present*, 2nd edn. New York: HaperCollins Publishers, 1980.

Webber, Kerry. "In the Shadow of the Crescent: The Life & Times of Colonel Stewart Francis Newcombe, R.E., D.S.O—Soldier, Explorer, Surveyor, Adventurer and Loyal Friend to Lawrence of Arabia," http://shadowofthecrescent.blogspot.co.uk/p/sf-newcombe-short-biography.html, accessed 4 September 1916.

Wells, H.G. *The War That Will End War*. London: Frank & Cecil Palmer, 1914.

Westrate, Bruce. *The Arab Bureau: British Policy in the Middle East 1916–1920*. University Park, PA: Pennsylvania State University Press, 1992.

Wilson, Trevor, ed. *The Political Diaries of C.P. Scott 1911–1928: A unique record of the Lloyd George Years by the Great Editor of the Manchester Guardian*. Ithaca, NY: Cornell University Press, 1970.

Williams, Michael. "Sykes–Picot drew lines in the Middle East sands that blood is washing away," *Reuters, Analysis & Opinion/ The Great Debate, 24 October 2014*, http://blogs.reuters.com/great-debate/2014/10/24/sykes-picot-drew-lines-in-the-middle-easts-sand-that-blood-is-washing-away/, accessed 2 September 2016.

Winstone, H.V.F. *Captain Shakespear: A Portrait*. New York: Quartet Books, 1978.

—— *Gertrude Bell*. London: Barzan Publishing, 2004.

—— *The Illicit Adventure: The Story of Political and Military Intelligence in the Middle East from 1898–1926*. London: Jonathan Cape, 1982.

—— *Leachman: O.C. Desert*. London: Quartet Books, 1982.

Wolpert, Stanley. *A New History of India*, 4th edn. New York: Oxford University Press, 1993.

Woodward, David R. *Field Marshal Sir William Robertson*. Westport CT: Praeger, 1998.

Wrench, John Evelyn. *Geoffrey Dawson and Our Times*. Foreword by the Earl of Halifax. London: Hutchinson, 1955.

Zeine, Zeine N. *The Struggle for Arab Independence: Western Diplomacy & the Rise and Fall of Faisal's Kingdom in Syria*. Beirut: Khayat's, 1960.

INDEX

Plate 1 Lt. Col. Sir Mark Sykes, Bart, MP, Commanding Officer, 5th Battalion Yorkshire "Green Howards" Regiment (c. 1914) – *Hull Daily Mail*

Plate 2 Sir Mark Sykes (c. 1916) – Unknown

Plate 3 Winston Churchill (c. 1916) – Churchill College Archives Centre, Churchill Press Photographs, Master, Fellows and Scholars of Churchill College, Cambridge

Plate 4 Lord Herbert Horatio Kitchener (c. 1914) – Imperial War Museum Photograph Archives

Plate 5 Lord Robert Cecil (c. 1916) – National Portrait Gallery

Plate 6 Sir Maurice Hankey (c. 1916) – National Photo Company Collection, Prints & Photographs Division, United States Library of Congress

Plate 7   Sir Reginald Wingate – Unknown

Plate 8   Sir Gilbert Clayton – National Portrait Gallery

Plate 9 T.E. Lawrence – Lowell Thomas Papers, Arabia, Graphic Materials, James A. Cannavino Library, Archives & Special Collections, Marist College, USA

Plate 10   Sir Mark Sykes, Unknown and François Georges-Picot, Unknown

Plate 11 Sharif/King Husayn – Lowell Thomas Papers, Arabia, Graphic Materials, James A. Cannavino Library, Archives & Special Collections, Marist College, USA

Plate 12   Lord Curzon – George Grantham Bain Collection, Prints & Photographs Division, United States Library of Congress

Plate 13 Chaim Weizmann – National Portrait Gallery

Plate 14 Nahum Sokolow – George Grantham Bain Collection, Prints & Photographs Division, United States Library of Congress

Plate 15 Arthur J. Balfour – Imperial War Museum Photograph Archives

9 781838 604677